Numerical Analysis

Numerical Analysis

L. Ridgway Scott

PRINCETON UNIVERSITY PRESS
PRINCETON AND OXFORD

Published by Princeton University Press, 41 William Street,
Princeton, New Jersey 08540
In the United Kingdom: Princeton University Press, 6 Oxford Street,
Woodstock, Oxfordshire OX20 1TW
press.princeton.edu

Library of Congress Control Number: 2010943322

ISBN: 978-0-691-14686-7
British Library Cataloging-in-Publication Data is available

The publisher would like to acknowledge the author of this volume for type-
setting this book using LATEX and Dr. Janet Englund and Peter Scott for
providing the cover photograph

Printed on acid-free paper ∞

Printed in the United States of America

10 9 8 7 6 5 4 3 2 1

Dedication

To the memory of Ed Conway[1] who, along with his colleagues at Tulane University, provided a stable, adaptive, and inspirational starting point for my career.

---

[1]Edward Daire Conway, III (1937–1985) was a student of Eberhard Friedrich Ferdinand Hopf at the University of Indiana. Hopf was a student of Erhard Schmidt and Issai Schur.

Contents

Preface

> "...by faith and faith alone, embrace, believing where we
> cannot prove," from *In Memoriam* by Alfred Lord Ten-
> nyson, a memorial to Arthur Hallum.

Numerical analysis provides the foundations for a major paradigm shift in what we understand as an acceptable "answer" to a scientific or technical question. In classical calculus we look for answers like $\sqrt{\sin x}$, that is, answers composed of combinations of names of functions that are familiar. This presumes we can evaluate such an expression as needed, and indeed numerical analysis has enabled the development of pocket calculators and computer software to make this routine. But numerical analysis has done much more than this. We will see that far more complex functions, defined, e.g., only implicitly, can be evaluated just as easily and with the same technology. This makes the search for answers in classical calculus obsolete in many cases. This new paradigm comes at a cost: developing stable, convergent algorithms to evaluate functions is often more difficult than more classical analysis of these functions. For this reason, the subject is still being actively developed. However, it is possible to present many important ideas at an elementary level, as is done here.

Today there are many good books on numerical analysis at the graduate level, including general texts [47, 134] as well as more specialized texts. We reference many of the latter at the ends of chapters where we suggest further reading in particular areas. At a more introductory level, the recent trend has been to provide texts accessible to a wide audience. The book by Burden and Faires [28] has been extremely successful. It is a tribute to the importance of the field of numerical analysis that such books and others [131] are so popular. However, such books intentionally diminish the role of advanced mathematics in the subject of numerical analysis. As a result, numerical analysis is frequently presented as an elementary subject. As a corollary, most students miss exposure to numerical analysis as a mathematical subject. We hope to provide an alternative.

Several books written some decades ago addressed specifically a mathematical audience, e.g., [80, 84, 86]. These books remain valuable references, but the subject has changed substantially in the meantime.

We have intentionally introduced concepts from various parts of mathematics as they arise naturally. In this sense, this book is an invitation to study more deeply advanced topics in mathematics. It may require a short detour to understand completely what is being said regarding operator theory in infinite-dimensional vector spaces or regarding algebraic concepts like tensors and flags. Numerical analysis provides, in a way that is accessible to advanced undergraduates, an introduction to many of the advanced concepts of modern analysis.

We have assumed that the general style of a course using this book will be to prove theorems. Indeed, we have attempted to facilitate a "Moore[2] method" style of learning by providing a sequence of steps to be verified as exercises. This has also guided the set of topics to some degree. We have tried to hit the interesting points, and we have kept the list of topics covered as short as possible. Completeness is left to graduate level courses using the texts we mention at the end of many chapters.

The prerequisites for the course are not demanding. We assume a sophisticated understanding of real numbers, including compactness arguments. We also assume some familiarity with concepts of linear algebra, but we include derivations of most results as a review. We have attempted to make the book self-contained. Solutions of many of the exercises are provided.

About the name: the term "numerical" analysis is fairly recent. A classic book [170] on the topic changed names between editions, adopting the "numerical analysis" title in a later edition [171]. The origins of the part of mathematics we now call analysis were all numerical, so for millennia the name "numerical analysis" would have been redundant. But analysis later developed conceptual (non-numerical) paradigms, and it became useful to specify the different areas by names.

There are many areas of analysis in addition to numerical, including complex, convex, functional, harmonic, and real. Some areas, which might have been given such a name, have their own names (such as probability, instead of random analysis). There is not a line of demarcation between the different areas of analysis. For example, much of harmonic analysis might be characterized as real or complex analysis, with functional analysis playing a role in modern theories. The same is true of numerical analysis, and it can be viewed in part as providing motivation for further study in all areas of analysis.

The subject of numerical analysis has ancient roots, and it has had periods of intense development followed by long periods of consolidation. In many cases, the new developments have coincided with the introduction of new forms of computing machines. For example, many of the basic theorems about computing solutions of ordinary differential equations were proved soon after desktop adding machines became common at the turn of the 20th century. The emergence of the digital computer in the mid-20th century spurred interest in solving partial differential equations and large systems of linear equations, as well as many other topics. The advent of parallel com-

---

[2]Robert Lee Moore (1882–1974) was born in Dallas, Texas, and did undergraduate work at the University of Texas in Austin where he took courses from L. E. Dickson. He got his Ph.D. in 1905 at the University of Chicago, studying with E. H. Moore and Oswald Veblen, and eventually returned to Austin where he continued to teach until his 87th year.

puters similarly stimulated research on new classes of algorithms. However, many fundamental questions remain open, and the subject is an active area of research today.

All of analysis is about evaluating limits. In this sense, it is about infinite objects, unlike, say, some parts of algebra or discrete mathematics. Often a key step is to provide uniform bounds on infinite objects, such as operators on vector spaces. In numerical analysis, the infinite objects are often sets of algorithms which are themselves finite in every instance. The objective is often to show that the algorithms are well-behaved uniformly and provide, in some limit, predictable results.

In numerical analysis there is sometimes a cultural divide between courses that emphasize theory and ones that emphasize computation. Ideally, both should be intertwined, as numerical analysis could well be called computational analysis because it is the analysis of computational algorithms involving real numbers. We present many computational algorithms and encourage computational exploration. However, we do not address the subject of software development (a.k.a., programming). Strictly speaking, programming is not required to appreciate the material in the book. However, we encourage mathematics students to develop some experience in this direction, as writing a computer program is quite similar to proving a theorem. Computer systems are quite adept at finding flaws in one's reasoning, and the organization required to make software readable provides a useful model to follow in making complex mathematical arguments understandable to others.

There are several important groups this text can serve. It is very common today for people in many fields to study mathematics through the beginning of real analysis, as might be characterized by the extremely popular "little Rudin" book [141]. Our book is intended to be at a comparable level of difficulty with little Rudin and can provide valuable reinforcement of the ideas and techniques covered there by applying them in a new domain. In this way, it is easily accessible to advanced undergraduates. It provides an option to study more analysis without raising the level of difficulty as occurs in a graduate course on measure theory.

People who go on to graduate work with a substantial computational component often need to progress further in analysis, including a study of measure theory and the Lebesgue integral. This is often done in a course at the "big Rudin" [142] level. Although the direct progression from little to big Rudin is a natural one, this book provides a way to interpolate between these levels while at the same time introducing ideas not found in [141] or [142] (or comparable texts [108, 121]). Thus the book is also appropriate as a course for graduate students interested in computational mathematics but with a background in analysis only at the level of [141].

We have included quotes at the beginning of each chapter and frequent footnotes giving historical information. These are intended to be entertaining and perhaps provocative, but no attempt has been made to be historically complete. However, we give references to several works on the history of mathematics that we recommend for a more complete picture. We indicate several connections among various mathematicians to give a sense of the personal interactions of the era. We use the terms "student" and "advisor" to describe general mentoring relationships which were sometimes different from what the terms might connote today. Although this may not be historically accurate in some cases, it would be tedious to use more precise terms to describe the relationships in each case in the various periods. We have used the MacTutor History of Mathematics archive extensively as an initial source of information but have also endeavored to refer to archival literature whenever possible.

In practice, numerical computation remains as much an art as it is a science. We focus on the part of the subject that is a science. A continuing challenge of current research is to transform numerical art into numerical analysis, as well as extending the power and reach of the art of numerical computation. Recent decades have witnessed a dramatic improvement in our understanding of many topics in numerical computation, and there is reason to expect that this trend will continue. Techniques that are supported only by heuristics tend to lose favor over time to ones that are understood rigorously. One of the great joys of the subject is when a heuristic idea succumbs to a rigorous analysis that reveals its secrets and extends its influence. It is hoped that this book will attract some new participants in this process.

Acknowledgments

I have gotten suggestions from many people regarding topics in this book, and my memory is not to be trusted to remember all of them. However, above all, Todd Dupont provided the most input regarding the book, including draft material, suggestions for additional topics, exercises, and overall conceptual advice. He regularly attended the fall 2009 class at the University of Chicago in which the book was given a trial run. I also thank all the students from that class for their influence on the final version.

Randy Bank, Carl de Boor and Nick Trefethen suggested novel approaches to particular topics. Although I cannot claim that I did exactly what they intended, their suggestions did influence what was presented in a substantial way.

Numerical Analysis

Chapter One

Numerical Algorithms

> The word "algorithm" derives from the name of the Persian mathematician (Abu Ja'far Muhammad ibn Musa) Al-Khwarizmi who lived from about 790 CE to about 840 CE. He wrote a book, *Hisab al-jabr w'al-muqabala*, that also named the subject "algebra."

Numerical analysis is the subject which studies algorithms for computing expressions defined with real numbers. The square-root $\sqrt{y}$ is an example of such an expression; we evaluate this today on a calculator or in a computer program as if it were as simple as y^2. It is numerical analysis that has made this possible, and we will study how this is done. But in doing so, we will see that the same approach applies broadly to include functions that cannot be named, and it even changes the nature of fundamental questions in mathematics, such as the impossibility of finding expressions for roots of order higher than 4.

There are two different phases to address in numerical analysis:

- the development of algorithms and

- the analysis of algorithms.

These are in principle independent activities, but in reality the development of an algorithm is often guided by the analysis of the algorithm, or of a simpler algorithm that computes the same thing or something similar.

There are three characteristics of algorithms using real numbers that are in conflict to some extent:

- the *accuracy* (or *consistency*) of the algorithm,

- the *stability* of the algorithm, and

- the effects of finite-precision arithmetic (a.k.a. round-off error).

The first of these just means that the algorithm approximates the desired quantity to any required accuracy under suitable restrictions. The second means that the behavior of the algorithm is continuous with respect to the parameters of the algorithm. The third topic is still not well understood at the most basic level, in the sense that there is not a well-established mathematical model for finite-precision arithmetic. Instead, we are forced to use crude upper bounds for the behavior of finite-precision arithmetic that often lead to overly pessimistic predictions about its effects in actual computations.

We will see that in trying to improve the accuracy or efficiency of a stable algorithm, one is often led to consider algorithms that turn out to be unstable and therefore of minimal (if any) value. These various aspects of numerical analysis are often intertwined, as ultimately we want an algorithm that we can analyze rigorously to ensure it is effective when using computer arithmetic.

The *efficiency* of an algorithm is a more complicated concept but is often the bottom line in choosing one algorithm over another. It can be related to all of the above characteristics, as well as to the *complexity* of the algorithm in terms of computational work or memory references required in its implementation.

Another central theme in numerical analysis is *adaptivity*. This means that the computational algorithm adapts itself to the data of the problem being solved as a way to improve efficiency and/or stability. Some adaptive algorithms are quite remarkable in their ability to elicit information automatically about a problem that is required for more efficient solution.

We begin with a problem from antiquity to illustrate each of these components of numerical analysis in an elementary context. We will not always disentangle the different issues, but we hope that the differing components will be evident.

1.1 FINDING ROOTS

People have been computing roots for millennia. Evidence exists [64] that the Babylonians, who used base-60 arithmetic, were able to approximate

$$\sqrt{2} \approx 1 + \frac{24}{60} + \frac{51}{60^2} + \frac{10}{60^3} \tag{1.1}$$

nearly 4000 years ago. By the time of Heron[1] a method to compute square-roots was established [26] that we recognize now as the Newton-Raphson-Simpson method (see section 2.2.1) and takes the form of a repeated iteration

$$x \leftarrow \tfrac{1}{2}(x + y/x), \tag{1.2}$$

where the backwards arrow $\leftarrow$ means *assignment* in algorithms. That is, once the computation of the expression on the right-hand side of the arrow has been completed, a new value is assigned to the variable x. Once that assignment is completed, the computation on the right-hand side can be redone with the new x.

The algorithm (1.2) is an example of what is known as *fixed-point iteration*, in which one hopes to find a fixed point, that is, an x where the iteration quits changing. A *fixed point* is thus a point x where

$$x = \tfrac{1}{2}(x + y/x). \tag{1.3}$$

---

[1] A.k.a. Hero, of Alexandria, who lived in the 1st century CE.

More precisely, x is a fixed point $x = f(x)$ of the function

$$f(x) = \tfrac{1}{2}(x + y/x), \tag{1.4}$$

defined, say, for $x \neq 0$. If we rearrange terms in (1.3), we find $x = y/x$, or $x^2 = y$. Thus a fixed point as defined in (1.3) is a solution of $x^2 = y$, so that $x = \pm\sqrt{y}$.

To describe actual implementations of these algorithms, we choose the scripting syntax implemented in the system *octave*. As a programming language, this has some limitations, but its use is extremely widespread. In addition to the public domain implementation of octave, a commercial interpreter (which predates octave) called Matlab is available. However, all computations presented here were done in octave.

We can implement (1.2) in octave in two steps as follows. First, we define the function (1.4) via the code

```
function x=heron(x,y)
x=.5*(x+y/x);
```

To use this function, you need to start with some initial guess, say, $x = 1$, which is written simply as

```
x=1
```

(Writing an expression with and without a semicolon at the end controls whether the interpreter prints the result or not.) But then you simply iterate:

```
x=heron(x,y)
```

until x (or the part you care about) quits changing. The results of doing so are given in table 1.1.

We can examine the accuracy by a simple code

```
function x=errheron(x,y)
for i=1:5
    x=heron(x,y);
    errheron=x-sqrt(y)
end
```

We show in table 1.1 the results of these computations in the case $y = 2$. This algorithm seems to "home in" on the solution. We will see that the accuracy doubles at each step.

1.1.1 Relative versus absolute error

We can require the accuracy of an algorithm to be based on the size of the answer. For example, we might want the approximation $\hat{x}$ of a root x to be small relative to the size of x:

$$\frac{\hat{x}}{x} = 1 + \delta, \tag{1.5}$$

$\sqrt{2}$ approximation	absolute error
1.50000000000000	8.5786e-02
1.41666666666667	2.4531e-03
1.41421568627451	2.1239e-06
1.41421356237469	1.5947e-12
1.41421356237309	-2.2204e-16

Table 1.1 Results of experiments with the Heron algorithm applied to approximate $\sqrt{2}$ using the algorithm (1.2) starting with $x = 1$. The boldface indicates the leading incorrect digit. Note that the number of correct digits essentially doubles at each step.

where δ satisfies some fixed tolerance, e.g., $|\delta| \leq \epsilon$. Such a requirement is in keeping with the model we will adopt for floating-point operations (see (1.31) and section 18.1).

    We can examine the relative accuracy by the simple code

```
function x=relerrher(x,y)
for i=1:6
    x=heron(x,y);
    errheron=(x/sqrt(y))-1
end
```

We leave as exercise 1.2 comparison of the results produced by the above code `relerrher` with the absolute errors presented in table 1.1.

1.1.2 Scaling Heron's algorithm

Before we analyze how Heron's algorithm (1.2) works, let us enhance it by a prescaling. To begin with, we can suppose that the number y whose square root we seek lies in the interval $[\frac{1}{2}, 2]$. If $y < \frac{1}{2}$ or $y > 2$, then we make the transformation

$$\tilde{y} = 4^k y \qquad (1.6)$$

to get $\tilde{y} \in [\frac{1}{2}, 2]$, for some integer k. And of course $\sqrt{\tilde{y}} = 2^k \sqrt{y}$. By scaling y in this way, we limit the range of inputs that the algorithm must deal with.

    In table 1.1, we showed the absolute error for approximating $\sqrt{2}$, and in exercise 1.2 the relative errors for approximating $\sqrt{2}$ and $\sqrt{\frac{1}{2}}$ are explored. It turns out that the maximum errors for the interval $[\frac{1}{2}, 2]$ occur at the ends of the interval (exercise 1.3). Thus five iterations of Heron, preceded by the scaling (1.6), are sufficient to compute $\sqrt{y}$ to 16 decimal places.

    Scaling provides a simple example of adaptivity for algorithms for finding roots. Without scaling, the global performance (section 1.2.2) would be quite different.

1.2 ANALYZING HERON'S ALGORITHM

As the name implies, a major objective of numerical analysis is to analyze
the behavior of algorithms such as Heron's iteration (1.2). There are two
questions one can ask in this regard. First, we may be interested in the local
behavior of the algorithm assuming that we have a reasonable start near
the desired root. We will see that this can be done quite completely, both
in the case of Heron's iteration and in general for algorithms of this type
(in chapter 2). Second, we may wonder about the global behavior of the
algorithm, that is, how it will respond with arbitrary starting points. With
the Heron algorithm we can give a fairly complete answer, but in general
it is more complicated. Our point of view is that the global behavior is
really a different subject, e.g., a study in dynamical systems. We will see
that techniques like scaling (section 1.1.2) provide a basis to turn the local
analysis into a convergence theory.

1.2.1 Local error analysis

Since Heron's iteration (1.2) is recursive in nature, it it natural to expect that
the errors can be expressed recursively as well. We can write an algebraic
expression for Heron's iteration (1.2) linking the error at one iteration to the
error at the next. Thus define

$$x_{n+1} = \tfrac{1}{2}(x_n + y/x_n), \tag{1.7}$$

and let $e_n = x_n - x = x_n - \sqrt{y}$. Then by (1.7) and (1.3),

$$
\begin{aligned}
e_{n+1} = x_{n+1} - x &= \tfrac{1}{2}(x_n + y/x_n) - \tfrac{1}{2}(x + y/x) \\
&= \tfrac{1}{2}(e_n + y/x_n - y/x) = \tfrac{1}{2}\left(e_n + \frac{y(x - x_n)}{xx_n}\right) \\
&= \tfrac{1}{2}\left(e_n - \frac{xe_n}{x_n}\right) = \tfrac{1}{2}e_n\left(1 - \frac{x}{x_n}\right) = \tfrac{1}{2}\frac{e_n^2}{x_n}.
\end{aligned}
\tag{1.8}
$$

If we are interested in the relative error,

$$\hat{e}_n = \frac{e_n}{x} = \frac{x_n - x}{x} = \frac{x_n}{x} - 1, \tag{1.9}$$

then (1.8) becomes

$$\hat{e}_{n+1} = \tfrac{1}{2}\frac{x\hat{e}_n^2}{x_n} = \tfrac{1}{2}\left(1 + \hat{e}_n\right)^{-1}\hat{e}_n^2. \tag{1.10}$$

Thus we see that

> *the error at each step is proportional to*
> *the square of the error at the previous step;*

for the relative error, the constant of proportionality tends rapidly to $\tfrac{1}{2}$. In
(2.20), we will see that this same result can be derived by a general technique.

1.2.2 Global error analysis

In addition, (1.10) implies a limited type of *global convergence* property, at least for $x_n > x = \sqrt{y}$. In that case, (1.10) gives

$$|\hat{e}_{n+1}| = \tfrac{1}{2}\frac{\hat{e}_n^2}{|1 + \hat{e}_n|} = \tfrac{1}{2}\frac{\hat{e}_n^2}{1 + \hat{e}_n} \le \tfrac{1}{2}\hat{e}_n. \tag{1.11}$$

Thus the relative error is reduced by a factor smaller than $\tfrac{1}{2}$ at each iteration, no matter how large the initial error may be. Unfortunately, this type of global convergence property does not hold for many algorithms. We can illustrate what can go wrong in the case of the Heron algorithm when $x_n < x = \sqrt{y}$.

Suppose for simplicity that $y = 1$, so that also $x = 1$, so that the relative error is $\hat{e}_n = x_n - 1$, and therefore (1.10) implies that

$$\hat{e}_{n+1} = \tfrac{1}{2}\frac{(1 - x_n)^2}{x_n}. \tag{1.12}$$

As $x_n \to 0$, $\hat{e}_{n+1} \to \infty$, even though $|\hat{e}_n| < 1$. Therefore, convergence is not truly global for the Heron algorithm.

What happens if we start with x_0 near zero? We obtain x_1 near ∞. From then on, the iterations satisfy $x_n > \sqrt{y}$, so the iteration is ultimately convergent. But the number of iterations required to reduce the error below a fixed error tolerance can be arbitrarily large depending on how small x_0 is. By the same token, we cannot bound the number of required iterations for arbitrarily large x_0. Fortunately, we will see that it is possible to choose good starting values for Heron's method to avoid this potential bad behavior.

1.3 WHERE TO START

With any iterative algorithm, we have to start the iteration somewhere, and this choice can be an interesting problem in its own right. Just like the initial scaling described in section 1.1.2, this can affect the performance of the overall algorithm substantially.

For the Heron algorithm, there are various possibilities. The simplest is just to take $x_0 = 1$, in which case

$$\hat{e}_0 = \frac{1}{x} - 1 = \frac{1}{\sqrt{y}} - 1. \tag{1.13}$$

This gives

$$\hat{e}_1 = \tfrac{1}{2}x\hat{e}_0^2 = \tfrac{1}{2}x\left(\frac{1}{x} - 1\right)^2 = \tfrac{1}{2}\frac{(x - 1)^2}{x}. \tag{1.14}$$

We can use (1.14) as a formula for $\hat{e}_1$ as a function of x (it is by definition a function of $y = x^2$); then we see that

$$\hat{e}_1(x) = \hat{e}_1(1/x) \tag{1.15}$$

by comparing the rightmost two terms in (1.14). Note that the maximum of $\hat{e}_1(x)$ on $[2^{-1/2}, 2^{1/2}]$ occurs at the ends of the interval, and

$$\hat{e}_1(\sqrt{2}) = \tfrac{1}{2}\frac{(\sqrt{2}-1)^2}{\sqrt{2}} = \tfrac{3}{4}\sqrt{2} - 1 \approx 0.060660. \qquad (1.16)$$

Thus the simple starting value $x_0 = 1$ is remarkably effective. Nevertheless, let us see if we can do better.

1.3.1 Another start

Another idea to start the iteration is to make an approximation to the square-root function given the fact that we always have $y \in [\tfrac{1}{2}, 2]$ (section 1.1.2). Since this means that y is near 1, we can write $y = 1 + t$ (i.e., $t = y - 1$), and we have

$$\begin{aligned} x = \sqrt{y} = \sqrt{1+t} &= 1 + \tfrac{1}{2}t + \mathcal{O}(t^2) \\ &= 1 + \tfrac{1}{2}(y-1) + \mathcal{O}(t^2) = \tfrac{1}{2}(y+1) + \mathcal{O}(t^2). \end{aligned} \qquad (1.17)$$

Thus we get the approximation $x \approx \tfrac{1}{2}(y+1)$ as a possible starting guess:

$$x_0 = \tfrac{1}{2}(y+1). \qquad (1.18)$$

But this is the same as x_1 if we had started with $x_0 = 1$. Thus we have not really found anything new.

1.3.2 The best start

Our first attempt (1.18) based on a linear approximation to the square-root did not produce a new concept since it gives the same result as starting with a constant guess after one iteration. The approximation (1.18) corresponds to the tangent line of the graph of $\sqrt{y}$ at $y = 1$, but this may not be the best affine approximation to a function on an interval. So let us ask the question, What is the best approximation to $\sqrt{y}$ on the interval $[\tfrac{1}{2}, 2]$ by a linear polynomial? This problem is a miniature of the questions we will address in chapter 12.

The general linear polynomial is of the form

$$f(y) = a + by. \qquad (1.19)$$

If we take $x_0 = f(y)$, then the relative error $\hat{e}_0 = \hat{e}_0(y)$ is

$$\hat{e}_0(y) = \frac{x_0 - \sqrt{y}}{\sqrt{y}} = \frac{a + by - \sqrt{y}}{\sqrt{y}} = \frac{a}{\sqrt{y}} + b\sqrt{y} - 1. \qquad (1.20)$$

Let us write $e_{ab}(y) = \hat{e}_0(y)$ to be precise. We seek a and b such that the maximum of $|e_{ab}(y)|$ over $y \in [\tfrac{1}{2}, 2]$ is minimized.

Fortunately, the functions

$$e_{ab}(y) = \frac{a}{\sqrt{y}} + b\sqrt{y} - 1 \qquad (1.21)$$

have a simple structure. As always, it is helpful to compute the derivative:

$$e'_{ab}(y) = -\tfrac{1}{2}ay^{-3/2} + \tfrac{1}{2}by^{-1/2} = \tfrac{1}{2}(-a + by)y^{-3/2}. \qquad (1.22)$$

Thus $e'_{ab}(y) = 0$ for $y = a/b$; further, $e'_{ab}(y) > 0$ for $y > a/b$, and $e'_{ab}(y) < 0$ for $y < a/b$. Therefore, e_{ab} has a minimum at $y = a/b$ and is strictly increasing as we move away from that point in either direction. Thus we have proved that

$$\min e_{ab} = \min e_{ba} = e_{ab}(a/b) = 2\sqrt{ab} - 1. \qquad (1.23)$$

Thus the maximum values of $|e_{ab}|$ on $[\frac{1}{2}, 2]$ will be at the ends of the interval or at $y = a/b$ if $a/b \in [\frac{1}{2}, 2]$. Moreover, the best value of $e_{ab}(a/b)$ will be negative (exercise 1.10). Thus we consider the three values

$$e_{ab}(2) = \frac{a}{\sqrt{2}} + b\sqrt{2} - 1$$

$$e_{ab}(\tfrac{1}{2}) = a\sqrt{2} + \frac{b}{\sqrt{2}} - 1 \qquad (1.24)$$

$$-e_{ab}(a/b) = 1 - 2\sqrt{ab}.$$

Note that $e_{ab}(2) = e_{ba}(1/2)$. Therefore, the optimal values of a and b must be the same: $a = b$ (exercise 1.11). Moreover, the minimum value of e_{ab} must be minus the maximum value on the interval (exercise 1.12). Thus the optimal value of $a = b$ is characterized by

$$a\tfrac{3}{2}\sqrt{2} - 1 = 1 - 2a \implies a = \left(\tfrac{3}{4}\sqrt{2} + 1\right)^{-1}. \qquad (1.25)$$

Recall that the simple idea of starting the Heron algorithm with $x_0 = 1$ yielded an error

$$|\hat{e}_1| \le \gamma = \tfrac{3}{4}\sqrt{2} - 1, \qquad (1.26)$$

and that this was equivalent to choosing $a = \frac{1}{2}$ in the current scheme. Note that the optimal $a = 1/(\gamma + 2)$, only slightly less than $\frac{1}{2}$, and the resulting minimum value of the maximum of $|e_{aa}|$ is

$$1 - 2a = 1 - \frac{2}{\gamma + 2} = \frac{\gamma}{\gamma + 2}. \qquad (1.27)$$

Thus the optimal value of a reduces the previous error of γ (for $a = \frac{1}{2}$) by nearly a factor of $\frac{1}{2}$, despite the fact that the change in a is quite small. The benefit of using the better initial guess is of course squared at each iteration, so the reduced error is nearly smaller by a factor of 2^{-2^k} after k iterations of Heron. We leave as exercise 1.13 the investigation of the effect of using this optimal starting place in the Heron algorithm.

1.4 AN UNSTABLE ALGORITHM

Heron's algorithm has one drawback in that it requires division. One can imagine that a simpler algorithm might be possible such as

$$x \leftarrow x + x^2 - y. \qquad (1.28)$$

n	0	1	2	3	4	5
x_n	1.5	1.75	2.81	8.72	82.8	6937.9
n	6	7	8	9	10	11
x_n	5×10^7	2×10^{15}	5×10^{30}	3×10^{61}	8×10^{122}	7×10^{245}

Table 1.2 Unstable behavior of the iteration (1.28) for computing $\sqrt{2}$.

Before experimenting with this algorithm, we note that a fixed point

$$x = x + x^2 - y \qquad (1.29)$$

does have the property that $x^2 = y$, as desired. Thus we can assert the *accuracy* of the algorithm (1.28), in the sense that any fixed point will solve the desired problem. However, it is easy to see that the algorithm is not *stable*, in the sense that if we start with an initial guess with any sort of error, the algorithm fails. table 1.2 shows the results of applying (1.28) starting with $x_0 = 1.5$. What we see is a rapid movement *away* from the solution, followed by a catastrophic blowup (which eventually causes failure in a fixed-precision arithmetic system, or causes the computer to run out of memory in a variable-precision system). The error is again being squared, as with the Heron algorithm, but since the error is getting bigger rather than smaller, the algorithm is useless. In section 2.1 we will see how to diagnose instability (or rather how to guarantee stability) for iterations like (1.28).

1.5 GENERAL ROOTS: EFFECTS OF FLOATING-POINT

So far, we have seen no adverse effects related to finite-precision arithmetic. This is common for (stable) iterative methods like the Heron algorithm. But now we consider a more complex problem in which rounding plays a dominant role.

Suppose we want to compute the roots of a general quadratic equation $x^2 + 2bx + c = 0$, where $b < 0$, and we chose the algorithm

$$x \leftarrow -b + \sqrt{b^2 - c}. \qquad (1.30)$$

Note that we have assumed that we can compute the square-root function as part of this algorithm, say, by Heron's method.

Unfortunately, the simple algorithm in (1.30) fails if we have $c = \epsilon^2 b^2$ (it returns $x = 0$) as soon as $\epsilon^2 = c/b^2$ is small enough that the floating-point representation of $1 - \epsilon^2$ is 1. For any (fixed) finite representation of real numbers, this will occur for some $\epsilon > 0$.

We will consider floating-point arithmetic in more detail in section 18.1, but the simple model we adopt says that the result of computing a binary operator $\oplus$ such as $+$, $-$, $/$, or $*$ has the property that

$$f\ell(a \oplus b) = (a \oplus b)(1 + \delta), \qquad (1.31)$$

where $|\delta| \le \epsilon$, where $\epsilon > 0$ is a parameter of the model.[2] However, this means that a collection of operations could lead to catastrophic cancellation, e.g., $f\ell(f\ell(1 + \frac{1}{2}\epsilon) - 1) = 0$ and not $\frac{1}{2}\epsilon$.

We can see the behavior in some simple codes. But first, let us simplify the problem further so that we have just one parameter to deal with. Suppose that the equation to be solved is of the form

$$x^2 - 2bx + 1 = 0. \qquad (1.32)$$

That is, we switch b to $-b$ and set $c = 1$. In this case, the two roots are multiplicative inverses of each other. Define

$$x_\pm = b \pm \sqrt{b^2 - 1}. \qquad (1.33)$$

Then $x_- = 1/x_+$.

There are various possible algorithms. We could use one of the two formulas $x_\pm = b \pm \sqrt{b^2 - 1}$ directly. More precisely, let us write $\tilde{x}_\pm \approx b \pm \sqrt{b^2 - 1}$ to indicate that we implement this in floating-point. Correspondingly, there is another pair of algorithms that start by computing $\tilde{x}_\mp$ and then define, say, $\hat{x}_+ \approx 1/\tilde{x}_-$. A similar algorithm could determine $\hat{x}_- \approx 1/\tilde{x}_+$.

All four of these algorithms will have different behaviors. We expect that the behaviors of the algorithms for computing $\tilde{x}_-$ and $\hat{x}_-$ will be dual in some way to those for computing $\tilde{x}_+$ and $\hat{x}_+$, so we consider only the first pair.

First, the function minus implements the $\tilde{x}_-$ square-root algorithm:

```
function x=minus(b)
% solving = 1-2bx +x^2
x=b-sqrt(b^2-1);
```

To know if it is getting the right answer, we need another function to check the answer:

```
function error=check(b,x)
error = 1-2*b*x +x^2;
```

To automate the process, we put the two together:

```
function error=chekminus(b)
x=minus(b);
error=check(b,x)
```

For example, when $b = 10^6$, we find the error is -7.6×10^{-6}. As b increases further, the error increases, ultimately leading to complete nonsense. For this reason, we consider an alternative algorithm suitable for large b.

The algorithm for $\hat{x}_-$ is given by

---

[2]The notation $f\ell$ is somewhat informal. It would be more precise to write $a \,\widehat{\oplus}\, b$ instead of $f\ell(a \oplus b)$ since the operator is modified by the effect of rounding.

```
function x=plusinv(b)
% solving = 1-2bx +x^2
y=b+sqrt(b^2-1);
x=1/y;
```

Similarly, we can check the accuracy of this computation by the code

```
function error=chekplusinv(b)
x=plusinv(b);
error=check(b,x)
```

Now when $b = 10^6$, we find the error is -2.2×10^{-17}. And the bigger b becomes, the more accurate it becomes.

Here we have seen that algorithms can have data-dependent behavior with regard to the effects of finite-precision arithmetic. We will see that there are many algorithms in numerical analysis with this property, but suitable analysis will establish conditions on the data that guarantee success.

1.6 EXERCISES

Exercise 1.1 *How accurate is the approximation (1.1) if it is expressed as a decimal approximation (how many digits are correct)?*

Exercise 1.2 *Run the code* `relerrher` *starting with $x = 1$ and $y = 2$ to approximate $\sqrt{2}$. Compare the results with table 1.1. Also run the code with $x = 1$ and $y = \frac{1}{2}$ and compare the results with the previous case. Explain what you find.*

Exercise 1.3 *Show that the maximum relative error in Heron's algorithm for approximating $\sqrt{y}$ for $y \in [1/M, M]$, for a fixed number of iterations and starting with $x_0 = 1$, occurs at the ends of the interval: $y = 1/M$ and $y = M$. (Hint: consider (1.10) and (1.14) and show that the function*

$$\phi(x) = \tfrac{1}{2}(1 + x)^{-1}x^2 \tag{1.34}$$

plays a role in each. Show that ϕ is increasing on the interval $[0, \infty[.)$

Exercise 1.4 *It is sometimes easier to demonstrate the relative accuracy of an approximation $\hat{x}$ to x by showing that*

$$|x - \hat{x}| \leq \epsilon'|\hat{x}| \tag{1.35}$$

instead of verifying (1.5) directly. Show that if (1.35) holds, then (1.5) holds with $\epsilon = \epsilon'/(1 - \epsilon')$.

Exercise 1.5 *There is a simple generalization to Heron's algorithm for finding kth roots as follows:*

$$x \leftarrow \frac{1}{k}((k - 1)x + y/x^{k-1}). \tag{1.36}$$

Show that, if this converges, it converges to a solution of $x^k = y$. Examine the speed of convergence both computationally and by estimating the error algebraically.

Exercise 1.6 *Show that the error in Heron's algorithm for approximating* $\sqrt{y}$ *satisfies*

$$\frac{x_n - \sqrt{y}}{x_n + \sqrt{y}} = \left(\frac{x_0 - \sqrt{y}}{x_0 + \sqrt{y}}\right)^{2^n} \tag{1.37}$$

for $n \geq 1$. *Note that the denominator on the left-hand side of (1.37) converges rapidly to* $2\sqrt{y}$.

Exercise 1.7 *We have implicitly been assuming that we were attempting to compute a positive square-root with Heron's algorithm, and thus we always started with a positive initial guess. If we give zero as an initial guess, there is immediate failure because of division by zero. But what happens if we start with a negative initial guess? (Hint: there are usually two roots to* $x^2 = y$, *one of which is negative.)*

Exercise 1.8 *Consider the iteration*

$$x \leftarrow 2x - yx^2 \tag{1.38}$$

and show that, if it converges, it converges to $x = 1/y$. *Note that the algorithm does not require a division. Determine the range of starting values* x_0 *for which this will converge. What sort of scaling (cf. section 1.1.2) would be appropriate for computing* $1/y$ *before starting the iteration?*

Exercise 1.9 *Consider the iteration*

$$x \leftarrow \tfrac{3}{2}x - \tfrac{1}{2}yx^3 \tag{1.39}$$

and show that, if this converges, it converges to $x = 1/\sqrt{y}$. *Note that this algorithm does not require a division. The computation of* $1/\sqrt{y}$ *appears in the Cholesky algorithm in (4.12).*

Exercise 1.10 *Suppose that* $a + by$ *is the best linear approximation to* $\sqrt{y}$ *in terms of relative error on* $[\tfrac{1}{2}, 2]$. *Prove that the error expression* e_{ab} *has to be negative at its minimum. (Hint: if not, you can always decrease* a *to make* $e_{ab}(2)$ *and* $e_{ab}(\tfrac{1}{2})$ *smaller without increasing the maximum value of* $|e_{ab}|$.)

Exercise 1.11 *Suppose that* $a + by$ *is the best linear approximation to* $\sqrt{y}$ *in terms of relative error on* $[\tfrac{1}{2}, 2]$. *Prove that* $a = b$.

Exercise 1.12 *Suppose that* $a + ay$ *is the best linear approximation to* $\sqrt{y}$ *in terms of relative error on* $[\tfrac{1}{2}, 2]$. *Prove that the error expression*

$$e_{aa}(1) = -e_{aa}(2). \tag{1.40}$$

(Hint: if not, you can always decrease a *to make* $e_{aa}(2)$ *and* $e_{aa}(\tfrac{1}{2})$ *smaller without increasing the maximum value of* $|e_{ab}|$.)

Exercise 1.13 *Consider the effect of the best starting value of* a *in (1.25) on the Heron algorithm. How many iterations are required to get 16 digits of accuracy? And to obtain 32 digits of accuracy?*

Exercise 1.14 *Change the function* minus *for computing* $\tilde{x}_-$ *and the function* plusinv *for computing* $\hat{x}_-$ *to functions for computing* $\tilde{x}_+$ *(call that function* plus*) and* $\hat{x}_+$ *(call that function* minusinv*). Use the* check *function to see where they work well and where they fail. Compare that with the corresponding behavior for* minus *and* plusinv.

Exercise 1.15 *The iteration (1.28) can be implemented via the function*

```
function y =sosimpl(x,a)
y=x+x^2-a;
```

Use this to verify that sosimpl(1,1) *is indeed 1, but if we start with*

```
x=1.000000000001
```

and then repeatedly apply x=sosimpl(x,1), *the result ultimately diverges.*

1.7 SOLUTIONS

Solution of Exercise 1.3. The function $\phi(x) = \frac{1}{2}(1+x)^{-1}x^2$ is increasing on the interval $[0, \infty[$ since

$$\phi'(x) = \frac{1}{2}\frac{2x(1+x) - x^2}{(1+x)^2} = \frac{1}{2}\frac{2x + x^2}{(1+x)^2} > 0 \tag{1.41}$$

for $x > 0$. The expression (1.10) says that

$$\hat{e}_{n+1} = \phi(\hat{e}_n), \tag{1.42}$$

and (1.14) says that

$$\hat{e}_1 = \phi(x - 1). \tag{1.43}$$

Thus

$$\hat{e}_2 = \phi(\phi(x - 1)). \tag{1.44}$$

By induction, define

$$\phi^{[n+1]}(t) = \phi(\phi^{[n]}(t)), \tag{1.45}$$

where $\phi^{[1]}(t) = \phi(t)$ for all t. Then, by induction,

$$\hat{e}_n = \phi^{[n]}(x - 1) \tag{1.46}$$

for all $n \geq 1$. Since the composition of increasing functions is increasing, each $\phi^{[n]}$ is increasing, by induction. Thus $\hat{e}_n$ is maximized when x is maximized, at least for $x > 1$. Note that

$$\phi(x - 1) = \phi((1/x) - 1), \tag{1.47}$$

so we may also write

$$\hat{e}_n = \phi^{[n]}((1/x) - 1). \tag{1.48}$$

Thus the error is symmetric via the relation

$$\hat{e}_n(x) = \hat{e}_n(1/x).$$ (1.49)

Thus the maximal error on an interval $[1/M, M]$ occurs simultaneously at $1/M$ and M.

Solution of Exercise 1.6. Define $d_n = x_n + x$. Then (1.37) in exercise 1.6 is equivalent to the statement that

$$\frac{e_n}{d_n} = \left(\frac{e_0}{d_0}\right)^{2^n}.$$ (1.50)

Thus we compute

$$d_{n+1} = x_{n+1} + x = \tfrac{1}{2}(x_n + y/x_n) + \tfrac{1}{2}(x + y/x) = \tfrac{1}{2}(d_n + y/x_n + y/x)$$

$$= \tfrac{1}{2}\left(d_n + \frac{y(x + x_n)}{xx_n}\right) = \tfrac{1}{2}\left(d_n + \frac{yd_n}{xx_n}\right) = \tfrac{1}{2}\left(d_n + \frac{xd_n}{x_n}\right)$$

$$= \tfrac{1}{2}d_n\left(1 + \frac{x}{x_n}\right) = \tfrac{1}{2}d_n\left(\frac{x_n + x}{x_n}\right) = \tfrac{1}{2}\frac{d_n^2}{x_n}.$$ (1.51)

Recall that (1.8) says that $e_{n+1} = \tfrac{1}{2}e_n^2/x_n$, so dividing by (1.51) yields

$$\frac{e_{n+1}}{d_{n+1}} = \left(\frac{e_n}{d_n}\right)^2$$ (1.52)

for any $n \geq 0$. A simple induction on n yields (1.50), as required.

Chapter Two

Nonlinear Equations

"A method algebraically equivalent to Newton's method was known to the 12th century algebraist Sharaf al-Din al-Tusi ... and the 15th century Arabic mathematician Al-Kashi used a form of it in solving $x^p - N = 0$ to find roots of N" [174].

Kepler's discovery that the orbits of the planets are elliptical introduced a mathematical challenge via his equation

$$x - \mathcal{E}\sin x = \tau, \qquad (2.1)$$

which defines a function $\phi(\tau) = x$. Here $\mathcal{E} = \sqrt{1 - b^2/a^2}$ is the eccentricity of the elliptical orbit, where a and b are the major and minor axis lengths of the ellipse and τ is proportional to time. See figure 2.1 regarding the notation [148]. Much effort has been expended in trying to find a simple representation of this function ϕ, but we will see that it can be viewed as just like the square-root function from the numerical point of view. Newton[1] proposed an iterative solution to Kepler's equation [174]:

$$x_{n+1} = x_n + \frac{\tau - x_n + \mathcal{E}\sin x_n}{1 - \mathcal{E}\cos x_n}. \qquad (2.2)$$

We will see that this iteration can be viewed as a special case of a general iterative technique now known as Newton's method.

We will also see that the method introduced in (1.2) as Heron's method, namely,

$$x_{n+1} = \tfrac{1}{2}\left(x_n + \frac{y}{x_n} \right), \qquad (2.3)$$

can be viewed as Newton's method for computing $\sqrt{y}$. Newton's method provides a general paradigm for solving nonlinear equations iteratively and changes qualitatively the notion of "solution" for a problem. Thus we see that Kepler's equation (2.1) is itself the solution, just as if it had turned out that the function $\phi(\tau) = x$ was a familiar function like square root or logarithm. If we need a particular value of x for a given τ, then we know there is a machine available to produce it, just as in computing $\sqrt{y}$ on a calculator.

---

[1] Isaac Newton (1643–1727) was one of the greatest and best known scientists of all time, to the point of being a central figure in popular literature [150].

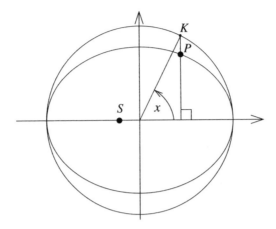

Figure 2.1 The notation for Kepler's equation. The sun is at S (one of the foci of the elliptical orbit), the planet is at P, and the point K lies on the indicated circle that encloses the ellipse of the orbit; the horizontal coordinates of P and K are the same, by definition. The angle x is between the principal axis of the ellipse and the point K.

First, we develop a general framework for iterative solution methods, and then we show how this leads to Newton's method and other iterative techniques. We begin with one equation in one variable and later extend to systems in chapter 7.

2.1 FIXED-POINT ITERATION

This goes by many names, including *functional iteration*, but we prefer the term *fixed-point iteration* because it seeks to find a fixed point

$$\alpha = g(\alpha) \tag{2.4}$$

for a continuous function g. Fixed-point iteration

$$x_{n+1} = g(x_n) \tag{2.5}$$

has the important property that, *if it converges, it converges to a fixed point* (2.4) (assuming only that g is continuous). This result is so simple (see exercise 2.1) that we hesitate to call it a theorem. But it is really the key fact about fixed-point iteration.

We now see that Heron's algorithm (2.3) may be written in this notation with

$$g(x) = \tfrac{1}{2}\left(x + \frac{y}{x}\right). \tag{2.6}$$

Similarly, the method (2.2) proposed by Newton to solve Kepler's equation (2.1) can be written as

$$g(x) = x + \frac{\tau - x + \mathcal{E}\sin x}{1 - \mathcal{E}\cos x}. \tag{2.7}$$

n	x_n from (2.8)	x_n from (2.2)
0	**1.00**	**1.00**
1	1.0**8**4147098480790	1.088**9**53263837373
2	1.088**3**90486229308	1.0885977**5**8269552
3	1.088588**1**38978555	1.088597752397894
4	1.088597**3**06592452	1.088597752397894
5	1.088597**7**31724630	1.088597752397894
6	1.0885977**5**1439216	1.088597752397894
7	1.08859775**2**353437	1.088597752397894
8	1.08859775239**5**832	1.088597752397894
9	1.0885977523977**9**8	1.088597752397894

Table 2.1 Computations of solutions to Kepler's equation (2.1) for $\mathcal{E} = 0.1$ and $\tau = 1$ via Newton's method (2.2) (third column) and by the fixed-point iteration (2.8). The boldface indicates the leading incorrect digit. Note that the number of correct digits essentially doubles at each step for Newton's method but increases only by about 1 at each step of the fixed-point iteration (2.8).

The choice of g is not at all unique. One could as well approximate the solution of Kepler's equation (2.1) via

$$g(x) = \tau + \mathcal{E} \sin x. \qquad (2.8)$$

In table 2.1, the methods (2.8) and (2.7) are compared. We find that Newton's method converges much faster, comparable to the way that Heron's method does, in that the number of correct digits doubles at each step.

The rest of the story about fixed-point iteration is then to figure out when and how fast it converges. For example, if g is Lipschitz[2]-continuous with constant $\lambda < 1$, that is,

$$|g(x) - g(y)| \leq \lambda |x - y|, \qquad (2.9)$$

then convergence will happen if we start close enough to α. This is easily proved by defining, as we did for Heron's method, $e_n = x_n - \alpha$ and estimating

$$|e_{n+1}| = |g(x_n) - g(\alpha)| \leq \lambda |e_n|, \qquad (2.10)$$

where the equality results from subtracting (2.4) from (2.5). Thus, by induction,

$$|e_n| \leq \lambda^n |e_0| \qquad (2.11)$$

for all $n \geq 1$. Thus we have proved the following.

Theorem 2.1 *Suppose that $\alpha = g(\alpha)$ and that the Lipschitz estimate (2.9) holds with $\lambda < 1$ for all $x, y \in [\alpha - A, \alpha + A]$ for some $A > 0$. Suppose that $|x_0 - \alpha| \leq A$. Then the fixed-point iteration defined in (2.5) converges according to (2.11).*

---

[2]Rudolf Otto Sigismund Lipschitz (1832–1903) had only one student, but that was Felix Klein.

Proof. The only small point to be sure about is that all the iterates stay in the interval $[\alpha - A, \alpha + A]$, but this follows from the estimate (2.11) once we know that $|e_0| \leq A$, as we have assumed.   QED

2.1.1 Verifying the Lipschitz condition

A Lipschitz-continuous function need not be C^1, but when a function is C^1, its derivative gives a good estimate of the Lipschitz constant. We formalize this simple result to highlight the idea.

Lemma 2.2 *Suppose $g \in C^1$ in an interval around an arbitrary point α. Then for any $\epsilon > 0$, there is an $A > 0$ such that g satisfies (2.9) in the interval $[\alpha - A, \alpha + A]$ with $\lambda \leq |g'(\alpha)| + \epsilon$.*

Proof. By the continuity of g', we can pick $A > 0$ such that $|g'(t) - g'(\alpha)| < \epsilon$ for all $t \in [\alpha - A, \alpha + A]$. Therefore,

$$|g'(t)| \leq |g'(\alpha)| + |g'(t) - g'(\alpha)| < |g'(\alpha)| + \epsilon \tag{2.12}$$

for all $t \in [\alpha - A, \alpha + A]$. Let $x, y \in [\alpha - A, \alpha + A]$, with $x \neq y$. Then

$$\left| \frac{g(x) - g(y)}{x - y} \right| = \left| \frac{1}{x - y} \int_y^x g'(t)\, dt \right|$$
$$\leq \max \left\{ |g'(t)| \mid t \in [\alpha - A, \alpha + A] \right\} \tag{2.13}$$
$$\leq |g'(\alpha)| + \epsilon,$$

by using (2.12).   QED

As a result, we conclude that the condition $|g'(\alpha)| < 1$ is sufficient to guarantee convergence of fixed-point iteration, as long as we start close enough to the root $\alpha = g(\alpha)$ (cf. exercise 2.2).

On the other hand, the Lipschitz constant λ in (2.9) also gives an upper bound for the derivative:

$$|g'(\alpha)| \leq \lambda \tag{2.14}$$

(cf. exercise 2.3). Thus if $|g'(\alpha)| > 1$, fixed-point iteration will likely not converge since the Lipschitz constant for g will be greater than 1 in any such interval. If we recall the iteration function $g(x) = x + x^2 - y$ in (1.28), we see that $g'(\sqrt{y}) = 1 + 2\sqrt{y} > 1$. Thus the divergence of that algorithm is not surprising.

It is not very useful to develop a general theory of divergence for fixed-point iteration, but we can clarify this by example. It is instructive to consider the simple case

$$g(x) := \alpha + \lambda(x - \alpha), \tag{2.15}$$

where for simplicity we take $\lambda > 0$. Then for all n we have

$$|x_n - \alpha| = |g(x_{n-1}) - \alpha| = \lambda |x_{n-1} - \alpha|, \tag{2.16}$$

and by induction

$$|x_n - \alpha| = \lambda^n |x_0 - \alpha|, \tag{2.17}$$

where x_0 is our starting value. If $\lambda < 1$, this converges, but if $\lambda > 1$, this diverges.

The affine example (2.15) not only gives an example of divergence when $|g'(\alpha)| > 1$ but also suggests the asymptotic behavior of fixed-point iteration. When $0 < |g'(\alpha)| < 1$, the asymptotic behavior of fixed-point iteration is (cf. exercise 2.4) given by

$$|x_n - \alpha| \approx C|g'(\alpha)|^n \tag{2.18}$$

as $n \to \infty$, where C is a constant that depends on g and the initial guess.

2.1.2 Second-order iterations

What happens if $g'(\alpha) = 0$? By Taylor's theorem,

$$g(x) - \alpha = \tfrac{1}{2}(x - \alpha)^2 g''(\xi) \tag{2.19}$$

for some ξ between x and α, and thus the error is squared at each iteration:

$$e^n = \tfrac{1}{2}(e^{n-1})^2 g''(\xi_n), \tag{2.20}$$

where $\xi_n \to \alpha$ if the iteration converges. Of course, squaring the error is not a good thing if it is too large initially (with regard to the size of g'').

We can now see why Heron's method converges so rapidly. Recall that $g(x) = \tfrac{1}{2}(x + y/x)$, so that $g'(x) = \tfrac{1}{2}(1 - y/x^2) = 0$ when $x^2 = y$. Moreover, $g''(x) = y/x^3$, so we could derive a result analogous to (1.8) from (2.20).

2.1.3 Higher-order iterations

It is possible to have even higher-order iterations. If $g(\alpha) = \alpha$ and $g'(\alpha) = g''(\alpha) = 0$, then Taylor's theorem implies that

$$g(x) - \alpha = \mathcal{O}((x - \alpha)^3). \tag{2.21}$$

In principle, any order of convergence could be obtained [8]. However, while there is a qualitative change from geometric convergence to quadratic convergence, all higher-order methods behave essentially the same. For example, given a second-order method, we can always create one that is fourth-order just by taking two steps and calling them one. That is, we define

$$x_{n+1} = g(g(x_n)). \tag{2.22}$$

We could view this as introducing a "half-step"

$$x_{n+1/2} = g(x_n) \quad \text{and} \quad x_{n+1} = g(x_{n+1/2}). \tag{2.23}$$

Applying (2.20) twice, we see that $x_{n+1} - \alpha = C(x_n - \alpha)^4$. We can also verify this by defining $G(x) = g(g(x))$ and evaluating derivatives of G:

$$\begin{aligned} G'(x) &= g'(g(x))g'(x) \\ G''(x) &= g''(g(x))g'(x)^2 + g'(g(x))g''(x) \\ G'''(x) &= g'''(g(x))g'(x)^3 + 3g''(g(x))g'(x)g''(x) + g'(g(x))g'''(x). \end{aligned} \tag{2.24}$$

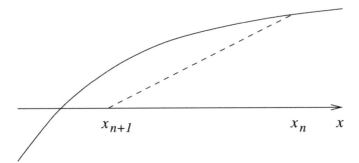

Figure 2.2 The geometric, or chord, method for approximating the solution of nonlinear equations. The slope of the dashed line is s.

Using the facts that $g(\alpha) = \alpha$ and $g'(\alpha) = 0$, we see that $G'(\alpha) = G''(\alpha) = G'''(\alpha) = 0$. Thus, if ϵ is the initial error, then the sequence of errors in a quadratic method (suitably scaled) is $\epsilon, \epsilon^2, \epsilon^4, \epsilon^8, \epsilon^{16}, \ldots$, whereas for a fourth-order method the sequence of errors (suitably scaled) is $\epsilon, \epsilon^4, \epsilon^{16}, \ldots$. That is, the sequence of errors for the fourth-order method is just a simple subsequence (omit every other term) of the sequence of errors for the quadratic method.

What is not so clear at this point is that it is possible to have fractional orders of convergence. In section 2.2.4 we will introduce a method with this property to illustrate how this is possible.

2.2 PARTICULAR METHODS

Now we consider solving a general nonlinear equation of the form

$$f(\alpha) = 0. \tag{2.25}$$

Several methods use a geometric technique, known as the *chord method*, designed to point to a good place to look for the root:

$$x_{n+1} = x_n - \frac{f(x_n)}{s}, \tag{2.26}$$

where s is the slope of a line drawn from the point $(x_n, f(x_n))$ to the next iterate x_{n+1}, as depicted in figure 2.2. This line is intended to intersect the x-axis at, or near, the root of f. This is based on the idea that the linear function ℓ with slope s which is equal to $f(x_n)$ at x_n vanishes at x_{n+1} (so $\ell(x) = s(x_{n+1} - x)$).

The simplest fixed-point iteration might be to choose $g(x) = x - f(x)$. The geometric method can be viewed as a damped version of this, where we take instead $g(x) = x - f(x)/s$. You can think of s as an adjustment parameter to help with convergence of the standard fixed-point iteration. The convergence of the geometric method is thus determined by the value of

$$g'(\alpha) = 1 - f'(\alpha)/s. \qquad (2.27)$$

Ideally, we would simply pick $s = f'(\alpha)$ if we knew how to compute it. We now consider various ways to approximate this value of s. These can all be considered different adaptive techniques for approximating $s \approx f'(\alpha)$.

2.2.1 Newton's method

The method in question was a joint effort of many people, including Newton and his contemporary Joseph Raphson[3] [32, 156]. In addition, Simpson[4] was the first (in 1740) to introduce it for systems of equations [99, 174]. Although Newton presented solutions most commonly for polynomial equations, he did suggest the method in (2.2). Newton's method for polynomials is different from what we describe here, but perhaps it should just be viewed as an additional method, slightly different from the one suggested by Raphson. Both because Newton is better known and because the full name is a bit long, we tend to abbreviate it by dropping Raphson's name, but for now let us retain it. We also add Simpson's name as suggested in [174].

The Newton-Raphson-Simpson method chooses the slope adaptively at each stage:

$$s = f'(x_n). \qquad (2.28)$$

The geometric method can be viewed as a type of difference approximation to this since we choose

$$s = \frac{0 - f(x_n)}{x_{n+1} - x_n}, \qquad (2.29)$$

and we are making the approximation $f(x_{n+1}) \approx 0$ in defining the difference quotient.

The Newton-Raphson-Simpson method is sufficiently important that we should write out the iteration in detail:

$$\boxed{x_{n+1} = x_n - \frac{f(x_n)}{f'(x_n)}.} \qquad (2.30)$$

This is fixed-point iteration with the iteration function $g = \mathcal{N}f$ defined by

$$g(x) = \mathcal{N}f(x) = x - \frac{f(x)}{f'(x)}. \qquad (2.31)$$

We can think of $\mathcal{N}$ as mapping the set of functions

$$V(I) = \{f \in C^{k+1}(I) \mid f'(x) \neq 0 \quad \forall x \in I\} \qquad (2.32)$$

to $C^k(I)$ for a given interval I and any integer $k \geq 0$. More generally, we can think of Newton's method as mapping problems of the form "find a root

---

[3]According to the mathematical historian Florian Cajori [32], the approximate dates for the life of Joseph Raphson are 1648–1715, but surprisingly little is known about his personal life [156].

[4]Thomas Simpson (1710–1761); see the quote on page 97.

of $f(x) = 0$" to algorithms using fixed-point iteration with $g = \mathcal{N}f$. We will not try to formalize such a space of problems or the space of such algorithms, but it is easy to see that Newton's method operates at a high level to solve a very general set of problems.

If $x_n \to \alpha$, then $f'(x_n) \to f'(\alpha)$, and so (2.27) should imply that the method is second-order convergent. Again, the second-order convergence of Newton's method is sufficiently important that it requires an independent computation:

$$
\begin{aligned}
g'(x) &= 1 - \frac{f'(x)^2 - f(x)f''(x)}{f'(x)^2} \\
&= \frac{f(x)f''(x)}{f'(x)^2}.
\end{aligned}
\tag{2.33}
$$

We conclude that $f(\alpha) = 0$ implies $g'(\alpha) = 0$, provided that $f'(\alpha) \neq 0$. Thus Newton's method is second-order convergent provided $f'(\alpha) \neq 0$ at the root α of $f(\alpha) = 0$.

To estimate the convergence rate, we simply need to calculate g'':

$$
\begin{aligned}
g''(x) &= \frac{d}{dx}\left(\frac{f(x)f''(x)}{f'(x)^2}\right) \\
&= \frac{f'(x)^3 f''(x) + f(x)f'(x)^2 f^{(3)}(x) - 2f(x)f'(x)f''(x)^2}{f'(x)^4} \\
&= \frac{f''(x)}{f'(x)} + \frac{f(x)}{f'(x)^3}\left(f'(x)f^{(3)}(x) - 2f''(x)^2\right).
\end{aligned}
\tag{2.34}
$$

Assuming $f'(\alpha) \neq 0$, this simplifies for $x = \alpha$:

$$
g''(\alpha) = \frac{f''(\alpha)}{f'(\alpha)}.
\tag{2.35}
$$

From (2.20), we expect that Newton's method converges asymptotically like

$$
\boxed{e_{n+1} \approx \frac{1}{2}\frac{f''(\alpha)}{f'(\alpha)}e_n^2,}
\tag{2.36}
$$

where we recall that $e_n = x_n - \alpha$.

We can see that Heron's method is the same as Newton's method if we take $f(x) = x^2 - y$. We have

$$
g(x) = x - \frac{f(x)}{f'(x)} = x - \frac{x^2 - y}{2x} = \frac{1}{2}x - \frac{y}{2x}.
\tag{2.37}
$$

Recall that $g'(x) = \frac{1}{2}(1 - y/x^2)$, so that $g''(x) = y/x^3 = 1/\sqrt{y}$ when $x = \sqrt{y}$. Thus we can assert that Heron's method is precisely second-order.

2.2.2 Stability of Newton's method

The mapping $\mathcal{N}f(x) = x - \frac{f(x)}{f'(x)}$ specified in (2.31) is not well-defined when $f'(x) = 0$. When this occurs at a root of f ($f(x) = 0$) it destroys the second-order convergence of Newton's method (see exercise 2.5). But it can cause a more serious defect if it occurs away from a root of f. For example, consider $f(x) = x - \cos x$ (cf. exercise 2.6). The root where $x = \cos x$ does not have $f'(x) = 0$, but $f'(x) = 1 + \sin x = 0$ for an infinite number of values. Just by drawing the graph corresponding to figure 2.2, we see that if we start near one of these roots of $f'(x) = 0$, then the next step of Newton's method can be arbitrarily large in magnitude (both negative and positive values are possible). Contrast this behavior with that of fixed-point iteration (cf. exercise 2.6).

2.2.3 Other second-order methods

The Steffensen iteration uses an adaptive difference:

$$s = \frac{f(x_n + f(x_n)) - f(x_n)}{f(x_n)} \tag{2.38}$$

in which the usual $\Delta x = f(x_n)$ (which will go to zero: very clever). The iteration thus takes the form

$$x_{n+1} = x_n - \frac{f(x_n)^2}{f(x_n + f(x_n)) - f(x_n)}. \tag{2.39}$$

We leave as exercise 2.10 verification that the iteration (2.39) is second-order convergent.

Steffensen's method is of the same order as Newton's method, but it has the advantage that it does not require evaluation of the derivative. If the derivative of f is hard to evaluate, this can be an advantage. On the other hand, it does require two function evaluations each iteration, which could make it comparable to Newton's method, depending on whether it is easier or harder to evaluate f' versus f. Unfortunately, Steffensen's method does not generalize to higher dimensions, whereas Newton's method does (section 7.1.4).

2.2.4 Secant method

The *secant method* approximates the slope by a difference method:

$$s = \frac{f(x_n) - f(x_{n-1})}{x_n - x_{n-1}}. \tag{2.40}$$

The error behavior is neither first- nor second-order but rather something in between. Let us derive an expression for the sequence of errors.

First, consider the method in the usual fixed-point form:

$$x_{n+1} = x_n - \frac{(x_n - x_{n-1})\, f(x_n)}{f(x_n) - f(x_{n-1})}. \tag{2.41}$$

Subtracting α from both sides and inserting $\alpha - \alpha$ in the numerator on the right-hand side and expanding, we find

$$
\begin{aligned}
e_{n+1} &= e_n - \frac{(e_n - e_{n-1}) f(x_n)}{f(x_n) - f(x_{n-1})} \\
&= \frac{-e_n f(x_{n-1}) + e_{n-1} f(x_n)}{f(x_n) - f(x_{n-1})} \\
&= \frac{x_n - x_{n-1}}{f(x_n) - f(x_{n-1})} \frac{-e_n f(x_{n-1}) + e_{n-1} f(x_n)}{x_n - x_{n-1}} \\
&= \frac{x_n - x_{n-1}}{f(x_n) - f(x_{n-1})} \frac{e_n \left(f(\alpha) - f(x_{n-1})\right) + e_{n-1} \left(f(x_n) - f(\alpha)\right)}{x_n - x_{n-1}} \\
&= \frac{x_n - x_{n-1}}{f(x_n) - f(x_{n-1})} \frac{e_n e_{n-1}}{x_n - x_{n-1}} \left(\frac{f(\alpha) - f(x_{n-1})}{e_{n-1}} - \frac{f(\alpha) - f(x_n)}{e_n} \right).
\end{aligned}
\tag{2.42}
$$

By Taylor's Theorem, we can estimate the expression (exercise 2.11)

$$
\frac{f(x_n) - f(x_{n-1})}{x_n - x_{n-1}} \approx f'(\alpha).
\tag{2.43}
$$

In section 10.2.3, we will formally define this approximation as the first divided difference $f[x_{n-1}, x_n]$, and we will also identify the expression

$$
\begin{aligned}
&\frac{1}{x_n - x_{n-1}} \left(\frac{f(\alpha) - f(x_{n-1})}{e_{n-1}} - \frac{f(\alpha) - f(x_n)}{e_n} \right) \\
&= \frac{1}{x_n - x_{n-1}} \left(-\frac{f(\alpha) - f(x_{n-1})}{\alpha - x_{n-1}} + \frac{f(\alpha) - f(x_n)}{\alpha - x_n} \right)
\end{aligned}
\tag{2.44}
$$

as the second divided difference $f[\alpha, x_{n-1}, x_n] \approx \frac{1}{2} f''(\alpha)$ (cf. (10.27)).

Rather than get involved in all the details, let us just use these approximations to see what is going on. Thus we find

$$
\boxed{e_{n+1} \approx \frac{1}{2} \frac{f''(\alpha)}{f'(\alpha)} e_n e_{n-1}.}
\tag{2.45}
$$

Thus the error is quadratic in the previous errors, but it is not exactly the square of the previous error. Instead, it is a more complicated combination.

Let M be an upper bound for $\frac{1}{2} f''/f'$ in an interval containing all the iterates. Then, analogous to (2.45), one can prove (see exercise 2.13) that

$$
|e_{n+1}| \le M |e_n| |e_{n-1}|.
\tag{2.46}
$$

To understand how the error is behaving, define a scaled error by $\epsilon_n = M|e_n|$. Then (2.46) means that

$$
\epsilon_{n+1} \le \epsilon_n \epsilon_{n-1}
\tag{2.47}
$$

for all n. If $\delta = \max\{\epsilon_0, \epsilon_1\}$, then (2.47) means that $\epsilon_2 \le \delta^2$, $\epsilon_3 \le \delta^3$, $\epsilon_4 \le \delta^5$, and so forth. In general, $\epsilon_n \le \delta^{f_n}$, where f_n is the Fibonacci sequence defined by

$$
f_{n+1} = f_n + f_{n-1},
\tag{2.48}
$$

with $f_0 = f_1 = 1$. Quadratic convergence would mean that $\epsilon_n \leq \delta^{2^n}$, but the Fibonacci sequence grows more slowly than 2^n. However, it is possible to determine the asymptotic growth exactly. In fact, we can write (exercise 2.14)

$$f_{n-1} = \frac{1}{\sqrt{5}}\left(r_+^n - r_-^n\right), \quad r_{\pm} = \frac{1 \pm \sqrt{5}}{2} \approx \begin{cases} 1.6180 & (+) \\ -0.6180 & (-). \end{cases} \quad (2.49)$$

Since $r_-^n \to 0$ as $n \to \infty$, $f_n \approx Cr_+^n$. Thus the errors for the secant method go to zero like (exercise 2.15)

$$e_{n+1} \approx Ce_n^{r_+}. \quad (2.50)$$

One iteration of the secant method requires only one function evaluation, so it can be more efficient than second-order methods. Two iterations of the secant method often require work comparable to one iteration of the Newton method, and thus a method in which one iteration is two iterations of the secant method has a faster convergence rate since $2r_+ > 2$.

2.3 COMPLEX ROOTS

Let us consider the situation when there are complex roots of equations. For example, what happens when we apply Heron's algorithm with $y < 0$? Let us write $y = -t$ $(t > 0)$ to clarify things, so that

$$x_{n+1} = \tfrac{1}{2}(x_n - t/x_n). \quad (2.51)$$

Unfortunately, for real values of x_0, the sequence generated by (2.51) does not converge to anything. On the other hand, if we take $x_0 = i\rho$, where $i = \sqrt{-1}$ and ρ is real, then

$$x_1 = \tfrac{1}{2}(x_0 - t/x_0) = \tfrac{1}{2}(i\rho - t/(i\rho)) = \tfrac{1}{2}i(\rho + t/\rho) \quad (2.52)$$

since $1/i = -i$. By induction, if $x_n = i\rho_n$, where ρ_n is real, then $x_{n+1} = i\rho_{n+1}$, where ρ_{n+1} is also real. More precisely,

$$x_{n+1} = \tfrac{1}{2}(x_n - t/x_n) = \tfrac{1}{2}(i\rho_n - t/(i\rho_n)) = \tfrac{1}{2}i(\rho_n + t/\rho_n), \quad (2.53)$$

so that

$$\rho_{n+1} = \tfrac{1}{2}(\rho_n + t/\rho_n). \quad (2.54)$$

We see that (2.54) is just the Heron iteration for approximating $\rho = \sqrt{t}$. Thus convergence is assured as long as we start with a nonzero value for ρ (cf. exercise 1.7).

The fact that Heron's method does not converge to an imaginary root given a real starting value is not spurious. Indeed, it is easy to see that Heron's method for a real root also does not converge if we start with a pure imaginary starting value. The set of values for which an iterative method converges is a valid study in dynamics, but here we are mostly interested in the local behavior, that is, convergence given a suitable starting guess. It is not hard to see that reasonable methods converge when starting within some open neighborhood of the root.

For a general complex y and a general starting guess x_0, it is not hard to see how Heron's algorithm will behave. Write $z_n = x_n/\sqrt{y}$. Then

$$z_{n+1} = x_{n+1}/\sqrt{y} = \frac{1}{2\sqrt{y}}(x_n + y/x_n) = \tfrac{1}{2}(z_n + 1/z_n). \tag{2.55}$$

Thus to study the behavior of Heron's method for complex roots, it suffices to study its behavior in approximating the square-root of one with a complex initial guess (exercise 2.16).

2.4 ERROR PROPAGATION

Suppose that the function g is not computed exactly. What can happen to the algorithm? Again, the affine function in (2.15) provides a good guide. Let us suppose that our computed function $\hat{g}$ satisfies

$$\hat{g}(x) = g(x) + \delta(x) = \alpha + \lambda(x - \alpha) + \delta(x) \tag{2.56}$$

for some error function $\delta(x)$. If, for example, we have $\delta(x) = \delta > 0$ for all x, then $\hat{g}(\hat{\alpha}) = \hat{\alpha}$ implies that

$$\hat{\alpha} = \hat{g}(\hat{\alpha}) = \alpha + \lambda(\hat{\alpha} - \alpha) + \delta = \alpha(1 - \lambda) + \lambda\hat{\alpha} + \delta, \tag{2.57}$$

so that

$$\hat{\alpha} = \alpha + \frac{\delta}{1 - \lambda}. \tag{2.58}$$

Thus the accuracy can degrade if λ is very close to one, but for a second-order method this is not an issue. A general theory can be found in [86] (see Theorem 3 on page 92 and equation (18) there), and (2.58) shows that the results there can be sharp.

The only problem with functional iteration in floating-point is that the function may not really be continuous in floating-point. Therefore it may be impossible to achieve true convergence in floating-point, in the sense that $fl(x) = \hat{g}(fl(x))$, where by $\hat{g}$ here we mean the computer implementation of the true function g in floating-point. Thus iterations should be terminated when the differences between successive iterates is on the order of the floating-point accuracy.

The effect of this kind of error on Newton's method is not so severe. Suppose that what we really compute is $\hat{f} = f + \delta$. Then Newton's method converges to a root $\hat{f}(\hat{\alpha}) = 0$, and doing a Taylor expansion of f around α, we find that

$$\alpha - \hat{\alpha} \approx \frac{\delta}{f'(\alpha)}. \tag{2.59}$$

2.5 MORE READING

Techniques for iteratively computing functions have been essential for providing software (and firmware) to compute square-roots, reciprocals, and other basic mathematical operations. For further reading, see the books [36, 61, 78, 113]. The book [86] relates Steffensen's method to a general acceleration process due to Aitken[5] to accelerate convergence of sequences. The method of *false position*, or *regula falsi*, is a slight modification of the secant method in which $x_{\nu-1}$ is replaced by x_k, where k is the last value where $f(x_k)$ has a sign opposite $f(x_\nu)$ [52]. For other generalizations of the secant method, see [22, 102].

2.6 EXERCISES

Exercise 2.1 *Suppose that g is a continuous function. Prove that, if fixed-point iteration (2.5) converges to some α, then α is a fixed point, i.e., it satisfies (2.4).*

Exercise 2.2 *Suppose that g is a C^1 function such that $\alpha = g(\alpha)$ and with the property that $|g'(\alpha)| < 1$. Prove that fixed-point iteration (2.5) converges if x_0 is sufficiently close to α. (Hint: note by lemma 2.2 that g is Lipschitz-continuous with a constant $\lambda < 1$ in an interval $[\alpha - A, \alpha + A]$ for some $A > 0$.)*

Exercise 2.3 *Suppose that g is a C^1 function such that the Lipschitz estimate (2.9) holds an interval $[\alpha - A, \alpha + A]$ for some $A > 0$. Prove that $|g'(\alpha)| \leq \lambda$. (Hint: consider the difference quotients used to define $g'(\alpha)$.)*

Exercise 2.4 *Suppose that g is a C^2 function such that $\alpha = g(\alpha)$ and with the property that $|g'(\alpha)| < 1$. Prove that fixed-point iteration (2.5) converges asymptotically according to (2.18), that is,*

$$\lim_{n \to \infty} \frac{|x_n - \alpha|}{|g'(\alpha)|^n} = C, \qquad (2.60)$$

where C is a constant depending only on g and the initial guess x_0.

Exercise 2.5 *Consider Newton's method for solving $f(\alpha) = 0$ in the case that $f'(\alpha) = 0$. Show that second-order convergence is lost. In particular, if p is the smallest positive integer such that $f^{(p)}(\alpha) \neq 0$, show that the*

---

[5] Alexander Craig Aitken (1895–1967) was born in New Zealand and studied at the University of Edinburgh, where his thesis was considered so impressive that he was both appointed to a faculty position there and elected a fellow of the Royal Society of Edinburgh, in 1925, before being awarded a D.Sc. degree in 1926. He was elected to the Royal Society of London in 1936 for his work on statistics, algebra, and numerical analysis. Aitken was reputedly one of the best mental calculators known [60, 85]. He was an accomplished writer, being elected to the Royal Society of Literature in 1964 in response to the publication of his war memoirs [4].

convergence is geometric with rate $1 - 1/p$. *(Hint: expand both* f *and* f' *in Taylor series around* α, *which both start with the terms involving* $f^{(p)}(\alpha)$ *as the first nonzero term.)*

Exercise 2.6 *Consider fixed-point iteration to compute the solution of*

$$\cos \alpha = \alpha,$$

using $g(x) = \cos x$. *Prove that this converges for any starting guess. Compute a few iterations to see what the approximate value of* α *is.*

Exercise 2.7 *A function* f *is said to be Hölder-continuous of exponent* $\alpha > 0$ *provided that*

$$|f(x) - f(y)| \le \lambda |x - y|^\alpha \tag{2.61}$$

for all x *and* y *in some interval. Show that the result (2.60) still holds as long as* g' *is Hölder-continuous of exponent* $\alpha > 0$.

Exercise 2.8 *Consider fixed-point iteration* $x \leftarrow y/x$ *for computing*

$$x = \sqrt{y}.$$

Explain its behavior. Why does it not contradict the result in exercise 2.2? (Hint: define $g(x) = y/x$ *and verify that a fixed point* $x = g(x)$ *must satisfy* $x^2 = y$. *Perform a few iterations of* $x \leftarrow y/x$ *and describe what you see. Evaluate* g' *at the fixed point.)*

Exercise 2.9 *Prove that the iteration (2.2) is Newton's method as defined in (2.30) for approximating the solution of Kepler's equation (2.1).*

Exercise 2.10 *Prove that Steffensen's iteration (2.39) is second-order convergent provided that* $f'(\alpha) \ne 0$ *at the root* $f(\alpha) = 0$ *and* $f \in C^2$ *near the root. (Hint: write Steffensen's iteration as a fixed-point iteration* $x_{\nu+1} = g(x_\nu)$ *and show that* $g'(\alpha) = 0$ *at the fixed point* $\alpha = g(\alpha)$.)

Exercise 2.11 *Verify the approximation (2.43). (Hint: write two Taylor expansions around* α, *one for* $f(x_n)$ *and one for* $f(x_{n-1})$, *and subtract them.)*

Exercise 2.12 *Investigate the fixed point(s) of the function*

$$g(x) = \frac{1}{e^{-x} - 1} + \frac{1}{e^x - 1} + 1. \tag{2.62}$$

What is the value of g' *at the fixed point(s)? (Hint: find a common denominator and simplify; see the solution to exercise 13.16 for an application of this important function.)*

Exercise 2.13 *Prove that (2.46) holds.*

Exercise 2.14 *Prove that the Fibonacci numbers satisfy (2.49). (Hint: see section 17.2.4; relate (2.48) with (17.7).)*

Exercise 2.15 *Prove that the error in the secant method behaves as predicted in (2.50). (Hint: first prove (2.45).)*

Exercise 2.16 *Investigate the behavior of the iteration (2.55) for various starting values of z_0. For what values of z_0 does the iteration converge? For what values of z_0 does the iteration not converge?*

Exercise 2.17 *Develop algorithms to solve $x^5 - x + b = 0$ for arbitrary $b \in \mathbb{R}$ (cf. (14.35)).*

Exercise 2.18 *The expression*

$$f(x) = \sqrt{1+x} - 1$$

arises in many computations. Unfortunately, for x small, round-off makes the obvious algorithm inaccurate; note that $f(x) \approx \frac{1}{2}x$ for x small. Develop an algorithm to compute f that is accurate for $|x| \leq \frac{1}{2}$. (Hint: write $t = \sqrt{1+x} - 1$ and observe that $(1+t)^2 - 1 - x = 0$. Try Newton's method starting with a good initial guess for $t = f(x)$.)

Exercise 2.19 *Consider the function*

$$f(x) = \frac{1}{x} \left(\sqrt{x^2 + 1} - |x - 1| \right). \tag{2.63}$$

Develop an algorithm to compute $f(x)$ with uniform accuracy for all $0 < x < \infty$. You may make reasonable assumptions about the accuracy of computing $\sqrt{y}$ but be explicit about them. (Hint: show that $f(1/x) = xf(x)$ and that $f(x) \approx 1 + \frac{1}{2}x + \mathcal{O}(x^3)$ for x small.)

Exercise 2.20 *Consider fixed-point iteration $x_{n+1} = g(x_n)$ for finding a fixed point $\alpha = g(\alpha)$. Suppose that the initial starting point $x_0 > \alpha$ and that $g'(x) > 0$ for $\alpha < x < x_0$. Prove that $x_n > \alpha$ for all $n \geq 0$. (Hint: write*

$$x_{n+1} - \alpha = g(x_n) - g(\alpha) = \int_\alpha^{x_n} g'(t)\, dt$$

and use induction.)

Exercise 2.21 *In many applications in which roots $f(x) = 0$ are sought, the cost of computing f (and f') is very large. You can simulate this via a loop*

```
function y = f(x,n)
for i=1:n
    t = exp(x);
    y = log(t);
end
y=y*y-2;
```

This computes the function $f(x) = x^2 - 2$ but can take arbitrarily long to do so. Use an example like this to compare the efficiency of the Steffensen and secant methods for various values of n.

2.7 SOLUTIONS

Solution of Exercise 2.4. We have

$$x_{n+1} - \alpha = g(x_n) - g(\alpha) = g'(\xi_n)(x_n - \alpha) \tag{2.64}$$

for some ξ_n between x_n and α. By induction, we thus find that

$$x_{n+1} - \alpha = \left(\prod_{i=0}^{n} g'(\xi_i) \right) (x_0 - \alpha). \tag{2.65}$$

Define $r_i = |g'(\xi_i)/g'(\alpha)|$. Then

$$\frac{|x_{n+1} - \alpha|}{|g'(\alpha)|^n} = \left(\prod_{i=0}^{n} r_i \right) |x_0 - \alpha|. \tag{2.66}$$

By (2.11), we know that $|\xi_n - \alpha| \leq \lambda^n |e_0|$. Therefore,

$$|g'(\xi_n) - g'(\alpha)| = |g''(\hat{\xi}_n)(\xi_n - \alpha)| \leq |g''(\hat{\xi}_n)|\lambda^n |e_0|. \tag{2.67}$$

for some $\hat{\xi}_n$ between ξ_n and α, and therefore between x_n and α. Therefore, $|r_n - 1| \leq \hat{C}\lambda^n$, where $\hat{C}$ is chosen to be larger than $|e_0 g''(\hat{\xi}_n)/g'(\alpha)|$ for all n. Since $r_n \to 1$ as $n \to \infty$, we can be assured that $r_n > 0$ for n sufficiently large. If $r_n = 0$ for some value of n, then we have $x_{n+i} = \alpha$ for all $i \geq 1$, so we can take the constant $\hat{C}$ in (2.60) to be zero. If all $r_n > 0$, then we can take the logarithm of the product in (2.66) to get

$$\log \left(\prod_{i=0}^{n} r_i \right) = \sum_{i=0}^{n} \log r_i. \tag{2.68}$$

Thus it suffices to prove that the sum on the right-hand side of (2.68) converges. Since we have an estimate on $r_i - 1$, we write $r_i = 1 + \epsilon_i$, where $|\epsilon_i| \leq \hat{C}\lambda^i$. Note that $\log(1 + x) \leq x$ for all $x > 0$. To get such a bound for $x < 0$, set $t = -x$ and write

$$|\log(1 - t)| = -\log(1 - t) = \int_{1-t}^{1} \frac{dx}{x} \leq \frac{t}{1 - t}. \tag{2.69}$$

Thus $|\log(1 + x)| \leq 2|x|$ for $|x| \leq \frac{1}{2}$. Therefore, $|\log r_i| \leq 2|1 - r_i| \leq \hat{C}\lambda^i$ for i sufficiently large, and thus the sum on the right-hand side of (2.68) converges to some value γ. Define $C = e^\gamma |x_0 - \alpha|$.

Solution of Exercise 2.6. Define $g(x) = \cos x$. Then a solution to $x = g(x)$ is what we are seeking, and we can apply fixed-point iteration $x_{n+1} = g(x_n)$. This can be computed in octave by simply repeating the command x=cos(x), having started with, say, $x = 1$. After about 40 iterations, it converges to $\alpha = 0.73909$. We have $g'(\alpha) = -\sin \alpha \approx -0.67362$. Thus it is clear that fixed-point iteration is locally convergent.

The set of values x_n generated by fixed-point iteration always lie in $[-1, 1]$, the range of $g(\cdot) = \cos \cdot$. But we cannot easily assert global convergence because the values of g' also extend over $[-1, 1]$. However, g maps $[-1, 0]$ to $[\beta, 1]$, where $\beta = \cos(-1) \approx 0.54030$. Similarly, g maps $[0, 1]$ to $[\beta, 1]$, but with the order reversed. Thus, regardless of the starting value x_0, $x_1 \in [\beta, 1]$. Moreover, this argument shows further that all subsequent $x_n \in [\beta, 1]$ as well. The maximum value of $|g'(x)| = \sin x$ on the interval $[\beta, 1]$ occurs at $x = 1$, since sin is strictly increasing on $[\beta, 1]$, and $\sin 1 = 0.84147$. Thus we conclude that fixed-point iteration converges at least as fast as 0.84147^n.

Solution of Exercise 2.18. We recall that

$$t = f(x) = \sqrt{1 + x} - 1. \tag{2.70}$$

We seek an algorithm that outputs $\hat{t}$ with the property that

$$|t - \hat{t}| = \left| \sqrt{1 + x} - 1 - \hat{t} \right| \leq \tfrac{1}{2} \epsilon |t|, \tag{2.71}$$

where $\epsilon > 0$ is a prescribed accuracy that we require to be sufficiently small. To begin with, we establish some inequalities of use later.

Adding 1 to both sides of (2.70) and squaring, we find that

$$(t + 1)^2 = 1 + x,$$

and thus

$$x = (t + 1)^2 - 1 = 2t + t^2 = t(2 + t). \tag{2.72}$$

Therefore,

$$\frac{x}{f(x)} = \frac{x}{t} = 2 + t = 1 + \sqrt{1 + x} \tag{2.73}$$

is a strictly increasing function of x. In particular,

$$|x/t| = |x/f(x)| \leq 1 + \sqrt{3/2} < 2.3 \tag{2.74}$$

for $|x| \leq \tfrac{1}{2}$. Similarly, $f(x)/x$ is a strictly decreasing function of x. We thus have

$$|t| = |f(x)| \leq \frac{1}{1 + \sqrt{1/2}} |x| < 0.6 |x| \tag{2.75}$$

for $|x| \leq \tfrac{1}{2}$. Therefore,

$$|t| \leq 0.3 \tag{2.76}$$

for $|x| \leq \tfrac{1}{2}$. Now let us define $\hat{t}$. Taylor's theorem shows that

$$\left| \sqrt{1 + x} - 1 - \tfrac{1}{2} x + \tfrac{1}{8} x^2 \right| \leq 0.1 \left| x^3 \right| \tag{2.77}$$

for $x \in [-0.1, 0.1]$. This means that for $|x| \leq \sqrt{\epsilon}$ and $\epsilon \leq 1/100$, we can simply define

$$\hat{t} = \tfrac{1}{2} x - \tfrac{1}{8} x^2. \tag{2.78}$$

Then (2.77) implies that

$$\left| \sqrt{1 + x} - 1 - \hat{t} \right| \leq 0.1 |x| \epsilon \leq \tfrac{1}{4} |t| \epsilon, \tag{2.79}$$

by (2.74). This proves (2.71) when $|x| \leq \sqrt{\epsilon}$.

Thus we can turn our attention to the case where $|x| > \sqrt{\epsilon}$. In view of (2.72), the function

$$\phi(\tau) = (1 + \tau)^2 - 1 - x = 2\tau + \tau^2 - x = 0$$

when $\tau = t$. Thus $t = f(x)$ is the solution to $\phi(t) = 0$, and we can consider using Newton's method to find t. Differentiating, we have $\phi'(\tau) = 2 + 2\tau$. Thus Newton's method is

$$t_{\nu+1} = t_\nu - \frac{2t_\nu + t_\nu^2 - x}{2 + 2t_\nu} = \frac{t_\nu^2 + x}{2 + 2t_\nu} =: g(t_\nu). \tag{2.80}$$

Differentiating again, we find that

$$g'(\tau) = \frac{2\tau + \tau^2 - x}{2(1 + \tau)^2} = \frac{\phi(\tau)}{2(1 + \tau)^2} = \frac{1}{2} - \frac{x + 1}{2(1 + \tau)^2}. \tag{2.81}$$

Therefore, $\phi(t) = 0$ implies $g'(t) = 0$, and differentiating yet again, we have

$$g^{(2)}(\tau) = -(x + 1)(1 + \tau)^{-3}. \tag{2.82}$$

Now let us consider using $t_0 = f\ell\sqrt{f\ell(1 + x)} - 1$ as a starting guess. More precisely, we define $a = f\ell(1 + x)$, $b = f\ell\sqrt{a}$, and $t_0 = f\ell(b - 1)$. We assume an estimate like (2.71) holds for all floating-point operations, namely, we assume that $a = (1 + x)(1 + \delta_1)$, $b = \sqrt{a}(1 + \delta_2)$, and $t_0 = (b - 1)(1 + \delta_3)$, where $|\delta_i| \le \epsilon$. For simplicity, we write $1 + \hat{\delta}_1 = (1 + \hat{\delta}_1)^2$; that is,

$$\hat{\delta}_1 = \sqrt{1 + \delta_1} - 1 = f(\delta) \approx \tfrac{1}{2}\delta_1.$$

In particular, $|\hat{\delta}_1| \le 0.6\delta_1$ as long as $\epsilon \le \frac{1}{2}$, by (2.76).

Therefore, $\sqrt{a} = \sqrt{1 + x}(1 + \hat{\delta}_1)$ and

$$b = \sqrt{1 + x}(1 + \hat{\delta}_1)(1 + \delta_2) = \sqrt{1 + x}\left(1 + \hat{\delta}_1 + \delta_2 + \hat{\delta}_1\delta_2\right). \tag{2.83}$$

Since $t = \sqrt{1 + x} - 1$, we find

$$b - 1 - t = \sqrt{1 + x}\left(\hat{\delta}_1 + \delta_2 + \hat{\delta}_1\delta_2\right). \tag{2.84}$$

Recall that $t_0 = (b - 1)(1 + \delta_3)$, so that $t_0 - (b - 1) = (b - 1)\delta_3$, and thus

$$\begin{aligned}
t_0 - t &= t_0 - (b - 1) + (b - 1) - t \\
&= (b - 1)\delta_3 + \sqrt{1 + x}\left(\hat{\delta}_1 + \delta_2 + \hat{\delta}_1\delta_2\right) \\
&= t\delta_3 + \sqrt{1 + x}\left(\hat{\delta}_1 + \delta_2 + \hat{\delta}_1\delta_2\right)(1 + \delta_3) \\
&= t\delta_3 + (t + 1)\left(\hat{\delta}_1 + \delta_2 + \hat{\delta}_1\delta_2\right)(1 + \delta_3).
\end{aligned} \tag{2.85}$$

In particular, (2.85) shows why t_0 cannot be used as a uniform approximation to t since the second term is multiplied by $t + 1$ and not t. However, if $\epsilon \le 10^{-2}$ (as we assumed above), then (2.85) implies that

$$|t_0 - t| \le 2.41\epsilon. \tag{2.86}$$

If we define t_1 by taking one Newton step (2.80), then (2.19) and (2.82) imply that

$$|t_1 - t| \leq 3\epsilon^2(x+1)(1+\xi)^{-3} = 3\epsilon^2(t+1)^2(1+\xi)^{-3}, \tag{2.87}$$

where ξ lies between t_0 and t. By (2.75), $t \geq -0.3$, and certainly $t_0 > -0.31$. Thus $1 + \xi \geq 0.69$, and applying (2.86) shows that

$$\frac{1+t}{1+\xi} = 1 + \frac{t-\xi}{1+\xi} \leq 1 + \frac{2.41\epsilon}{1+\xi} \leq 1 + \frac{2.41\epsilon}{0.69} \leq 1 + 3.5\epsilon. \tag{2.88}$$

Therefore,

$$\frac{(1+t)^2}{(1+\xi)^3} \leq (1+3.5\epsilon)^2 \frac{1}{1+\xi} \leq \frac{1.08}{0.69} < 1.57. \tag{2.89}$$

Applying this in (2.87) shows that (recall $|x| > \sqrt{\epsilon}$)

$$|t_1 - t| \leq 4.72\epsilon^2 \leq 4.72\epsilon^{3/2}|x| \leq 11\epsilon^{3/2}|t|, \tag{2.90}$$

by (2.73). This proves (2.71) provided $\epsilon \leq 1/484$, as we now require.

Note that for $|x| \approx \sqrt{\epsilon}$, (2.86) implies that t_0 is accurate to about half of the digits required, but only half. One Newton step then provides the other half of the required digits. A similar statement can be made about (2.78). The term $\frac{1}{2}x$ provides a substantial fraction of the required digits, and the correction term $-\frac{1}{8}x^2$ provides the rest.

Chapter Three

Linear Systems

> "The *Nine Chapters on the Mathematical Art* has played a central role in Oriental mathematics somewhat similar to Euclid's *Elements of Geometry* in the West. However, the *Nine Chapters* has always been more involved in the methods for finding an algorithm to solve a problem, so that its influence has been both pedagogical and practical. Instead of theorems ... the *Nine Chapters* provides algorithmic Rules." [92]

Most interesting problems involve more than one variable, so we now move to systems of equations. Before we turn to nonlinear systems, we look in detail at algorithms for solving linear systems for two reasons. First, the linear case is a prerequisite to the nonlinear case; we will reduce the solution of nonlinear problems to an iteration involving the solution of linear systems of equations. Second, this allows us to introduce ideas from linear algebra that we need for subsequent developments.

This chapter formalizes the familiar method used to solve systems of equations by elimination. The basic elimination method is associated with the name of Gauss[1] [7], although the method was in use well before he lived. The *fangcheng* methods were known in China hundreds of years before the birth of Gauss [92]. The method is now found in the middle-school curriculum, so we review it quickly with a view to establishing notation that will be useful to derive its fundamental properties.

Gaussian elimination can be applied to equations involving entities in any field $\mathbb{F}$. Our focus here will be limited to the fields of real ($\mathbb{R}$) and complex ($\mathbb{C}$) numbers. We might further limit our scope just to real numbers, but later we will want to consider methods for finding eigenvalues and eigenvectors of matrices. Matrices with real entries can have complex eigenvalues and eigenvectors, so we are forced to consider complex numbers in working with them.

Linear systems of n equations in n unknowns arise in many applications. Problems of size $n \approx 10^5$ occur [56] even when the matrices are dense (i.e, have no substantial amount of entries known to be zero). We will see that with $n = 10^5$, the amount of computation required to solve such a system is near a *petaflop* (10^{15} floating-point operations). A typical (single) processor

---

[1] Johann Carl Friedrich Gauss (1777–1855) was one of the greatest scientists of all time and is celebrated both in scientific and popular media [93].

today can perform roughly one gigaflops (10^9 floating-point operations per second). Thus problems of this size often require parallel computation [144].

3.1 GAUSSIAN ELIMINATION

A system of n linear equations in n unknowns can be written as

$$
\begin{aligned}
a_{11}x_1 + a_{12}x_2 + \cdots + a_{1n}x_n &= f_1 \\
a_{21}x_1 + a_{22}x_2 + \cdots + a_{2n}x_n &= f_2 \\
\vdots \qquad\qquad \vdots \\
a_{n1}x_1 + a_{n2}x_2 + \cdots + a_{nn}x_n &= f_n.
\end{aligned}
\tag{3.1}
$$

Here all entities a_{ij}, f_i, and x_i are in some field $\mathbb{F}$, which we may assume for simplicity is either $\mathbb{R}$ or $\mathbb{C}$. The principle of elimination is to subtract suitable multiples of the first equation from the remaining equations in such a way as to eliminate x_1 from the later equations. More precisely, we subtract

$$
\frac{a_{i1}}{a_{11}}\left(a_{11}x_1 + a_{12}x_2 + \cdots + a_{1n}x_n = f_1\right)
\tag{3.2}
$$

from the ith equation for each $i = 2, \ldots, n$. This converts the original tableau of expressions to

$$
\begin{aligned}
a_{11}x_1 + a_{12}x_2 + \cdots + a_{1n}x_n &= f_1 \\
\hat{a}_{22}x_2 + \cdots + \hat{a}_{2n}x_n &= \hat{f}_2 \\
\vdots \qquad\qquad \vdots \\
\hat{a}_{n2}x_2 + \cdots + \hat{a}_{nn}x_n &= \hat{f}_n,
\end{aligned}
\tag{3.3}
$$

where the hatted coefficients result from the appropriate subtractions.

The key point is that the original system of n equations in n unknowns is converted to one with only $n-1$ equations in $n-1$ unknowns. Continuing in this way, we eventually arrive at a simple equation for x_n, and then this value can be used in the previous equation involving x_n and x_{n-1} to solve for x_{n-1}, and so on. Let us now write down all this as concise algorithms.

3.1.1 Elimination algorithm

First, we can express the elimination algorithm as

$$
\begin{aligned}
a_{ij}^{k+1} &= a_{ij}^k - m_{ik}a_{kj}^k \quad \forall i = k+1, \ldots, n, \ j = k+1, \ldots, n, \\
f_i^{k+1} &= f_i^k - m_{ik}f_k^k \quad \forall i = k+1, \ldots, n,
\end{aligned}
\tag{3.4}
$$

where the *multipliers* m_{ik} are defined by

$$
m_{ik} = \frac{a_{ik}^k}{a_{kk}^k} \quad \forall i = k+1, \ldots, n, \ k = 1, \ldots, n.
\tag{3.5}
$$

Here $a_{ij}^1 = a_{ij}$ and $f_{ij}^1 = f_{ij}$. Note that we have defined a_{ij}^k only for $i \geq k$ and $j \geq k$. More precisely, we have defined $\hat{A}^k$ to be this $(n+1-k) \times (n+1-k)$ matrix. If we want to complete the $n \times n$ system, we set $a_{ij}^k = a_{ij}^{k-1}$ for other values of i and j and then set

$$a_{ik}^{k+1} = 0 \quad \text{for } i > k. \tag{3.6}$$

The tableau (3.3) displays the case involving the terms a_{ij}^2, which define $\hat{A}^2$ and A^2. We have glossed over the possibility that $a_{kk}^k = 0$, which can certainly happen, but let us suppose for the moment that a_{kk}^k does not vanish.

We can formalize the algorithm of elimination by saying that the input is an $n \times n$ matrix, A, and the output is an $(n-1) \times (n-1)$ matrix, $\hat{A}$, together with a column vector m of length $n-1$. We iteratively apply this algorithm to $\hat{A}$ until $n = 1$.

After applying (3.4) for $k = 1, \ldots, n-1$, we have a triangular system of equations

$$
\begin{aligned}
a_{11}^1 x_1 + a_{12}^1 x_2 + \cdots + a_{1n}^1 x_n &= f_1^1 \\
a_{22}^2 x_2 + \cdots + a_{2n}^2 x_n &= f_2^2 \\
&\vdots \qquad \vdots \\
a_{nn}^n x_n &= f_n^n.
\end{aligned}
\tag{3.7}
$$

This can be proved rigorously by observing that the elimination algorithm is applied inductively to the matrices $\hat{A}^k$ for $k = 1, 2, \ldots, n-1$, which are of size $(n+1-k) \times (n+1-k)$, with $\hat{A}^1 = A$. The induction step is the combination of (3.4) and (3.5). This will be covered in more detail in section 3.2.1.

It is helpful to use matrix notation at this point. The original set of equations (3.1) can be written in matrix-vector form as $AX = F$, where A is the matrix with entries (a_{ij}), X is the vector with entries (x_j), and F is the vector with entries (f_i). Define the matrix U via

$$u_{ij} = a_{ij}^i \quad \forall i = 1, \ldots, n \text{ and } j = i, \ldots, n \tag{3.8}$$

(and zero elsewhere). Then (3.7) becomes $UX = G$, where G is the vector with entries (f_i^i). By construction, U is an *upper-triangular matrix*.

Definition 3.1 *A matrix $B = (b_{ij})$ is upper-triangular (respectively, lower-triangular) if $b_{ij} = 0$ for all $j > i$ (respectively, $i > j$).*

3.1.2 Backward substitution

The key reason for reducing $AX = F$ to the triangular form $UX = G$ is that triangular systems are easier to solve. In particular, upper-triangular systems can be solved by the *backsubstitution* algorithm, which we write in equational form as

$$x_n = g_n / u_{nn}$$

$$x_{n-i} = \left(g_{n-i} - \sum_{j=n-i+1}^{n} u_{n-i,j} x_j \right) / u_{n-i,n-i} \quad \forall i = 1, \ldots, n-1. \quad (3.9)$$

A similar algorithm can be used to solve lower-triangular systems (exercise 3.9), which is called *forward substitution*.

Two corollaries of the algorithm (3.9) are as follows.

Corollary 3.2 *A triangular matrix is singular if and only if one of its diagonal entries is zero.*

Proof. If none of the diagonal entries of U is zero, then (3.9) provides a way to determine a solution X for an arbitrary right-hand side G. Thus U is invertible. The converse is left as exercise 3.1.   QED

Corollary 3.3 *The diagonal entries of a triangular matrix are its eigenvalues.*

Proof. The eigenvalues of T are the values λ such that $T - \lambda I$ is singular. But $T - \lambda I$ is also triangular and so is singular only if λ is equal to one of the diagonal entries of T.   QED

Corollary 3.4 *The determinant of a triangular matrix is equal to the product of the diagonal entries of the triangular matrix.*

Proof. It is a result of linear algebra that the product of eigenvalues of a matrix is equal to the determinant, so the result follows from corollary 3.4. But this result from linear algebra need not be invoked. Instead, the fact that $\det U = \prod_{i=1}^{n} u_{ii}$ may be derived by a direct proof using the fact that U is upper-triangular (exercise 3.2).   QED

We will explore triangular matrices in more detail in section 3.3.

3.2 FACTORIZATION

Let us form the lower-triangular matrix L via

$$L = \begin{pmatrix} 1 & 0 & \cdots & 0 & 0 \\ m_{21} & 1 & \cdots & 0 & 0 \\ & & \vdots & & \\ m_{n1} & m_{n2} & \cdots & m_{n,n-1} & 1 \end{pmatrix}, \quad (3.10)$$

where the multipliers m_{ik} are defined in (3.5).

Theorem 3.5 *The matrices L and U defined via the elimination process form a multiplicative factorization of $A = LU$.*

3.2.1 Proof of the factorization

The theorem can be proved by multiplying the factors (exercise 3.3), but we prefer to give a more modern treatment that also can be used to generate code automatically [164]. This approach also has the added benefit that it can be easily modified to identify a permutation matrix that corresponds to pivoting (section 3.4 and exercise 3.4).

We will prove this by induction on n. The case $n = 1$ being trivial and not very instructive, we consider the case $n = 2$; in this case, the statement is that

$$A = \begin{pmatrix} a_{11} & a_{12} \\ a_{21} & a_{22} \end{pmatrix} = \begin{pmatrix} 1 & 0 \\ a_{21}/a_{11} & 1 \end{pmatrix} \begin{pmatrix} a_{11} & a_{12} \\ 0 & a_{22} - a_{12}a_{21}/a_{11} \end{pmatrix}, \qquad (3.11)$$

which is easily verified by multiplying the factors on the right-hand side. The induction step is very similar to this as well.

Suppose that A is an $n \times n$ matrix that we write in the form

$$A = \begin{pmatrix} a_{11} & a_1^{\mathrm{T}} \\ a^1 & B \end{pmatrix}, \qquad (3.12)$$

where a_1 and a^1 are (column) vectors of length $n - 1$, with $(a_1)_j = a_{1j}$ and $(a^1)_j = a_{j1}$, and B is an $(n-1) \times (n-1)$ matrix; we assume $a_{11} \neq 0$. Then the first step of Gaussian elimination may be interpreted as forming the factorization

$$A = \begin{pmatrix} 1 & 0 \\ a_{11}^{-1}a^1 & I_{n-1} \end{pmatrix} \begin{pmatrix} a_{11} & a_1^{\mathrm{T}} \\ 0 & B - a_{11}^{-1}a^1 a_1^{\mathrm{T}} \end{pmatrix}, \qquad (3.13)$$

where I_{n-1} denotes the $(n-1) \times (n-1)$ identity matrix and $(a^1)_j = a_{j1}$. That is, we first prove that the product of matrices on the right-hand side of (3.13) equals A (exercise 3.6). This takes some care in interpretation since the matrices have been represented in block form, but it is elementary to verify that matrix multiplication follows these block rules (exercise 3.5). Then by inspection we see that the second factor is the matrix we have identified in the elimination process as A^2:

$$A^2 = \begin{pmatrix} a_{11} & a_1^{\mathrm{T}} \\ 0 & B - a_{11}^{-1}a^1 a_1^{\mathrm{T}} \end{pmatrix} = \begin{pmatrix} a_{11} & a_1^{\mathrm{T}} \\ 0 & \hat{A}^2 \end{pmatrix}. \qquad (3.14)$$

By induction, we can now assume that $\hat{A}^2 = L^2 U^2$, with L^2 defined in terms of the corresponding multipliers and U^2 defined by the elimination process. By induction in the definition of the elimination process, we know that the multipliers (3.5) for $\hat{A}^2$ are the same as the ones for A, and similarly for U^2. Thus we know that

$$\begin{aligned} A &= \begin{pmatrix} 1 & 0 \\ a_{11}^{-1}a^1 & I_{n-1} \end{pmatrix} \begin{pmatrix} a_{11} & a_1^{\mathrm{T}} \\ 0 & L^2 U^2 \end{pmatrix} \\ &= \begin{pmatrix} 1 & 0 \\ a_{11}^{-1}a^1 & I_{n-1} \end{pmatrix} \begin{pmatrix} 1 & 0 \\ 0 & L^2 \end{pmatrix} \begin{pmatrix} a_{11} & a_1^{\mathrm{T}} \\ 0 & U^2 \end{pmatrix} \\ &= \begin{pmatrix} 1 & 0 \\ a_{11}^{-1}a^1 & L^2 \end{pmatrix} \begin{pmatrix} a_{11} & a_1^{\mathrm{T}} \\ 0 & U^2 \end{pmatrix}, \end{aligned} \qquad (3.15)$$

where the last two equalities are left as exercise 3.7.   QED

The inductive nature of the proof also corresponds to a recursive definition of the algorithm. The representation (3.12) provides the base case, and (3.13) provides the reduction to a smaller problem. We leave as an exercise the development of a factorization in this way (exercise 3.8).

The above argument directly establishes the existence of an LU factorization by induction. The argument is complicated by the fact that we also want to identify the entities generated in the elimination process as the ingredients in the factorization. If one were willing to forget about the elimination process, the factorization could be derived more easily. We will see that the factors are the proper focus and indeed that there are other algorithms for determining factors that are more efficient than the elimination process (section 4.1.1). The introduction of Gaussian elimination is mainly done to connect the concept of the factorization with the approach that one uses by hand to solve small systems of linear equations.

3.2.2 Using the factors

There are several uses for the factorization of a matrix as a product of triangular matrices. For one thing, we can derive a simple formula for the determinant:

$$\det A = \det L \ \det U = \prod_{i=1}^{n} u_{ii}. \tag{3.16}$$

Here we have used corollary 3.4 together with the fact from linear algebra that the determinant of a product of matrices is the product of the determinants of the two matrices separately.

The factorization can also provide an alternate way to solve equations. The original equation $AX = F$ can now be written as $L(UX) = F$, so if we write $G = UX$, then G must solve the lower-triangular system $LG = F$. Moreover, lower-triangular systems can be solved by an algorithm similar to (3.9) (see exercise 3.9).

Suppose that we had already performed elimination and had kept a copy of the multipliers. Then we can solve for X in two steps:

$$\begin{aligned} &\text{first solve } LG = F; \\ &\text{then solve } UX = G. \end{aligned} \tag{3.17}$$

3.2.3 Operation estimates

One reason that the algorithm (3.17) is of interest is that it allows us to split the overall task of solving linear systems into parts that have different *operation estimates*. Such estimates give upper bounds on the number of floating-point operations and memory references, which in turn can be used to model the performance of an algorithm. For this reason, such estimates are often called *work estimates*.

Operation estimates are done by counting the number of basic steps in an algorithm. Let us write the first line of (3.4) in a more algorithmic form as

$$\text{For } k = 1, \ldots, n-1,$$
$$\text{For } i = k+1, \ldots, n \text{ and } j = k+1, \ldots, n, \qquad (3.18)$$
$$a_{ij} \leftarrow a_{ij} - m_{ik} * a_{kj}.$$

Here the leftward-facing arrow $\leftarrow$ means *assignment*. That is, the expression on the right side of $\leftarrow$ is computed and then deposited in the memory location allocated for a_{ij}. Note that we are assuming that all the values a_{ij}^k are stored in the same location.

In the algorithm (3.18) for updating a_{ij}, there is one subtraction and one multiplication. In addition, three items have to be read from memory, and one written to memory at the end. We will discuss memory models in more detail in section 4.1.2. For now we simply focus on floating-point operations.

The innermost loop in (3.18) involves repeating the a_{ij} computation $(n - k)^2$ times since it is done for $i = k+1, \ldots, n$ and $j = k+1, \ldots, n$; this involves $n - k$ values of i and j, independently. Since this is done (in the outer loop) $n - 1$ times for different values of k, the total number of multiplications (and subtractions) is

$$\sum_{k=1}^{n-1} (n-k)^2 = \sum_{j=1}^{n-1} j^2 = \tfrac{1}{3}n^3 - \tfrac{1}{2}n^2 + \tfrac{1}{6}n \qquad (3.19)$$

(see exercise 3.10).

The backsubstitution algorithm (3.9) can be analyzed similarly. For each i, there are i product terms $u_{ij}x_j$ in the summation, and computing the sum requires at least $i - 1$ additions. If we include the subtraction in the count of additions (it is a reasonable assumption that the costs of additions and subtractions are roughly the same), then we find a total of i multiplications and additions in each case. Thus the total amount of multiplications and additions is

$$\sum_{i=1}^{n-1} i = \tfrac{1}{2}n(n-1) = \tfrac{1}{2}n^2 - \tfrac{1}{2}n. \qquad (3.20)$$

In addition, there are n divisions.

The main conclusion from this exercise is that the amount of work required to factor an $n \times n$ matrix is $\mathcal{O}(n^3)$, whereas the amount of work required to perform backsubstitution is only $\mathcal{O}(n^2)$. We leave as exercise 3.13 the estimation of the work involved in computing the multipliers (3.13). It is worth noting that in both the elimination and backsubstitution algorithms, multiplications come paired with additions. This fact has been exploited in some hardware systems which can perform a linked multiply-add pair faster than the sum of times required to do each operation separately. For these reasons, it is sometimes useful to think of the multiply-add pair as a single type of operation.

3.2.4 Multiple right-hand sides

Note that if we defined $a_{i,n+1}^k = f_i^k$, then the two equations in (3.4) could be written as one. Moreover, multiple right-hand sides $f_{i,j}$ could be treated similarly by writing $a_{i,n+j}^k = f_{i,j}^k$, in which case (3.4) becomes

$$a_{ij}^{k+1} = a_{ij}^k - m_{ik} a_{kj}^k \quad \forall i = k+1, \ldots, n; \ j = k+1, \ldots, n+m; \quad (3.21)$$

where m is the number of right-hand sides. Thus the elimination process can be used to reduce multiple systems to triangular form. However, using the algorithm (3.17), this can be presented more simply.

Let us write the multiple right-hand sides in vector form as $F^{(j)}$ and suppose that we want to solve $A X^{(j)} = F^{(j)}$ for $j = 1, \ldots, m$. We take advantage of the factorization $A = LU$ to factorize only once and solve multiple times via

$$\begin{aligned} \text{first solve } & L G^{(j)} = F^{(j)}, \\ \text{then solve } & U X^{(j)} = G^{(j)}, \end{aligned} \quad (3.22)$$

for $j = 1, \ldots, m$. See exercise 3.14 for an analysis of the operation counts.

3.2.5 Computing the inverse

One example with multiple right-hand sides arises in computation of the inverse of a matrix. The columns $X^{(j)}$ of the inverse of A satisfy $A X^{(j)} = E^{(j)}$, where the $E^{(j)}$'s are the standard basis vectors for $\mathbb{R}^n$: $e_i^{(j)} = \delta_{ij}$, where δ_{ij} denotes the Kronecker δ. Applying the algorithm (3.22), we see that the inverse can be computed in $(5/6)n^3 + \mathcal{O}(n^2)$ operations, i.e., multiply-add pairs (exercise 3.14).

Thus, asymptotically, the amount of work required to compute the inverse of a matrix is less than twice the work needed just to solve a single system of equations. However, computing the inverse is not a good general approach to solving equations. For example, the number of multiply-add pairs required to multiply the inverse times the right-hand side is n^2, whereas the back-substitution algorithm requires about half this much work. With banded matrices (section 4.3), the comparison can become much more dramatic.

3.3 TRIANGULAR MATRICES

The sets $\mathcal{U}$ and $\mathcal{L}$ of upper-triangular matrices and lower-triangular matrices, respectively, have a ring[2] structure, as follows. For the time being, let $\mathcal{T}$ stand for one of these sets, i.e., either $\mathcal{U}$ or $\mathcal{L}$. Then we can add any two matrices in $\mathcal{T}$ componentwise, i.e, $(T^1 + T^2)_{ij} = T_{ij}^1 + T_{ij}^2$. The fact that the sum is still triangular is obvious. What is somewhat surprising is that the product of two triangular matrices is also triangular.

---

[2] A ring is a set in which addition and multiplication are defined.

Suppose that A and B are upper-triangular matrices of the form

$$
A = \begin{pmatrix}
a_1 & a_{12} & a_{13} & \cdots & a_{1n} \\
0 & a_2 & a_{23} & \cdots & a_{2n} \\
\vdots & \vdots & \vdots & \vdots & \vdots \\
0 & 0 & 0 & \cdots & a_n
\end{pmatrix}, \quad
B = \begin{pmatrix}
b_1 & b_{12} & b_{13} & \cdots & b_{1n} \\
0 & b_2 & b_{23} & \cdots & b_{2n} \\
\vdots & \vdots & \vdots & \vdots & \vdots \\
0 & 0 & 0 & \cdots & b_n
\end{pmatrix}.
$$
$$\tag{3.23}$$

Then (exercise 3.15) the product AB is (a) also an upper-triangular matrix with the property (b) that

$$
AB = \begin{pmatrix}
a_1 b_1 & c_{12} & c_{13} & \cdots & c_{1n} \\
0 & a_2 b_2 & c_{23} & \cdots & c_{2n} \\
\vdots & \vdots & \vdots & \vdots & \vdots \\
0 & 0 & 0 & \cdots & a_n b_n
\end{pmatrix}
\tag{3.24}
$$

for some coefficients c_{ij}. That is,

the diagonal entries of the product are the products of the diagonal entries.

A similar formula holds for lower-triangular matrices.

It is natural to guess that, if none of the diagonal entries of a matrix A is zero, then

$$
A^{-1} = \begin{pmatrix}
a_1 & a_{12} & a_{13} & \cdots & a_{1n} \\
0 & a_2 & a_{23} & \cdots & a_{2n} \\
\vdots & \vdots & \vdots & \vdots & \vdots \\
0 & 0 & 0 & \cdots & a_n
\end{pmatrix}^{-1}
= \begin{pmatrix}
a_1^{-1} & b_{12} & b_{13} & \cdots & b_{1n} \\
0 & a_2^{-1} & b_{23} & \cdots & b_{2n} \\
\vdots & \vdots & \vdots & \vdots & \vdots \\
0 & 0 & 0 & \cdots & a_n^{-1}
\end{pmatrix};
\tag{3.25}
$$

that is, A^{-1} is also upper-triangular.

Lemma 3.6 *Suppose that A is an upper-triangular (respectively, a lower-triangular) matrix with nonzero diagonal entries. Then A is invertible, and A^{-1} is also upper-triangular (respectively, a lower-triangular) and (3.25) holds.*

Proof. The invertibility of a triangular matrix is determined by having nonzero diagonal entries in view of corollary 3.2. Thus all we need to show is that the inverse has the corresponding triangular structure. This is easily seen by examining the backsubstitution algorithm applied to the equations $A X^{(k)} = E^{(k)}$ for the columns of the inverse (cf. section 3.2.5). Since the diagonal entries of A are nonzero, the algorithm will produce solutions to $A X = F$ for all F. Since $e_j^{(k)} = 0$ for $j > k$, we have $x_j^{(k)} = 0$ for $j > k$. Thus the kth column of A^{-1} is zero for $j > k$, and A^{-1} is upper-triangular; that is, A^{-1} must be of the form

$$
A^{-1} = \begin{pmatrix}
b_1 & b_{12} & b_{13} & \cdots & b_{1n} \\
0 & b_2 & b_{23} & \cdots & b_{2n} \\
\vdots & \vdots & \vdots & \vdots & \vdots \\
0 & 0 & 0 & \cdots & b_n
\end{pmatrix}.
\tag{3.26}
$$

Using the product formula (3.24) completes the proof since $A^{-1}A = I$. QED

3.4 PIVOTING

By *pivoting*, we mean reordering of the unknowns in the system (3.1). The numbering scheme for the equations and unknowns does not affect the solutions, but it can affect the outcome of the elimination algorithm. In particular, it may happen at a certain stage that $a_{kk}^k = 0$, at which point the elimination algorithm breaks down since the multipliers (3.5) become undefined. At such a point, we can choose among the remaining rows and columns (numbered i and j, respectively, such that $i, j \geq k$) to find some value of i and j such that $a_{i,j}^k \neq 0$. We can either imagine moving the required row and column into the right position or just keep track of a new numbering scheme. The row and/or column exchanges in pivoting can be applied to the original matrix, which we denote by $\widehat{A}$, in which case Gaussian elimination with pivoting is just the original elimination algorithm applied to $\widehat{A}$ (with no pivoting).

The imagined movement of rows and columns can be thought of as pivoting them around an imaginary midpoint. For example, if we do only row pivoting, we swap row k with row i, pivoting around the row position $\frac{1}{2}(i+k)$ (which might be in between two rows if this is not an integer). If we do only row or column pivoting (but not both) then it is called *partial pivoting*. Doing both is called *full pivoting*.

Pivoting is an adaptive algorithm that attempts to keep elimination going when it might otherwise fail. More generally, it can improve the quality of the solution process by minimizing the size of the multipliers m_{ik} in (3.5). We will say more about this in section 18.2.3.

3.4.1 When no pivoting is needed

We now address the issue of what governs whether Gaussian elimination fails by having $a_{kk}^k = 0$ for some $k \geq 1$. Define A_k to be the $k \times k$ upper-left minor of A, that is,

$$A_k = \begin{pmatrix} a_{11} & \cdots & a_{1k} \\ \vdots & \vdots & \vdots \\ a_{k1} & \cdots & a_{kk} \end{pmatrix}. \tag{3.27}$$

It is easy to see that $(A_k)^k = (A^k)_k$, that is, doing Gaussian elimination on the $k \times k$ upper-left minor of A yields the same result as the $k \times k$ upper-left minor of the result of Gaussian elimination on A. We can refer to $(A_k)^k = (A^k)_k$ simply as A_k^k. If Gaussian elimination has proceeded to step k without producing a zero value a_{kk}^k, then we know that A_k must be invertible.

Suppose that k is the first index where $a_{kk}^k = 0$. We have a factorization $A_k = L_k U_k$, where $U_k = A_k^k$, since Gaussian elimination has proceeded successfully until this point. But the upper-triangular matrix A_k^k is singular since there is a zero on the diagonal. Therefore, A_k is singular for either of two reasons: (1) since A_k^k is obtained from A_k by elementary row operations, or (2) picking $X \neq 0$ such that $U_k X = A_k^k X = 0$ and noting that we must have $A_k X = L_k U_k X = 0$. Thus we have proved the following result.

Lemma 3.7 *Gaussian elimination using no pivoting proceeds with nonzero pivotal elements to produce nonsingular factors $A = LU$ if and only if each $k \times k$ upper-left minor A_k is nonsingular for $k = 1, \ldots, n$.*

One corollary of lemma 3.7 is that Gaussian elimination always succeeds without pivoting for a positive definite matrix (exercise 3.16).

3.4.2 Partial pivoting

The following result shows that partial pivoting is sufficient to solve any invertible system.

Theorem 3.8 *Gaussian elimination with partial pivoting (either row or column) proceeds with nonzero pivotal elements to produce nonsingular factors $\widetilde{A} = LU$ if and only if A is nonsingular.*

theorem 3.8 refers to two algorithms involving partial pivoting: one in which row pivoting is done and the other in which column pivoting is done. It does not refer to some mixture of the two. The only issue in question is whether or not there are nonzero pivotal elements available.

Proof. If Gaussian elimination succeeds, then A is invertible because it is a permutation of LU, both of which are invertible, so we need to address only the converse.

If we perform column pivoting and find at the kth stage that there are no nonzero pivotal elements available in the kth row, then it means that the entire kth row of A^k is zero, and hence A must have been singular to start with. Similarly, if we perform row pivoting and fail at the kth stage to find a nonzero pivotal element, then the kth column of A^k is zero on and below the diagonal. We can write the upper part of the kth column as a linear combination of the first $k-1$ columns by solving a $(k-1) \times (k-1)$ triangular system using the upper $(k-1) \times (k-1)$ block of A^k, which we can assume is nonsingular by induction. Thus again A must be singular.   QED

3.4.3 Full pivoting and matrix rank

Since partial pivoting succeeds for any nonsingular matrix, we might wonder about the role of full pivoting. We will see that there are two applications, one algebraic, which we explore here, and the other analytic in nature, which we discuss in section 18.2.3. So let us assume that full pivoting is done, and at some point Gaussian elimination fails to find a nonzero pivotal element.

If there is no element $a_{i,j}^k \neq 0$, then this means that an entire $(n - k + 1) \times$ $(n - k + 1)$ subblock of A^k is zero. In particular, it means that $A^k x = 0$ for all x such that $x_j = 0$ for $j \geq k$. Since we have performed only row operations on A to define A^k, this means that the kernel of A is (at least) $(n - k + 1)$-dimensional. Assuming this is the first time this has happened in the elimination algorithm, this shows that the rank of A is precisely $n - k + 1$. Moreover, there are solutions $Ax = f$ if and only if $f_j^k = 0$ for $j \geq k$.

Although we will not make substantial use of these facts, we see that Gaussian elimination allows us to find the rank of a matrix and to determine whether f satisfies the compatibility conditions required to have a solution. Compare this with research on computing the rank of a general tensor (e.g., SIAM News, Volume 37, Number 9, November 2004).

We will see later that there are other benefits to full pivoting (cf. section 18.2.3) in addition to the fact that it allows solution of rank-deficient systems (in exact arithmetic). Thus one might wonder why full pivoting is not the default approach. One reason is the increased work, since it requires a comparison of $(n - k)^2$ floating-point numbers, for $k = 1, \ldots, n - 1$, and this essentially doubles the work. Partial pivoting reduces the extra work to comparing only $n - k$ floating-point numbers for each k, and this extra work is asymptotically hidden for large n.

3.4.4 The one-dimensional case

Let us work out some of the details in the simple case when the dimension of the null space of A is 1. Thus there is a solution $x \neq 0$ to $Ax = 0$, and all solutions y of $Ay = 0$ satisfy $y = \alpha x$ for some scalar α. In this case, the codimension of the range of A is also 1, and there is a vector g such that there is a solution y to

$$Ay = f \quad \text{if and only if} \quad g^* f = 0. \tag{3.28}$$

Let us show how the LU factorization can be used to compute g.

We can characterize g as a solution to $A^* g = 0$, as follows:

$$0 = g^* A y = \overline{y^* (A^* g)} \quad \forall y \in \mathbb{F}^n \tag{3.29}$$

if and only if $A^* g = 0$. Thus we need to see how to compute a null solution, something of interest in its own right.

If we do full pivoting, then Gaussian elimination proceeds to the end, with the only oddity being that $u_{nn} = 0$. But the matrix factor L is computed without incident; note that there is nothing to compute in the nth column of L (the only nonzero entry is $l_{nn} = 1$). Of course, $A = LU$ implies that $A^* = U^* L^*$, although now it is the upper-triangular factor (L^*) that is 1 on the diagonal. Transposing the algorithm (3.17), we first solve $U^* w = 0$ and then $L^* g = w$. Since L^* is always invertible, there is no obstruction to obtaining g from w. So the constraint in (3.28) has to involve w. The null space of U^* is easy to characterize: it is generated by the vector $E^n = (0, 0, \cdots, 0, 1)^\mathsf{T}$. Thus g is computed via

$$L^* g = E^n. \tag{3.30}$$

In a similar way, we can compute a generator x for the null space of A. Thus we seek x such that $L(Ux) = 0$, and since L is invertible, we must have $Ux = 0$. We can take $x_n = 1$ since we have $u_{nn} = 0$. Then the second line in (3.9) can be used to generate the remaining values of x_i, $i = n - 1, \ldots, 1$; cf. exercise 3.18.

3.4.5 Uniqueness of the factorization

Suppose we assume that the row and/or column exchanges have been incorporated into A, so that Gaussian elimination with no pivoting factors $A = LU$. We now ask whether such a factorization is unique. Of course, different row and/or column exchanges will give in general a different factorization, but we assume that these exchanges are fixed for this discussion.

If $\widetilde{L}$ and $\widetilde{U}$ are two other triangular factors of $A = \widetilde{L}\widetilde{U}$ with the property that the diagonal elements of $\widetilde{L}$ are also all 1's, then we claim that $L = \widetilde{L}$ and $U = \widetilde{U}$. To see this, write $LU = \widetilde{L}\widetilde{U}$ and multiply on the right by $\widetilde{U}^{-1}$ and on the left by L^{-1}. We find $U\widetilde{U}^{-1} = L^{-1}\widetilde{L}$. But $U\widetilde{U}^{-1}$ is upper-triangular and $L^{-1}\widetilde{L}$ is lower-triangular (cf. section 3.3). To be equal, they both must be diagonal. Since both L and $\widetilde{L}$ have only 1's on the diagonal, this must also hold for $L^{-1}\widetilde{L}$; cf. (3.24) and (3.25). Thus $U\widetilde{U}^{-1} = L^{-1}\widetilde{L} = I$.

There are other possible factorizations with different diagonal assignments. We see that $A = LDU$ gives a description of the general case, where both L and U have 1's on the diagonal and D is a diagonal matrix. More precisely, if $A = L\widehat{U}$ is the factorization provided by Gaussian elimination, define D to be the diagonal matrix whose entries are the diagonal elements of $\widehat{U}$. Then set $U = D^{-1}\widehat{U}$, yielding $A = LDU$. By (3.24), U has 1's on the diagonal.

A corollary of the uniqueness of the factorization is that if A is symmetric, then $A = LDL^T$.

3.5 MORE READING

There are excellent texts on numerical linear algebra, e.g., by Demmel [45] and by Trefethen and Bao [160]. The monograph by van de Geijn and Quintana-Ortí [164] develops the approach given in section 3.2.1. Parallel linear algebra is covered to a limited extent in [144], and references to further work can be found there.

3.6 EXERCISES

Exercise 3.1 *Show directly that a triangular matrix with a 0 diagonal entry must be singular. (Hint: suppose the ith diagonal entry of an lower-triangular matrix is 0. Show that the vector that is 1 in the ith position and 0 for indices less than i can be extended such that it is mapped to 0 by multiplication by the triangular matrix.)*

Exercise 3.2 *Prove that the product of the diagonal entries of a triangular matrix is equal to its determinant. (Hint: show that the determinant of a triangular matrix can be computed by a simple recursion.)*

Exercise 3.3 *Prove theorem 3.5 by multiplying the expressions for the factors L and U in terms of the individual coefficients.*

Exercise 3.4 *Prove that if A is invertible, then there is a permutation matrix P such that PA can be factored as $PA = LU$. (Hint: modify the proof of theorem 3.5 by introducing a permutation matrix at each step. Define $P^{(1)}$ to be a permutation matrix such that $P^{(1)}A$ can be written in the form (3.12) with $a_{11} \neq 0$. Then use (3.13) and apply, by induction, the result to the submatrix $B - a_{11}^{-1}a^1a_1{}^T$. Show that $P^{(1)}A$ is invertible if and only if $B - a_{11}^{-1}a^1a_1{}^T$ is invertible. First, prove that (3.13) is correct, by multiplying it out, and then compare the second factor with $A^{(2)}$. Note that $\det P^{(1)}A = a_{11} \det(B - a_{11}^{-1}a^1a_1{}^T).$)*

Exercise 3.5 *Suppose that α and β are complex scalars, $\tilde{a}$ and $\tilde{b}$ are complex column vectors of length $n - 1$, $a^\star$ and $b^\star$ are complex row vectors of length $n - 1$, and A and B are $(n - 1) \times (n - 1)$ complex matrices. Prove that the block matrix multiplication formula holds:*

$$\begin{pmatrix} \alpha & a^\star \\ \tilde{a} & A \end{pmatrix} \begin{pmatrix} \beta & b^\star \\ \tilde{b} & B \end{pmatrix} = \begin{pmatrix} \alpha\beta + a^\star\tilde{b} & \alpha b^\star + a^\star B \\ \beta a + Ab & \tilde{a}b^\star + AB \end{pmatrix}. \tag{3.31}$$

Note that $a^\star\tilde{b}$ is an inner product (resulting in a scalar) and $\tilde{a}b^\star$ is an outer product (resulting in an $(n - 1) \times (n - 1)$ matrix).

Exercise 3.6 *Prove (3.13) by multiplying out the factors in block form. (Hint: use exercise 3.5.)*

Exercise 3.7 *Prove the last two equalities in (3.15) by multiplying out the factors. (Hint: use exercise 3.5.)*

Exercise 3.8 *Using a programming language that supports recursion, develop a code to perform an LU factorization in which the representation (3.11) provides the base case and (3.13) provides the reduction to a smaller problem.*

Exercise 3.9 *Derive an algorithm for forward solution for lower-triangular systems $LY = F$. (Hint: the transpose of a lower-triangular matrix is upper-triangular.)*

Exercise 3.10 *Prove that $\sum_{j=1}^{n-1} j^2 = \frac{1}{3}n^3 + an^2 + bn$ and determine the constants a and b. (Hint: summing is like integrating; compare $\sum_{j=1}^{n-1} j^2$ with $\int_1^n x^2\, dx$.)*

Exercise 3.11 *Prove that, for $p = 0, 1, 2, \ldots, 8$,*

$$\sum_{j=0}^{n-1} j^p = \sum_{i=0}^{p} \frac{B_{p-i}}{i+1} \binom{p}{i} n^{i+1}, \tag{3.32}$$

where B_i is the ith Bernoulli[3] number, which are given by

$$[B_0, B_1, \ldots] = [1, -\tfrac{1}{2}, \tfrac{1}{6}, 0, -\tfrac{1}{30}, 0, \tfrac{1}{42}, 0, -\tfrac{1}{30}, \ldots]. \tag{3.33}$$

(Hint: see exercise 13.16.)

Exercise 3.12 *The definition of matrix-vector multiplication is*

$$(AV)_i = \sum_{j=1}^{n} a_{ij} v_j. \tag{3.34}$$

If B is another matrix, then prove that $B(AV) = (BA)V$, where the matrix BA is defined by

$$(BA)_{ij} = \sum_{k=1}^{n} B_{ik} A_{kj}. \tag{3.35}$$

(Hint: just apply (3.34) twice.)

Exercise 3.13 *Determine the number of floating-point operations required to determine the multipliers defined in (3.5) as a function of n.*

Exercise 3.14 *Consider an $n \times n$ matrix A and equations $AX^i = F^i$ with m right-hand sides F^i. Show that the number of floating-point operations required to solve the m systems of equations with the same matrix but with m different right-hand sides is $\frac{1}{3}n^3 + \frac{1}{2}mn^2$ to leading order. Show that the inverse of A can be computed in $(5/6)n^3$ operations to leading order.*

Exercise 3.15 *Verify the form (3.24) of the product of triangular matrices.*

Exercise 3.16 *Show that Gaussian elimination can proceed without pivoting for positive definite matrices.*

Exercise 3.17 *Suppose that A is an $n \times n$ matrix and there is an $\alpha > 0$ such that, for all $X \in \mathbb{R}^n$,*

$$X^T A X \geq \alpha X^T X. \tag{3.36}$$

Prove that, without pivoting, the pivotal elements a_{kk}^k are always at least as big as α. Show that pivoting is not necessary if the symmetric part $\frac{1}{2}(A + A^T)$ of A is positive definite.

Exercise 3.18 *Modify the backsubstitution algorithm (3.9) to solve for a nonzero solution to $Ux = 0$ in the case where $u_{nn} = 0$ but all other diagonal entries of the upper-triangular matrix U are nonzero. Take $x_n = 1$.*

---

[3]Jakob Bernoulli (1654–1705) was the brother of Johann Bernoulli (1667–1748), who was a tutor of Euler. Johann's son David Bernoulli (1700–1782) also interacted with Euler, as did his nephew Nicolaus Bernoulli (1687–1759). The Bernoulli family tree included as well several other mathematicians active in the 18th century.

3.7 SOLUTIONS

Solution of Exercise 3.3. We begin by using (3.8) and rewriting (3.4) as

$$m_{ik}u_{kj} = m_{ik}a_{kj}^k = a_{ij}^k - a_{ij}^{k+1} \quad \forall i = k+1, \ldots, n; \ j = k+1, \ldots, n. \quad (3.37)$$

This is valid for all k as long as $i > k$ and $j > k$, that is, for $k < \nu :=$ $\min\{i, j\}$. Summing (3.37), we find (for all $i, j = 1, \ldots, n$)

$$\sum_{k=1}^{\nu-1} m_{ik}u_{kj} = \sum_{k=1}^{\nu-1} \left(a_{ij}^k - a_{ij}^{k+1}\right) = a_{ij}^1 - a_{ij}^\nu \quad (3.38)$$

because the sum on the right-hand side of (3.38) telescopes. Suppose that $i \le j$, so that $\nu = i$. Then (3.38) simplifies to

$$\sum_{k=1}^{i-1} m_{ik}u_{kj} = a_{ij}^1 - a_{ij}^i = a_{ij} - u_{ij}. \quad (3.39)$$

Since $\ell_{ii} = 1$, we can write (3.39) as

$$\sum_{k=1}^{i} \ell_{ik}u_{kj} = a_{ij}, \quad (3.40)$$

which verifies the factorization for $i \le j$.

Now suppose that $i > j$, so that $\nu = j$. Then (3.38) simplifies to

$$\sum_{k=1}^{j-1} m_{ik}u_{kj} = a_{ij}^1 - a_{ij}^j = a_{ij} - m_{ij}a_{jj}^j = a_{ij} - m_{ij}u_{jj} \quad (3.41)$$

in view of the definition of the multipliers (3.5) and U (3.8). Therefore,

$$\sum_{k=1}^{j} \ell_{ik}u_{kj} = a_{ij}, \quad (3.42)$$

which verifies the factorization for $i > j$.

Solution of Exercise 3.10. We have

$$\int_k^{k+1} x^2 \, dx = \tfrac{1}{3}((k+1)^3 - k^3) = \tfrac{1}{3}(3k^2 + 3k + 1) = k^2 + k + \tfrac{1}{3}. \quad (3.43)$$

Summing, we find

$$\tfrac{1}{3}n^3 = \int_0^n x^2 \, dx = \sum_{k=0}^{n-1}(k^2 + k + \tfrac{1}{3}). \quad (3.44)$$

Thus

$$\sum_{k=1}^{n-1} k^2 = \sum_{k=0}^{n-1} k^2 = \tfrac{1}{3}n^3 - \sum_{k=0}^{n-1}(k + \tfrac{1}{3})$$
$$= \tfrac{1}{3}n^3 - \tfrac{1}{2}n(n-1) - \tfrac{1}{3}n = \tfrac{1}{3}n^3 - \tfrac{1}{2}n^2 + \tfrac{1}{6}n. \quad (3.45)$$

Solution of Exercise 3.16. Let $0 \ne x \in \mathbb{R}^k$ be arbitrary and let

$$X = (x_1, \ldots, x_k, 0, \ldots, 0) \in \mathbb{R}^n.$$

Then $AX = (A_k x, y)^T$ for some $y = (y_{k+1}, \ldots, y_n)$, and thus $X^T A X = x^T A_k x$. Therefore, if A is positive definite, then so is A_k for all $k = 1, \ldots, n$, and in particular, A_k is invertible.

Chapter Four

Direct Solvers

> There are two Jordans who appear in numerical analysis. Wilhelm Jordan (1842–1899) was German and is associated with the variant of the elimination method known as Gauss-Jordan [7]. Marie Ennemond Camille Jordan (1838–1922) was French and is associated with the decomposition leading to the Jordan canonical form, as well as many other important ideas in mathematics.

We now consider the problem of solving linear systems of equations more extensively. This chapter is not required for reading subsequent chapters, so it can be skipped without affecting the flow in later chapters. However, it provides a more algorithmic view of linear algebra that can be of interest in its own right. In particular, we will see that the Gaussian elimination algorithm is not optimal on current computers. We will see that there are other algorithms that can produce the same, or similar, factors that are more efficient. Moreover, we will see that factorizations, and the algorithms that produce them, can be tailored to particular properties of the linear system, such as symmetry of the corresponding matrix. In addition, we will see that these factorization methods preserve common patterns of *sparsity*, that is, systematic occurrences of coefficients in the linear system that are known in advance to be zero.

4.1 DIRECT FACTORIZATION

Now that we know that a factorization $A = LDU$ exists (section 3.4.5), it is reasonable to ask if there is a more direct way to derive the factors. There are several algorithms that are quite similar but deal with the diagonal in different ways. These schemes are often called *compact factorization* schemes; they can be written succinctly, and their memory reference patterns are more controlled.

One might ask what the benefit of a different algorithm might be. We will see that the number of floating-point operations is the same, and the memory usage is the also the same (e.g., can be done using only the storage allocated to A itself). However, the number of *memory references* is not the same, and this is what distinguishes compact factorization schemes from Gaussian elimination. Another advantage of compact schemes is that higher precision can be used for the intermediary accumulations; see section 18.1.3 for ways to compute equation (4.4) more accurately.

4.1.1 Doolittle factorization

Doolittle[1] factorization [158] produces the same factors as Gaussian elimination; that is, $A = LU$ with L always having 1's on the diagonal. We can derive the algorithm by simply writing the equation for the product and rearranging:

$$a_{ij} = \sum_{k=1}^{\min\{i,j\}} \ell_{ik} u_{kj}, \tag{4.1}$$

where the summation is limited because we know that L is lower-triangular and U is upper-triangular. If $i \leq j$, then $\min\{i,j\} = i$, and we can write (4.1) as

$$u_{ij} = a_{ij} - \sum_{k=1}^{i-1} \ell_{ik} u_{kj} \tag{4.2}$$

since $\ell_{ii} = 1$. If we use this with $i = 1$, we find $u_{1j} = a_{1j}$ for $j = 1, \ldots, n$, as we expect from Gaussian elimination. If $j \leq i$, then $\min\{i,j\} = j$, and we can write (4.1) as

$$\ell_{ij} = u_{jj}^{-1} \left(a_{ij} - \sum_{k=1}^{j-1} \ell_{ik} u_{kj} \right). \tag{4.3}$$

If we use this with $j = 1$, we find $\ell_{i1} = a_{i1}/u_{11}$ for $i = 2, \ldots, n$, as we also expect from Gaussian elimination. Once we have the ℓ_{i1}'s, we see that we can now use (4.2) for $i = 2$ since it involves only $k = 1$ if $i = 2$ and the u_{1j}'s are already known.

Thus we can alternate between (4.2) and (4.3), computing what we need for the next step. By reversing the index names in (4.3), we can collect these steps as a single algorithm: for $i = 1, \ldots, n$,

$$
\begin{aligned}
u_{ij} &\leftarrow a_{ij} - \sum_{k=1}^{i-1} \boxed{\ell_{ik}} u_{kj} \quad \forall j = i, \ldots, n \\
\ell_{ji} &\leftarrow \boxed{u_{ii}^{-1}} \left(a_{ji} - \sum_{k=1}^{i-1} \ell_{jk} \boxed{u_{ki}} \right) \quad \forall j = i+1, \ldots, n.
\end{aligned}
\tag{4.4}
$$

This computes the first row of U and then the first column of L, then the second row of U and the second column of L, and so forth. The boxes around terms in a given line indicate variables that get used multiple times for a given value of i. We will now see how this can lead to an improvement in performance.

---

[1] Myrick Hascall Doolittle (1830–1913) was a mathematician in the Computing Division of the U.S. Coast and Geodetic Survey [24, 59]. Although the basic algorithm [51] was known to Gauss, and Doolittle's "contribution seems to have been to design a tableau in which the terms were expeditiously recorded" [151], his work stands as one of the earliest American algorithms. Doolittle studied briefly with Benjamin Peirce at Harvard [72].

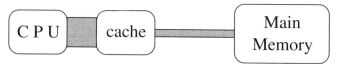

Figure 4.1 A simple model for a computer with a cache. The shaded "pipes" indicate the memory pathways; the larger and shorter pipe between the CPU and cache is much faster than the narrower and longer pipe to main memory.

There is some fine print relating to the initialization and finalization of (4.4). For $i = 1$, the summation in the first line is empty, so it corresponds to the simple assignment $u_{1j} \leftarrow a_{1j}$ for $j = 1, \ldots, n$. For $i = n$, there are no valid values of j in the last line, and there is no work to be done. Thus (4.4) really applies only for $i = 2, \ldots, n - 1$, with slightly different work to be done for $i = 1$ and $i = n$.

4.1.2 Memory references

To understand the impact of memory references, we need to have some model of memory. Indeed, if memory access were very fast compared to floating-point operations (as it was in early digital computers), we could ignore its effect. But modern computer architectures have very complex memory systems, with multiple levels of memory including different levels of cache as well as more conventional memory components. As computer designs have progressed, processor speeds have increased exponentially, but the speed of typical memory operations has not advanced as quickly as the speed of floating-point operations. In current computers, only carefully chosen memory operations proceed quickly. The trend is toward even more complex memory systems, with the ratio of the speed of a general memory operation to that of a floating-point operation increasing in the process. We consider a very simple model just to give the flavor of the issues and leave detailed analyses of different algorithms for various memory systems to a text devoted to computer architecture [126].

All computers today utilize the concept of a *cache*, which we depict in figure 4.1 in a very simplified model. The cache duplicates a portion of the main memory, and it typically does so using technology that allows faster access by the *central processing unit* (*CPU*) where floating-point arithmetic (and other operations) take place. If a certain variable is not available in cache, it is retrieved from main memory (and stored as well in cache) but in a less timely fashion. Moreover, the previous cache contents are overwritten in the process and are no longer available. This action is called a *cache miss*, and it can cost orders of magnitude more than a *cache hit*, which is a memory reference to something already in cache. Thus algorithms that can reuse the data stored in cache can perform much faster than ones that must repeatedly access main memory.

Figure 4.2 Residence of variables in cache and main memory in the compact scheme in the computation of line 1 (a) and line 2 (b) of (4.4).

With this model in mind, let us examine the memory reference pattern of the compact scheme (4.4) and compare it with the memory reference pattern of Gaussian elimination. For simplicity, we will count only memory references to main memory, that is, cache misses, essentially assuming that the cache is infinitely fast by comparison. Furthermore, we will assume n is such that the cache cannot hold n^2 numbers, but it can hold ρn floating-point numbers, where ρ is a small but fixed integer. We explore in exercise 4.1 the possibility of having ρ be a factor less than 1.

Let's do the numbers for memory references for Gaussian elimination first. For $k = 1, \ldots, n-1$, we have to read a block of memory of size $(n-k) \times (n-k)$ corresponding to the a_{ij}'s in (3.18) and then write it back to memory. In our model, at least for the beginning values of k, this requires two memory references for each i, j and thus a total of

$$2 \sum_{k=1}^{n-1} (n-k)^2 = 2 \sum_{m=1}^{n-1} m^2 = \tfrac{2}{3} n^3 + \mathcal{O}(n^2) \tag{4.5}$$

memory references for Gaussian elimination. Since each step requires accessing $(n-k)^2$ items, we assume that all these represent cache misses. In addition, for each k, we need to read $n - k$ a_{kj}'s and m_{ik}'s. However, this contribution to the overall memory reference count is of lower order, contributing only to the $\mathcal{O}(n^2)$ term, as the a_{kj}'s and m_{ik}'s can be reused effectively from cache.

Now let us examine the compact scheme (4.4). Fix i for the moment. We need to read $i - 1$ values of ℓ_{ik}, that is,

$$\ell_{i1}, \ell_{i2}, \ldots, \ell_{i,i-1} \quad [i - 1 \text{ cache misses}] \tag{4.6}$$

to compute the first line of (4.4), but these are reused for different values of j, so we may assume that they are stored in cache and that there is only one cache miss involved in acquiring each of them for a given i. Note that these variables are inside a box in (4.4), indicating that we assume they reside in cache; also see figure 4.2(a). For each j (of which there are $n + 1 - i$ values), we further need to read $i - 1$ values of u_{kj}, that is,

$$u_{1j}, u_{2j}, \ldots, u_{i-1,j}, \quad \forall j = i, \ldots, n \quad [(n+1-i) \times (i-1) \text{ cache misses}] \tag{4.7}$$

and one value of a_{ij} and to write u_{ij} to memory. All told, this amounts to $i + 1$ cache misses for each $j = i, \ldots, n$, and adding the cache misses for acquiring ℓ_{ik} at the beginning amounts to

$$i - 1 + (n + 1 - i)(i + 1) \tag{4.8}$$

cache misses for the first line of (4.4). When $i = 1$, there is no computation to be done; all we do is read a_{1j} and write u_{1j} for $j = 1, \ldots, n$, corresponding to $2n$ cache misses, but this agrees with (4.8) for the case $i = 1$. If by design the factors are being stored in the memory locations for the matrix a_{ij}, then these memory references can be avoided. However, we will not make this assumption for simplicity.

The second line is very similar but with the roles reversed. The variables inside a box in line 2 of (4.4), indicate that they reside in cache; also see figure 4.2(b). In particular, there is one more memory reference, to u_{ii}, but one less value of j, for a total of

$$i + (n - i)(i + 1) \tag{4.9}$$

cache misses for the second line of (4.4). Totaling the cache misses in both lines in (4.4) by adding (4.8) and (4.9), we obtain at most

$$2i - 1 + (2n - 2i + 1)(i + 1) = 2(n - i + 1)(i + 1) - 3 \tag{4.10}$$

cache misses. But recall that the second line is executed only for $i < n$. Summing the leading term in (4.10) over all $i < n$ gives ($\iota = i + 1$)

$$2 \sum_{i=1}^{n-1} (n - i + 1)(i + 1) = 2 \sum_{\iota=2}^{n} (n - \iota + 2)\iota = 2(n + 2) \sum_{\iota=2}^{n} \iota - 2 \sum_{\iota=2}^{n} \iota^2$$

$$= (n + 2)(n^2 + n - 2) - 2 \sum_{\iota=2}^{n} \iota^2 \tag{4.11}$$

$$= \tfrac{1}{3} n^3 + \mathcal{O}(n^2)$$

cache misses for the direct factorization method (see exercise 3.10 for the last step).

The factor of 2 reduction in memory references of (4.11) for the compact scheme over the number (4.5) for Gaussian elimination means a

factor of 2 *improvement in overall performance*
on contemporary computers.

We have not said how to ensure that the boxed variables remain in cache, only that there is enough room to hold them. We leave the question of how to program this algorithm to ensure that this happens to a more advanced reference [126] on software implementation. The key requirement of the computer system is that the cache be set associative with a replacement policy such as *least recently used* (LRU) [126].

4.1.3 Cholesky factorization and algorithm

For simplicity, we restrict attention to linear systems with real entries; for the complex case, see exercise 4.2. The Cholesky[2] algorithm was not published

---

[2] André-Louis Cholesky (1875–1918) attended classes given by Camille Jordan (see page 51) at École Polytechnique [23].

until after Cholesky's death, but it was later examined by Turing[3] [161] and Wilkinson[4] and others [65]. The *Cholesky factorization* is of the form $A = U^\mathrm{T}U$, where U is upper-triangular. Thus by definition, A must be symmetric. Cholesky's algorithm is appropriate only for symmetric, positive definite matrices, but it is quite important as it applies to such an important subclass of matrices. The Cholesky algorithm can be written by looping the following expression for $j = 1, \ldots, n$:

$$u_{ij} = \frac{1}{u_{ii}} \left(a_{ij} - \sum_{k=1}^{i-1} u_{ki} u_{kj} \right), \quad i = 1, \ldots, j-1$$

$$u_{jj} = \left(a_{jj} - \sum_{k=1}^{j-1} u_{kj}^2 \right)^{1/2}. \tag{4.12}$$

The first instance ($j = 1$) of this involves only the second equation: $u_{11} = (a_{11})^{1/2}$.

One use of this algorithm is to determine whether a symmetric matrix is positive definite or not.

Theorem 4.1 *Suppose that A is an $n \times n$ symmetric real matrix. Then the Cholesky algorithm (4.12) determines the factorization $A = U^\mathrm{T}U$ with the quantities*

$$a_{jj} - \sum_{k=1}^{j-1} u_{kj}^2 > 0 \tag{4.13}$$

for all $j = 1, \ldots, n$ if and only if A is positive definite.

Proof. First, if the algorithm (4.12) determines U with each u_{jj} a positive real number, then $X^\mathrm{T}AX = (UX)^\mathrm{T}UX = 0$ if and only if $UX = 0$. But since U is triangular with nonzero diagonal entries, $UX = 0$ if and only if $X = 0$. Now suppose that A is positive definite. Let $A = LDL^\mathrm{T}$ be the factorization provided by Gaussian elimination (cf. exercise 3.16). Then $X^\mathrm{T}AX = (LX)^\mathrm{T}DLX > 0$ for any nonzero X. Let Y be arbitrary and solve $LX = Y$. Then $Y^\mathrm{T}DY > 0$ for all nonzero Y. Thus $D > 0$ and the Cholesky algorithm must correctly produce this factorization. QED

Cholesky factorization uses the symmetry of A to write $A = \hat{L}D\hat{L}^\mathrm{T}$ and then takes the square root of D, defining $L = \hat{L}\sqrt{D}$.

4.2 CAUTION ABOUT FACTORIZATION

In our study of algorithms for solving linear systems, we have ignored the effects of finite-precision arithmetic so far. In section 18.2, we will consider

---

[3] Alan Turing (1912–1954) is known for many things in addition to numerical analysis, including his work on cryptography and the foundations of computer science, for which he is memorialized by the Turing award.

[4] James Hardy Wilkinson (1919–1986) worked with Turing and received one of the first Turing awards.

this in more detail, including the potential effect of the accumulation of round-off errors in very large systems. Here we consider by example something simpler, just related to the limits of representation of real numbers in finite precision. However, we will see that this places some severe limits on what can be achieved using the factorization algorithms considered so far. The Hilbert[5] matrix

$$H = \begin{pmatrix} 1 & \frac{1}{2} & \frac{1}{3} & \cdots & \frac{1}{n} \\ \frac{1}{2} & \frac{1}{3} & \frac{1}{4} & \cdots & \frac{1}{n+1} \\ \vdots & \vdots & \vdots & \cdots & \vdots \\ \frac{1}{n} & \frac{1}{n+1} & \frac{1}{n+2} & \cdots & \frac{1}{2n-1} \end{pmatrix} \qquad (4.14)$$

reveals limitations of even the best algorithms. More precisely,

$$h_{ij} = \frac{1}{i+j-1} \quad \text{for } i,j = 1,\ldots,n. \qquad (4.15)$$

The Hilbert matrix is clearly symmetric, and it can be verified (exercise 4.4) that it is positive definite by exhibiting the Cholesky factor

$$u_{ij} = \sqrt{2i-1}\frac{((j-1)!)^2}{(i+j-1)!(j-i)!} \quad \text{for } j = 1,\ldots,n \text{ and } i = 1,\ldots,j, \qquad (4.16)$$

that is, $H = U^T U$. Unfortunately, the diagonal terms

$$u_{jj} = \sqrt{2j-1}\frac{((j-1)!)^2}{(2j-1)!} = \sqrt{2j-1}\frac{(j-1)!}{j(j+1)\cdots(2j-1)} \qquad (4.17)$$

decrease exponentially as j increases (exercise 4.5), whereas $u_{1,n} = 1/n$. For example, $u_{16,16}$ is less than 1.2×10^{-9}. Thus the terms

$$a_{16,16} = \frac{1}{31} \quad \text{and} \quad \sum_{k=1}^{15} u_{kj}^2 \qquad (4.18)$$

in (4.12) must differ by less than 2×10^{-18} (all the terms are positive). Thus 16 digits are insufficient to resolve the difference. Indeed, the Cholesky algorithm implemented to this accuracy can indicate that the Hilbert matrix is not positive definite for n as small as 14. This is not an effect of the accumulation of round-off error (chapter 18) but rather a simple failure of representation. That is, without a way to represent more digits in the terms, there is no way to determine whether the condition (4.13), that is,

$$h_{jj} > \sum_{k=1}^{j-1} u_{kj}^2, \qquad (4.19)$$

holds or not. The factorization can be computed using extended precision arithmetic, but this allows the factorization to continue only for n proportional to the number of digits used in the representation. This does not make

---

[5]David Hilbert (1862–1943) was one of the most influential mathematicians of the 20th century, in part because of a set of 23 problems he posed at an international meeting in 1900, some of which remain unsolved, especially the sixth.

the computation intractable, but an adaptive approach in which different entries are represented to different levels of accuracy might be needed to be efficient. Many systems allow computation with rational coefficients, and this allows exact arithmetic. However, it does not eliminate the possibility of growth of the size of the representation. The Hilbert matrix (4.14) and its Cholesky factorization (4.17) have rational entries. We leave as exercise 4.6 the question of how large the denominator can become in the factors.

4.3 BANDED MATRICES

In many applications, most entries in a matrix are zero. Such a matrix is called *sparse*. One structured matrix of this type is called a *banded matrix*. We will see that the factorizations of a banded matrix retain the band structure.

Define the *bandwidth* w of a matrix A to be the smallest integer such that $a_{ij} = 0$ whenever $|i - j| \geq w$. A full matrix is a matrix with no significant zero structure and thus corresponds to $w = n$ for an $n \times n$ matrix. A diagonal matrix has bandwidth 1. The main fact of interest about banded matrices is the following.

Lemma 4.2 *Suppose A is an $n \times n$ matrix of bandwidth w such that Gaussian elimination can be performed on A without pivoting to produce the factors $A = LU$. Then the bandwidth of both L and U is at most w.*

The proof of this fact is obtained by exhibiting algorithms that provide the factorization and avoid involvement of terms above and below the nonzero band. The banded Cholesky algorithm (section 4.3.1) provides one example, and the general case is similar. For now, we give an example and illustrate the value of working with banded structures.

The matrix A that results from discretizing the second derivative is a tridiagonal ($w = 2$) matrix with 2 on the diagonal and -1 above and below the diagonal:

$$
A = \begin{pmatrix}
1 & -1 & 0 & 0 & \cdots & 0 & 0 & 0 \\
-1 & 2 & -1 & 0 & \cdots & 0 & 0 & 0 \\
0 & -1 & 2 & -1 & \cdots & 0 & 0 & 0 \\
\vdots & \vdots & \vdots & \vdots & \vdots & \vdots & \vdots & \vdots \\
0 & 0 & 0 & 0 & \cdots & -1 & 2 & -1 \\
0 & 0 & 0 & 0 & \cdots & 0 & -1 & 2
\end{pmatrix},
\tag{4.20}
$$

where a slight modification has been done in the first row.

The LU factors of A take a simple form: $U = L^{\mathsf{T}}$ with L given by

$$
L = \begin{pmatrix}
1 & 0 & 0 & \cdots & 0 & 0 \\
-1 & 1 & 0 & \cdots & 0 & 0 \\
0 & -1 & 1 & \cdots & 0 & 0 \\
\vdots & \vdots & \vdots & \vdots & \vdots & \vdots \\
0 & 0 & 0 & \cdots & -1 & 1
\end{pmatrix}.
\tag{4.21}
$$

Thus we see that we can solve linear systems involving the matrix A very effectively by LU factorization and backsubstitution (exercise 4.10). On the other hand, we can now see a real disadvantage to working with the inverse of a matrix.

Define a matrix M, where $m_{ij} = n + 1 - i$ for $j \leq i$ and $m_{ij} = n + 1 - j$ for $j \geq i$. We can visualize this matrix as

$$M = \begin{pmatrix} n & n-1 & n-2 & n-3 & \cdots & 3 & 2 & 1 \\ n-1 & n-1 & n-2 & n-3 & \cdots & 3 & 2 & 1 \\ n-2 & n-2 & n-2 & n-3 & \cdots & 3 & 2 & 1 \\ n-3 & n-3 & n-3 & n-3 & \cdots & 3 & 2 & 1 \\ \vdots & \vdots & \vdots & \vdots & \vdots & \vdots & \vdots & \vdots \\ 3 & 3 & 3 & 3 & \cdots & 3 & 2 & 1 \\ 2 & 2 & 2 & 2 & \cdots & 2 & 2 & 1 \\ 1 & 1 & 1 & 1 & \cdots & 1 & 1 & 1 \end{pmatrix}. \tag{4.22}$$

Then $AM = I$ (exercise 4.11), so $A^{-1} = M$. Thus we see that the inverse of a banded matrix can be full. Solving $AX = F$ for A given by (4.20) requires n^2 (multiply and add) operations using the formula $X = MF$, whereas computing and using the LU factors (section 3.2.2) require only $\mathcal{O}(n)$ operations (see section 4.3.2).

Pivoting increases the bandwidth, but bounds can clearly be made on how much the bandwidth increases. We refer the reader to [39] and [67] for more details.

4.3.1 Banded Cholesky

The Cholesky algorithm can then be written by looping the following expression for $j = 1, \ldots, n$:

$$u_{ij} = \frac{1}{u_{ii}} \left(a_{ij} - \sum_{k=\max\{1, i-w\}}^{i-1} u_{ki} u_{kj} \right), \quad i = \max\{1, j - w\}, \ldots, j - 1,$$

$$u_{jj} = \left(a_{jj} - \sum_{k=\max\{1, j-w\}}^{j-1} u_{kj}^2 \right)^{1/2}.$$

$$\tag{4.23}$$

The Doolittle scheme, Gaussian elimination, and other algorithms can also be written in banded form (see exercises 4.12 and 4.13).

4.3.2 Work estimates for banded algorithms

It is not hard to guess the leading terms in the work estimates for band factorization algorithms such as Cholesky (4.23). The amount of floating-point operations required to factor a matrix is cubic in n, and thus it is reasonable to assume that the banded version would involve either $w^2 n$ or $w n^2$ work (n^3 would mean no gain, and w^3 would be impossible because there are nw nonzero coefficients). Fortunately, the more optimistic $w^2 n$ is correct, as we can see specifically in the banded Cholesky algorithm (4.23), as follows.

Consider the work done for each $j = 1, \ldots, n$. Once $i > w$ and $j > w$, the first line in (4.23) takes the form

$$u_{ij} = \frac{1}{u_{ii}} \left(a_{ij} - \sum_{k=i-w}^{i-1} u_{ki} u_{kj} \right), \quad i = j - w, \ldots, j - 1, \qquad (4.24)$$

and involves exactly w products in the sum, and there are exactly w values of i for which it is computed. Thus the total work is $2w^2$ multiplies and additions and w divisions. The second line in (4.23) has a similar structure and for $j > w$ involves $2w$ multiplies and additions and one square-root. Ignoring the divisions and square-roots, we see that to leading order, the primary work is $2nw^2$ multiplies and additions, with the remaining terms of order nw and smaller.

Let us consider the implications of these results for banded matrices like the one in (4.20), in which $w = 2$. We can factor it using the banded Cholesky algorithm (see exercise 4.14) in an amount of work proportional to nw^2, and we can further use banded forward and backward solution (exercise 4.10) in $\mathcal{O}(wn)$ work. Thus the total amount of work needed to solve $A x = f$ is at most $\mathcal{O}(w^2 n)$. On the other hand, even if we know explicitly the formula for A^{-1} in (4.22), it would take $\mathcal{O}(n^2)$ work to use it to determine $x = A^{-1} f$. Thus if $w << \sqrt{n}$, it is much more efficient to use the factorization rather than the inverse.

4.4 MORE READING

In [144], several algorithms for matrix factorization on parallel computers are presented. This shows the diversity of algorithms available even for such a basic problem.

4.5 EXERCISES

Exercise 4.1 *Show that direct factorization methods can still be effective for the situation where the cache cannot hold an entire row of the matrix.*

Exercise 4.2 *Describe the Cholesky factorization $A = U^\star U$ for matrices with complex entries, where now we must assume that A is Hermitian: $A^\star = A$. How do the equations (4.12) change in this case? How is the condition (4.13) different?*

Exercise 4.3 *Implement the Cholesky factorization (4.12) using arithmetic with finite-precision and apply it to the Hilbert matrix (4.14). For what n does it fail?*

Exercise 4.4 *Verify computationally that (4.16) defines the Cholesky factorization of the Hilbert matrix (4.14), that is, $H = U^T U$.*

Exercise 4.5 *Prove that the diagonal terms of the Cholesky factorization (4.17) of the Hilbert matrix decay exponentially. (Hint: write (4.17) as*

$$\frac{\sqrt{2j-1}}{j} \frac{1}{j+1} \frac{2}{j+2} \cdots \frac{j-1}{2j-1} \tag{4.25}$$

and show that all the factors are less than $\frac{1}{2}$ for $j > 7$.)

Exercise 4.6 *Both the entries of the Hilbert matrix (4.14) and the entries of its Cholesky factorization (4.17) are rational numbers. Does this resolve the issue of representing the computation with finite precision? How large can the denominator of the diagonal entries in the factors become? How many bits are required to represent them in a binary expansion?*

Exercise 4.7 *Develop an algorithm, and determine the operation count, for the symmetric Gaussian elimination method. Compare it to the Cholesky method and to the regular Gaussian elimination method. First, you need to explain the algorithm in detail. (Hint: note the symmetry in the computation $a_{ij} - a_{ik}a_{kj}/a_{kk}$ when A is symmetric. Show that this allows you to work with only the upper-triangular part of A.)*

Exercise 4.8 *Count the number of memory references for symmetric Gaussian elimination (exercise 4.7) and compare it to the Cholesky method and to the regular Gaussian elimination method.*

Exercise 4.9 *The Crout[6] factorization $A = LU$ sets the diagonal of U to be all 1's [38], instead of having the diagonal of L all 1's as in Gaussian elimination. Derive an algorithm analogous to the Doolittle algorithm (4.4) for computing the Crout factorization.*

Exercise 4.10 *Derive the algorithm for banded forward and backward solution and estimate the number of arithmetic operations and memory references.*

Exercise 4.11 *Prove that $AM = I$, where A is defined in (4.20) and M is defined in (4.22).*

Exercise 4.12 *Derive the algorithm for banded Gaussian elimination and estimate the number of arithmetic operations and memory references.*

Exercise 4.13 *Derive the banded version of the Doolittle algorithm for direct factorization and estimate the number of arithmetic operations and memory references.*

---

[6]Prescott Durand Crout (1907–1984) was a professor of mathematics at MIT from 1934 to 1973 and a member of the Radiation Laboratory staff from 1941 to 1945.

Exercise 4.14 *Show that the banded matrix matrix A in (4.20) is positive definite. (Hint: use the factor L.)*

Exercise 4.15 *Let A be an invertible $n \times n$ matrix with an existing factorization $A = LU$. Let u and v be nonzero vectors of length n (that is, matrices of size $n \times 1$). The* Sherman-Morrison *formula states that*

$$(A - uv^T)^{-1} = A^{-1} + \alpha A^{-1} uv^T A^{-1}, \qquad (4.26)$$

where $\alpha = (1 - v^T A^{-1} u)^{-1}$, provided that $r = v^T A^{-1} u \neq 1$ as we now assume. First, prove that this formula is correct. Second, show that if $r = v^T A^{-1} u = 1$, then $A - uv^T$ is not invertible. Finally, suppose that $B = A - uv^T$ and that $r \neq 1$. Use the factorization $A = LU$ to compute the solution to $BX = F$ in only $\mathcal{O}(n^2)$ work. The matrix B is called a rank-one update of A since the matrix uv^T has rank one. (Hint: write $X = B^{-1}F = Y + \alpha W v^T Y$, where Y solves $AY = F$ and W solves $AW = u$. Don't forget to check the value of $r = v^T W$ before you compute $\alpha = (1 - r)^{-1}$.)

Exercise 4.16 *The first and last lines of the matrix A defined in (4.20) depend on the boundary conditions being employed in the definition of the difference operator. The top line corresponds to a Neumann, or derivative, boundary condition in which the derivative of the function value is set to zero; the bottom line corresponds to a Dirichlet[7] boundary condition in which the function value is set to zero. If we change the first line of A to*

$$\begin{pmatrix} 2 & -1 & 0 & 0 & \cdots & 0 & 0 & 0 \end{pmatrix}, \qquad (4.27)$$

then we get a new matrix $B = A - uv^T$ corresponding to having Dirichlet conditions at both ends. Use the Sherman-Morrison formula (exercise 4.15) to prove that B is invertible. (Hint: figure out what u and v need to be. To prove $r \neq 1$, you need to check only the sign of $v^T A^{-1} u$.) If we change the last line of A to

$$\begin{pmatrix} 0 & 0 & 0 & 0 & \cdots & 0 & -1 & 1 \end{pmatrix}, \qquad (4.28)$$

then we get a new matrix $\widetilde{B} = A - uv^T$ corresponding to having Neumann conditions at both ends. Show that this matrix is singular and examine what goes wrong with the Sherman-Morrison formula in this case. (Hint: apply $\widetilde{B}$ to the vector of all 1's.)

Exercise 4.17 *Use the Sherman-Morrison formula (exercise 4.15) and the factor L in (4.21) to define an algorithm to solve $BX = F$ that avoids factoring B, where B is defined in exercise 4.16. Give the explicit formulas, including those for u and v.*

---

[7]Gustav Peter Lejeune Dirichlet (1805–1859) studied at gymnasium with Georg Ohm (of Ohm's Law [129]) and then in Paris during 1823–1825 before returning to Germany, where he obtained a position in Berlin with help from Humboldt [93]. After the death of Gauss, Dirichlet succeeded him in Göttingen [40]. Dirichlet's students included Leopold Kronecker (of the δ symbol) and Rudolf Lipschitz (see page 17).

Exercise 4.18 *Let A and B be any $n \times n$ matrices. Prove that*

$$\det \begin{pmatrix} A & B \\ 0 & I \end{pmatrix} = \det \begin{pmatrix} A & 0 \\ B & I \end{pmatrix} = \det A, \qquad (4.29)$$

where I denotes the $n \times n$ identity matrix (cf. the more general case in [146]). (Hint: expand the determinant around the lower-right corner.)

Exercise 4.19 *Let A, B, C, D, K, L, M, and N be any $n \times n$ matrices. Prove the block multiplication formula*

$$\begin{pmatrix} A & B \\ C & D \end{pmatrix} \begin{pmatrix} K & L \\ M & N \end{pmatrix} = \begin{pmatrix} AK + BM & AL + BN \\ CK + DM & CL + DN \end{pmatrix}. \qquad (4.30)$$

(Hint: compare (3.31).)

Exercise 4.20 *Let A and B be any $n \times n$ matrices. Prove that $\det(A B - \lambda I) = \det(B A - \lambda I)$ for any $\lambda \in \mathbb{C}$, where I denotes the $n \times n$ identity matrix. Thus $A B$ and $B A$ have the same eigenvalues. (Hint: show that, for $\lambda \neq 0$,*

$$\begin{pmatrix} \lambda I - A B & A \\ 0 & I \end{pmatrix} \begin{pmatrix} I & 0 \\ B & \lambda I \end{pmatrix} = \begin{pmatrix} \lambda I & \lambda A \\ B & \lambda I \end{pmatrix}$$

$$= \begin{pmatrix} I & 0 \\ (1/\lambda)B & \lambda I - B A \end{pmatrix} \begin{pmatrix} \lambda I & \lambda A \\ 0 & I \end{pmatrix} \qquad (4.31)$$

using exercise 4.19 and then apply exercise 4.18. Use a continuation argument to include $\lambda = 0$.)

4.6 SOLUTIONS

Solution of Exercise 4.1. When $i = 1$ in (4.4), only assignment takes place, so we can assume that $1 < i \leq n$ in step 1. We segment the summation over k using a subdivision $1 < k_1 < \cdots < k_r = i - 1$ such that, say, $k_{m+1} - k_m \leq (\rho/3)n$. We write the first line of computation in (4.4) as

$$t_j \leftarrow a_{ij} - \sum_{k=1}^{k_1} \boxed{\ell_{ik}} u_{kj} \quad \forall j = i, \ldots, n$$

$$\text{for } m = 2, \ldots, r-1, \quad t_j \leftarrow t_j - \sum_{k=k_{m-1}}^{k_m} \boxed{\ell_{ik}} u_{kj} \quad \forall j = i, \ldots, n. \qquad (4.32)$$

$$u_{ij} \leftarrow t_j - \sum_{k=k_{r-1}}^{i-1} \boxed{\ell_{ik}} u_{kj} \quad \forall j = i, \ldots, n.$$

Here we introduced temporary variables t_j to indicate more clearly how the computation is blocked. There is no need for extra memory; the t_j's could be stored in the storage locations for the u_{ij}'s. The second line of computation in (4.4) is blocked in an analogous way for $i = 1, \ldots, n - 1$:

$$t_j \leftarrow a_{ji} - \sum_{k=1}^{k_1} \ell_{jk}\boxed{u_{ki}} \quad \forall j = i+1, \ldots, n.$$

$$\text{for } m = 2, \ldots, r-1, \quad t_j \leftarrow t_j - \sum_{k=k_{m-1}}^{k_m} \ell_{jk}\boxed{u_{ki}} \quad \forall j = i+1, \ldots, n. \quad (4.33)$$

$$\ell_{ji} \leftarrow \boxed{u_{ii}^{-1}}\left(t_j - \sum_{k=k_{r-1}}^{i-1} \ell_{jk}\boxed{u_{ki}}\right) \quad \forall j = i+1, \ldots, n.$$

Again, we must assume that there is some way to ensure that the variables in boxes are not removed from cache.

Now let us examine the memory references in (4.32). Fix i for the moment. We read $i-1$ values of ℓ_{ik} in blocks that fit within cache and these are reused for different values of j, so we assume there is only one cache miss involved in acquiring each of them for a given i. For each j, we further need to read $i-1$ values of u_{kj} and one value of a_{ij}, but now we also have to read u_{ij} to memory $r-1$ times and write u_{ij} to memory r times. This gives

$$i - 1 + (n - i + 1)(i - 1 + 2r) \quad (4.34)$$

cache misses for (4.32). Similarly, (4.33) requires

$$i + (n - i)(i - 1 + 2r) \quad (4.35)$$

cache misses for (4.33), for a total of

$$2i - 1 + 2(n - i + \tfrac{1}{2})(i - 1 + 2r) \quad (4.36)$$

cache misses. Summing the expression (4.36) over i still yields a total of at most $\frac{1}{3}n^3 + 2rn^2 + \mathcal{O}(n^2)$ cache misses. Thus as long as r is not too large, the performance will be similar.

Solution of Exercise 4.15. Let $C = uv^{\mathsf{T}}A^{-1}$ and $r = v^{\mathsf{T}}Au$. Then

$$(A - uv^{\mathsf{T}})(A^{-1} + \alpha A^{-1}uv^{\mathsf{T}}A^{-1}) = I - C + \alpha(C - C^2)$$
$$= I - C + \alpha(C - u(v^{\mathsf{T}}A^{-1}u)v^{\mathsf{T}}A^{-1}) = I - C + \alpha(C - rC) \quad (4.37)$$
$$= I - C(1 - \alpha(1 - r)) = I.$$

Now suppose that $r = 1$. Then $(A - uv^{\mathsf{T}})(A^{-1}u) = u - ru = 0$. Thus $A^{-1}u$ is a null vector of $A - uv^{\mathsf{T}}$ if $r = 1$. Note that $A^{-1}u = 0$ if and only if $u = 0$ since A is assumed to be invertible. Since $A^{-1}u \neq 0$ is a null vector of $B = A - uv^{\mathsf{T}}$, B cannot be invertible.

The hint explains most of the algorithm. Use LU factorization to solve for W and compute the scalar product $r = v^{\mathsf{T}}W$. If $r = 1$, then stop, noting that B is not invertible in this case. If $r \neq 1$, define $\alpha = (1 - r)^{-1}$ and use LU factorization to solve for Y. Finally, compute $X = Y + \alpha Wv^{\mathsf{T}}Y$. Note that $Wv^{\mathsf{T}}Y = (v^{\mathsf{T}}Y)W$ requires the computation of only a scalar product $v^{\mathsf{T}}Y$ and multiplication of this scalar times the vector W. If you computed in the opposite order, i.e., $(Wv^{\mathsf{T}})V$, forming the matrix Wv^{T} and then multiplying this matrix times the vector V, it would take much more work (and temporary storage).

Chapter Five

Vector Spaces

> In defending his thesis in 1913, S. N. Bernstein (see page 187) said, "Mathematicians for a long time have confined themselves to the finite or algebraic integration of differential equations, but after the solution of many interesting problems the equations that can be solved by these methods have to all intents and purposes been exhausted, and one must either give up all further progress or abandon the formal point of view and start on a new analytic path." [6]

So far, we have dealt with simple functions of a single variable with values that are also one-dimensional. But we want to consider multidimensional objects, and we need to establish some basic ideas. The first is a way to measure sizes of things. So far, the absolute value of a real number was sufficient, or the modulus of a complex number. But in higher dimensions the issue is more complicated. The concept of a *norm* on a vector space provides such a measure.

Suppose we are at point A and need to see something at point B. We imagine this takes place in a two-dimensional plane, as indicated in figure 5.1. How long it takes us to complete the task is context-dependent.

If we are in a typical urban center, we have to move along a grid defined by the streets and sidewalks. The time it takes is proportional to the so-called Manhattan distance, which we will see corresponds to the norm $\|A - B\|_1$

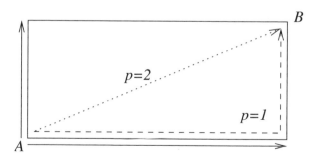

Figure 5.1 Three ways to see what is at point B when starting at point A. The norm $\|A - B\|_\infty$ is the maximum of the length of the two solid arrows. The dashed arrow corresponds to $\|A - B\|_1$, and the dotted arrow corresponds to $\|A - B\|_2$.

(see (5.5)). A typical path is indicated by the dashed line in figure 5.1, but such a path is not unique.

On the other hand, if we are in an empty (flat) field, then we could walk directly and the time it takes would be proportional to the Euclidean distance $\|A - B\|_2$ (see (5.5) again, or (5.4), and the dotted line in figure 5.1). We will see that $\|A - B\|_2 \leq \|A - B\|_1$, so the direct approach is faster when feasible.

Finally, we might not really want to go to B but just see something there. Suppose that A and B are at the corners of a rectangular forest. It is faster to walk around the forest until we can get a clear view of B but we may not know which way to go. The two possible paths are indicated as solid lines in figure 5.1. However, we know that the worst case is the long side of the rectangle, and this is $\|A - B\|_\infty$ (see (5.6)). We will see that $\|A - B\|_\infty$ is always smaller than the Euclidean distance $\|A - B\|_2$, so this strategy is better than walking through the forest.

All these norms are relevant in certain contexts, and none is more important than the others intrinsically. Other norms are of interest as well, and we will explore the concept of norms in general. We have already seen how three of them arise naturally in a context in which length relations among them were of interest. One key result in the chapter (section 5.3.2) is that all norms on a finite-dimensional vector space are equivalent in terms of estimating "distance" (for any two norms, there is a constant such that, for any vector, the first norm of the vector is no larger than that constant times the second norm of that vector).

Some norms (and vector spaces) support a geometric interpretation familiar in Euclidean spaces. Such *inner-product spaces* generalize $\mathbb{R}^n$ and $\mathbb{C}^n$ and allow many operations to be carried out abstractly. One such operation is the Gram-Schmidt orthonormalization process. We will see in section 12.3 that this can be used to construct orthogonal polynomials, which have many applications both theoretical and practical. In this chapter, we show how the Gram-Schmidt process leads to the important QR matrix factorization.

5.1 NORMED VECTOR SPACES

The main point of the section is to introduce ways to estimate accurately the size of things which have complicated forms. This is a simple generalization of Euclidean distance, but it can also apply to rather complicated objects such as operators on vector spaces.

Suppose that V is a (finite-dimensional) vector space such as $V = \mathbb{R}^n$. Then a norm $\| \cdot \|$ is a mapping from V to nonnegative real numbers such that three properties hold. First, it is nondegenerate: if $v \in V$ satisfies

$$\|v\| = 0, \tag{5.1}$$

then v must be the zero element of the vector space V. Second, it is homogeneous with respect to scalar multiplication:

$$\|sv\| = |s|\,\|v\| \tag{5.2}$$

for all scalars s and all $v \in V$. Third, and most important, the triangle inequality must hold:

$$\|v + w\| \leq \|v\| + \|w\| \tag{5.3}$$

for $v, w \in V$.

We have not yet identified the set $\mathbb{F}$ of scalars for our vector spaces. In general for a vector space, it can be any division ring,[1] but for simplicity we will restrict to the case where $\mathbb{F}$ is a field. Moreover, we further restrict to the case where $\mathbb{F}$ is the real or complex numbers. In the latter case, the expression $|s|$ means the complex modulus of s. The reason for this restriction is that *normed* linear spaces require $\mathbb{F}$ to have a norm itself, together with the Archimedian property that $|st| = |s|\,|t|$ for $s, t \in \mathbb{F}$. There are essentially only two such fields: $\mathbb{R}$ and $\mathbb{C}$ [5]. However, it should be noted that the quaternions provide an example of a division ring having an Archimedian norm.

5.1.1 Examples of norms

The Euclidean norm $\|\cdot\|_2$ is defined on $\mathbb{R}^n$ by

$$\|x\|_2 = \left(\sum_{i=1}^{n} x_i^2\right)^{1/2}. \tag{5.4}$$

More generally, for any p in the interval $1 \leq p < \infty$, we define

$$\boxed{\|x\|_p = \left(\sum_{i=1}^{n} |x_i|^p\right)^{1/p}} \tag{5.5}$$

for $x \in \mathbb{F}^n$, where $\mathbb{F} = \mathbb{R}$ or $\mathbb{C}$. Whenever our vector space is $\mathbb{F}^n$, it is understood that the field of scalars is $\mathbb{F}$. That is, the field of scalars for $\mathbb{R}^n$ is $\mathbb{R}$, and for $\mathbb{C}^n$ is $\mathbb{C}$. We use the notation $\mathbb{F}^n$ to avoid having to repeat things for both $\mathbb{R}^n$ and $\mathbb{C}^n$.

It is elementary to establish (5.1) and (5.2) for the p-norms (5.5) (see exercise 5.1). The triangle inequality is elementary for $p = 1$ (exercise 5.2). However, the triangle inequality is far less obvious for other values of p. We postpone the proof until section 5.2.

We want to think of $\mathbb{F}^n$ endowed with different norms as different (normed, linear) spaces. Thus the notation ℓ_p is used to denote $\mathbb{F}^n$ endowed with the p-norm. More precisely, we should write this as $\ell_p(\mathbb{F})$ or even $\ell_p(\mathbb{F}, n)$.

---

[1]A division ring is the same as a field, but the multiplication is not assumed to be commutative [87]. The quaternions are an important example.

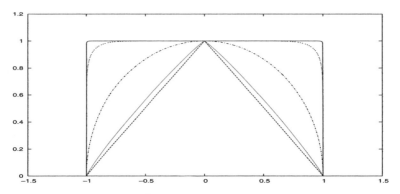

Figure 5.2 Unit "balls" for $\mathbb{R}^2$ with the norms $\|\cdot\|_p$ for $p = 1.01, 1.1, 2, 10, 100$.

5.1.2 Unit balls

It is useful to visualize the *unit ball* for a given norm, that is, the set of vectors of unit size with respect to the norm. Like sports balls, these sets come in different shapes. In $\mathbb{R}^n$, for the Euclidean norm, it is the unit sphere; for $p = 1$, it is a union of $(n-1)$-dimensional simplices. For example, for $n = 2$, it is a diamond-shaped parallelogram with vertices at $(0, \pm 1)$ and $(\pm 1, 0)$. The unit balls in $\mathbb{R}^2$ are depicted in figure 5.2 for various values of p.

It is clear from figure 5.2 that the unit ball for $\|\cdot\|_p$ approaches the square with corners $\pm(1, \pm 1)$ as $p \to \infty$. We have not characterized a norm yet with such a unit ball, but it is not hard to see that this is the unit ball for the max norm

$$\|x\|_\infty = \max_{i=1}^n |x_i|. \tag{5.6}$$

For any fixed $x \in \mathbb{F}^n$,

$$\|x\|_\infty = \lim_{p \to \infty} \|x\|_p, \tag{5.7}$$

which explains the ∞ subscript in the "max" norm. This is a consequence of the following two inequalities.

Since the the set of points on, or inside, the unit ball for the max norm contains all the points on the unit balls for the finite p-norms, we have

$$\|x\|_\infty \le \|x\|_p \tag{5.8}$$

(see exercise 5.3 for a less visual proof of (5.8)). All p-norms are dominated by the max norm:

$$\|x\|_p \le n^{1/p} \|x\|_\infty \quad \forall x \in \mathbb{F}^n \tag{5.9}$$

because each term in the sum (5.5) is bounded by $\|x\|_\infty$. The result (5.7) now follows from exercise 5.4.

5.1.3 Seminorms

The first condition (5.1) on the norm can be relaxed to obtain what is called a seminorm. By the other two properties, the set
$$K = \{v \in V \mid \|v\| = 0\} \tag{5.10}$$
(called the *kernel*) is a linear subspace of V (exercise 5.5). Thus one can define a norm on the quotient space V/K in a natural way (exercise 5.6). An example of a seminorm arises naturally in section 12.5.

5.2 PROVING THE TRIANGLE INEQUALITY

The key step in the proof is an inequality that is of interest in its own right; its proof was published by Rogers in 1888 and in the following year by Hölder[2] [115].

5.2.1 The Rogers-Hölder inequality

Suppose that p and q are positive real numbers related by
$$\frac{1}{p} + \frac{1}{q} = 1, \tag{5.11}$$
where $1 < p, q < \infty$. We augment these pairings with the cases $p = 1$ and $q = \infty$, or $p = \infty$ and $q = 1$, which correspond to the limiting cases. Then

$$\sum_{i=1}^{n} |x_i y_i| \leq \|x\|_p \|y\|_q \quad \forall x, y \in \mathbb{F}^n. \tag{5.12}$$

The case $p = 1$ and $q = \infty$ (or $p = \infty$ and $q = 1$) is elementary to verify (see exercise 5.11), so we focus on the case of finite p and q.

First, we prove (5.12) for all vectors satisfying
$$\|x\|_p = 1 \quad \text{and} \quad \|y\|_q = 1, \tag{5.13}$$
in which case (5.12) is just the statement that $\sum_{i=1}^{n} |x_i y_i| \leq 1$. Then the general case follows by a simple scaling, because of the homogeneity of (5.12) (see exercise 5.12). One inequality due to Young[3] states that

$$\sum_{i=1}^{n} |x_i y_i| \leq \frac{1}{p} \sum_{i=1}^{n} |x_i|^p + \frac{1}{q} \sum_{i=1}^{n} |y_i|^q, \tag{5.14}$$

---

[2]Otto Ludwig Hölder (1859–1937) was a student of Paul Du Bois-Reymond in Germany. Both Du Bois-Reymond and K. H. A. Schwarz (page 73) were students of Ernst Kummer, in Berlin, together with Cantor, Christoffel, and Fuchs. Although Hölder's famous inequality arose from work in analysis, he also worked extensively in algebra and was the advisor of Emil Artin.

[3]William Henry Young (1863–1942) was the husband of Grace Chisholm Young (1868–1944) and the father of Laurence Chisholm Young (1905–2000), both also mathematicians. L. C. Young is known for the concept of Young's measure, which provides an extended notion of solution for a partial differential equation, and he was also the father of a mathematician. A different William Henry Young was the father of the mathematician John Wesley Young (1879–1932), who was the brother-in-law of E. H. Moore.

provided that (5.11) holds. Given (5.14) and our assumption (5.13), (5.12) follows from Young's inequality in view of (5.11).

Now let us prove Young's inequality (5.14). It suffices to prove (5.14) for $n = 1$, in which case we drop the subscripts. It also suffices to assume that x and y are nonnegative. It is clear that $2xy \leq x^2 + y^2$ because this is just the statement $(y - x)^2 \geq 0$. The general case is just a consequence of the convexity of the exponential:

$$e^{(1/p)X + (1/q)Y} \leq (1/p)e^X + (1/q)e^Y \tag{5.15}$$

(exercise 5.13), provided that (5.11) holds. With $X = \log x^p$ and $Y = \log y^q$, we find

$$xy = e^{(1/p)\log x^p + (1/q)\log y^q} \leq (1/p)x^p + (1/q)y^q. \tag{5.16}$$

This completes the proof of Young's inequality and thus also of (5.12).

5.2.2 Minkowski's inequality

The Minkowski[4] inequality is just a name for the triangle inequality for $\|\cdot\|_p$, and it follows from (5.12), as we now show. For $p = 1$, Minkowski's inequality is elementary (exercise 5.2), so let us assume that $p > 1$; we write

$$\sum_{i=1}^n |x_i + y_i|^p = \sum_{i=1}^n |x_i + y_i| \, |x_i + y_i|^{p-1}$$
$$\leq \sum_{i=1}^n (|x_i| + |y_i|) \, |x_i + y_i|^{p-1}. \tag{5.17}$$

Applying (5.12), we find

$$\sum_{i=1}^n |x_i| |x_i + y_i|^{p-1} \leq \|x\|_p \left(\sum_{i=1}^n |x_i + y_i|^{(p-1)q} \right)^{1/q}$$
$$= \|x\|_p \left(\sum_{i=1}^n |x_i + y_i|^p \right)^{1-1/p} \tag{5.18}$$
$$= \|x\|_p \|x + y\|_p^{p-1},$$

where $q = (1 - 1/p)^{-1}$ since

$$(p-1)q = (p-1)\left(1 - \frac{1}{p}\right)^{-1} = (p-1)\left(\frac{p-1}{p}\right)^{-1} = p. \tag{5.19}$$

Applying (5.18) in (5.17), both as is and with the roles of x and y reversed, yields

$$\|x + y\|_p^p \leq (\|x\|_p + \|y\|_p) \, \|x + y\|_p^{p-1}. \tag{5.20}$$

Dividing by $\|x+y\|_p^{p-1}$ completes the proof (if by chance $\|x+y\|_p = 0$, there is nothing to prove).

---

[4]Hermann Minkowski (1864–1909) is also known for the concept of four-dimensional space time that formed the basis for special relativity.

5.3 RELATIONS BETWEEN NORMS

We saw at the beginning of the chapter that relationships among norms can provide valuable information, e.g., regarding optimal strategies for navigation. Here we consider such relationships for norms in general. In particular, it is elementary that

$$\|x\|_\infty \leq \|x\|_1 \leq n\|x\|_\infty \qquad (5.21)$$

for $x \in \mathbb{F}^n$; moreover, these constants are sharp (exercise 5.14). Following this approach, we can prove the following.

Lemma 5.1 *Let N be any norm on $\mathbb{F}^n$. Then there is a constant $K < \infty$ such that $N(x) \leq K\|x\|_\infty$ for all $x \in \mathbb{F}^n$.*

Proof. Let $e^1, \ldots, e^n$ be the standard basis for $\mathbb{F}^n$ and write $x = \sum_{j=1}^{n} x_j e^j$. Applying the triangle inequality $n - 1$ times, we find that

$$N(x) \leq \sum_{j=1}^{n} |x_j| N(e^j) \leq \|x\|_\infty \sum_{j=1}^{n} N(e^j). \qquad (5.22)$$

Define $K = \sum_{j=1}^{n} N(e^j)$. \hfill QED

5.3.1 Continuity of norms

Norms are Lipschitz-continuous in the following sense. By the triangle inequality,

$$\|x\| = \|(x - y) + y\| \leq \|x - y\| + \|y\|. \qquad (5.23)$$

Rearranging, we find

$$\|x\| - \|y\| \leq \|x - y\|. \qquad (5.24)$$

Reversing the names of x and y and using the fact that $\| - x\| = \|x\|$, we find

$$|\,\|x\| - \|y\|\,| \leq \|x - y\| \leq K\|x - y\|_\infty, \qquad (5.25)$$

where the last inequality is a consequence of lemma 5.1. In particular, this provides a way to prove ordinary continuity of the norm. That is, if $x \to y$, then $x - y \to 0$, and hence $\|x\| \to \|y\|$.

5.3.2 Norm equivalence

Definition 5.2 *Two norms N_1 and N_2 on a vector space V are said to be equivalent if there are positive constants C_i such that $N_1(x) \leq C_1 N_2(x)$ and $N_2(x) \leq C_2 N_1(x)$ for all $x \in V$.*

Theorem 5.3 *Any two norms on a finite-dimensional vector space are equivalent.*

Proof. It is sufficient to prove this for $V = \mathbb{F}^n$ (exercise 5.15). In view of lemma 5.1, it suffices to prove that there is a constant $C < \infty$ such that

$$\|x\|_\infty \leq CN(x) \quad \forall x \in \mathbb{F}^n. \tag{5.26}$$

This shows that any norm N is equivalent to $\|\cdot\|_\infty$, and hence all are equivalent to each other.

To prove (5.26), let $B = \{x \in \mathbb{F}^n \mid \|x\|_\infty = 1\}$ and define

$$\nu = \inf \{N(x) \mid x \in B\}. \tag{5.27}$$

If $\nu > 0$, set $C = 1/\nu$. Then we have proved (5.26) for all $x \in B$. For general $x \neq 0$, apply this result to $\|x\|_\infty^{-1}x$ and use the homogeneity of both norms to prove (5.26) for $x \neq 0$. It is obvious for $x = 0$.

Thus it suffices to show that we cannot have $\nu = 0$. But if $\nu = 0$, then there must be a point $x \in B$ where $N(x) = 0$ since a continuous function must attain its minimum on a compact set [141]. Since N is a norm, this implies that $x = 0$. But this contradicts the fact that $x \in B$, so we must have $\nu > 0$.   QED

The proof that $\nu > 0$ is nonconstructive and relies on the compactness of a closed, bounded set in $\mathbb{F}^n$ [141]. There are similar "compactness arguments" in infinite dimensions, but a finite dimension $(n < \infty)$ is essential in this particular argument.

5.4 INNER-PRODUCT SPACES

Inner products on a vector space V are conjugate-symmetric, nonnegative, bilinear forms; i.e., they satisfy the following conditions:

$$\begin{aligned}
(u, v) &= \overline{(v, u)} \\
(u, u) &> 0 \quad \text{for } u \neq 0 \\
(u + sv, w) &= (u, w) + s(v, w)
\end{aligned} \tag{5.28}$$

for all $u, v, w \in V$ and any scalar s. In addition, we assume that $(u, u) = 0$ implies $u \equiv 0$ for any $u \in V$. Note that the bilinearity implies that $(u, v) = 0$ if either u or v is zero. Also,

$$\begin{aligned}
(u, v + sw) &= \overline{(v + sw, u)} = \overline{(v, u)} + \overline{s(w, u)} \\
&= (u, v) + \overline{s}(u, w).
\end{aligned} \tag{5.29}$$

The canonical example of an inner-product space is $\mathbb{F}^n$, with

$$(x, y) = y^*x = \sum_{i=1}^n x_i \overline{y_i} \quad \text{for } x, y \in \mathbb{F}^n \tag{5.30}$$

(recall $\mathbb{F}^n = \mathbb{R}^n$ or $\mathbb{C}^n$). Another example of an inner-product space is any linear subspace V of $\mathbb{F}^n$. The critical difference between a normed linear space and an inner-product space is the Cartesian geometry that the latter inherits from the inner product. All (finite-dimensional) inner-product spaces can be viewed as copies of $\mathbb{F}^n$. For any x and y in V, we can consider the two-dimensional plane spanned by them, that is, the set

$$\left\{\alpha x + \beta y \mid (\alpha, \beta) \in \mathbb{F}^2\right\}. \tag{5.31}$$

The representation of this space on a blackboard is a faithful presentation of the geometry of the space (at least for $\mathbb{F} = \mathbb{R}$). When $(x, y) = 0$, the vectors are perpendicular (orthogonal) in that plane. This geometric interpretation is valid for all inner-product spaces.

Define a norm associated with the inner product by

$$\|f\|_2 = \sqrt{(f, f)}. \tag{5.32}$$

In the case $V = \mathbb{F}^n$, we have already seen that this is indeed a norm, but now we have to verify that it is in the case of a general inner product. The key step in proving the triangle inequality for $\|f\|_2$ in the case of $V = \mathbb{F}^n$ was the inequality (5.12) in the case $p = q = 2$:

$$|(x, y)| \leq \sqrt{(x, x)}\sqrt{(y, y)} \quad \forall x, y \in V. \tag{5.33}$$

This inequality is associated with the names of Cauchy,[5] Schwarz,[6] and sometimes [58] Bunyakovsky.[7] We leave the proof of (5.33) to exercise 5.18. Given (5.33), the triangle inequality is immediate:

$$\begin{aligned}
\|x + y\|_2^2 &= (x + y, x + y) = (x, x) + (x, y) + \overline{(x, y)} + (y, y) \\
&= (x, x) + 2|(x, y)| + (y, y) \\
&\leq (x, x) + 2\|x\|_2\|y\|_2 + (y, y) = (\|x\|_2 + \|y\|_2)^2.
\end{aligned} \tag{5.34}$$

Having an orthonormal basis set for a vector space V allows a simple correspondence between V and $\mathbb{F}^n$. These vectors form the axes in a Cartesian representation. We now show how a set of orthonormal vectors can be generated from an arbitrary linearly independent set.

5.4.1 Inductive orthonormalization

We construct orthonormal vectors inductively starting from an arbitrary linearly independent set $\{v^1, \ldots, v^k\}$ that spans a subspace $V_k \subset V$. That is, we construct a new set $\{p^1, \ldots, p^k\} \subset V_k$ such that

$$(p^i, p^j) = \delta_{ij}. \tag{5.35}$$

More precisely, we will have

$$p^i = t_i v^i + w^i, \tag{5.36}$$

---

[5]Augustin Louis Cauchy (1789–1857) was mentored by Lagrange (see page 152) early in life and "led what has been described as the first revolution of rigor in mathematics" [88].

[6]Karl Hermann Amandus Schwarz (1843–1921) is also known for his alternating method for approximating the solutions of partial differential equations [21], among many other things. He was a student of both Ernst Kummer and Karl Weierstrass.

[7]Victor Yakovlevich Bunyakovsky (1804–1889) obtained a doctorate from Paris in 1825 after working with Cauchy.

where $t_i \neq 0$ and $w^i \in V_{i-1}$ (and V_{i-1} denotes the space spanned by $\{v^1, \ldots, v^{i-1}\}$).

Notice that these conditions imply (exercise 5.19) that the p^i's are linearly independent. Thus the set $\{p^1, \ldots, p^k\}$ forms a basis for the space generated by $\{v^1, \ldots, v^k\}$. These conditions will be proved by induction. For $i = 1$, it is trivial: $p^1 = \|v^1\|_2^{-1} v^1$ (and $t_1 = \|v^1\|_2^{-1}$).

5.4.2 Orthogonal projections

Suppose that we have constructed a system satisfying the above for $i = 1, \ldots, k$. Then given any $f \in V$, define

$$L_k^S f = \sum_{i=1}^{k} (f, p^i) p^i. \tag{5.37}$$

The next result shows how simple the question of best approximation is in this context.

Theorem 5.4 *Given any* $f \in V$,

$$\|f - L_k^S f\|_2 = \min_{q \in V_k} \|f - q\|_2. \tag{5.38}$$

The key step in the proof of theorem 5.4 is an *orthogonality condition* that we state separately.

Lemma 5.5 *Given any* $f \in V$,

$$(f - L_k^S f, q) = 0 \tag{5.39}$$

for all $q \in V_k$.

Proof. To prove the lemma, note that (5.37) implies that

$$(L_k^S f, p^j) = \left(\sum_{i=1}^{k} (f, p^i) p^i, p^j \right) = \sum_{i=1}^{k} (f, p^i) (p^i, p^j)$$

$$= \sum_{i=1}^{k} (f, p^i) \delta_{ij} = (f, p^j), \tag{5.40}$$

where we used (5.28) to expand the inner product and the orthonormality (5.35). This verifies the lemma for $q = p^j$. The general result follows by writing q as a linear combination of the p^j's and expanding via (5.28). QED

Proof. The proof of theorem 5.4 is rather elementary. Let $p \in V_k$. Then expanding using the definition (5.32) of the norm and the properties (5.28), we find

$$\|f - L_k^S f + p\|_2^2 = (f - L_k^S f + p, f - L_k^S f + p)$$

$$= (f - L_k^S f, f - L_k^S f) + (f - L_k^S f, p) + (p, f - L_k^S f) + (p, p)$$

$$= (f - L_k^S f, f - L_k^S f) + (f - L_k^S f, p) + \overline{(f - L_k^S f, p)} + (p, p)$$

$$= (f - L_k^S f, f - L_k^S f) + (p, p), \tag{5.41}$$

using the orthogonality condition in lemma 5.5 in the last step. This says that

$$\|f - L_k^S f\|_2 < \|f - L_k^S f - p\|_2 \quad \forall\, 0 \neq p \in V_k. \tag{5.42}$$

Now let $q \in V_k$ and write $p = q - L_k^S f$. Then (5.42) implies that

$$\|f - L_k^S f\|_2 < \|f - L_k^S f - p\|_2 = \|f - q\|_2 \tag{5.43}$$

unless $q = L_k^S f$.   QED

There are some immediate corollaries. Suppose that $q \in V_k$. Then $q = L_k^S q$ (L_k^S is a projection) because we must have $\|q - L_k^S q\|_2 = 0$. Moreover,

$$\|f - L_k^S f\|_2 = 0 \tag{5.44}$$

if and only if $f \in V_k$.

5.4.3 Least squares

The basic idea of least squares is to find the best approximation in a given subspace V_k of some larger space V. That is, we start with a set of vectors $\{v^1, \ldots, v^k\}$ in V that span V_k, the approximation space. Given an arbitrary $v^{k+1} \in V$, we construct the closest element of V_k to v^{k+1}.

In general, we assume that we have $v^1, \ldots, v^{k+1}$ that are linearly independent. Thus $v^{k+1} \notin V_k$. Then we define

$$p^{k+1} = \frac{1}{\|v^{k+1} - L_k^S v^{k+1}\|_2} \left(v^{k+1} - L_k^S v^{k+1}\right). \tag{5.45}$$

We assume that the vectors $\{p^1, \ldots, p^k\}$ have already been constructed, so that allows use to define L_k^S via (5.37). The coefficient

$$t_{k+1} = 1/\|v^{k+1} - L_k^S v^{k+1}\|_2 \tag{5.46}$$

is well-defined (and nonzero) because we must have $v^{k+1} - L_k^S v^{k+1} \neq 0$ since $v^{k+1} \notin V_k$. The scaling ensures that $(p^{k+1}, p^{k+1}) = 1$, and the orthogonality $(p^{k+1}, p^j) = 0$ is a consequence of lemma 5.5. Note that $w^{k+1} = -t_{k+1} L_k^S v^{k+1}$ in (5.36).

We can write (5.45) algorithmically as

$$\begin{aligned}
e^{k+1} &= v^{k+1} - \sum_{j=1}^{k} (v^{k+1}, p^j) p^j = v^{k+1} - \sum_{j=1}^{k} r_{j,k+1} p^j \\
t_{k+1} &= 1/\|e^{k+1}\|_2 \\
p^{k+1} &= t_{k+1} e^{k+1},
\end{aligned} \tag{5.47}$$

where

$$r_{j,k+1} = (v^{k+1}, p^j), \; j = 1, \ldots, k. \tag{5.48}$$

The algorithm (5.47) is known as the Gram[8]-Schmidt[9] process. By analogy with (5.48), we define

$$r_{k+1,k+1} = (v^{k+1}, p^{k+1}) = (e^{k+1}, p^{k+1}) = \|e^{k+1}\|_2. \tag{5.49}$$

For completeness, define $r_{j,k+1} = 0$ for $j > k + 1$.

---

[8]Jørgen Pedersen Gram (1850–1916) was a Danish mathematician, statistician, and actuary [77].

[9]Erhard Schmidt (1876–1959) was a student of David Hilbert and, together with Issai Schur, was an advisor of the Brauer brothers (see page 229).

5.4.4 The QR decomposition

In the case where $\{v^1, \ldots, v^k\} \subset V = \mathbb{F}^n$, then we have generated orthonormal vectors $\{p^1, \ldots, p^k\} \subset \mathbb{F}^n$ with the property (5.36). Define a unitary matrix Q whose columns are the vectors p^i:

$$Q = \begin{bmatrix} p^1 & \cdots & p^k \end{bmatrix}, \tag{5.50}$$

that is, $Q_{li} = (p^i)_l$; and let A be the matrix with jth column v^j:

$$A = \begin{bmatrix} v^1 & \cdots & v^k \end{bmatrix}, \tag{5.51}$$

that is, $a_{lj} = (v^j)_l$. We can easily identify the upper-triangular matrix R, defined in (5.48), (5.49), and following, as the matrix corresponding to the change of basis from the p^j's to the v^j's:

$$v^j = \sum_{i=1}^{k} (v^j, p^i) p^i = \sum_{i=1}^{k} r_{ij} p^i, \tag{5.52}$$

which is equivalent to

$$a_{lj} = (v^j)_l = \sum_{i=1}^{k} (p^i)_l r_{ij} = \sum_{i=1}^{k} (Q)_{li} r_{ij} = (QR)_{lj}. \tag{5.53}$$

Thus we have shown that

$$A = QR. \tag{5.54}$$

The equation (5.54) is known as the QR *factorization* (or *decomposition*) of A. Note that we assume only that the columns of A are linearly independent. It need not be that A is square.

We can connect the QR factorization to the factorizations studied earlier. If you consider

$$A^\star A = R^\star Q^\star Q R = R^\star R, \tag{5.55}$$

we recognize that R is the Cholesky factor of $A^\star A$. The form $A^\star A$ occurs frequently, especially as *normal equations* in statistics [77], and the QR decomposition provides a way to determine the Cholesky factor without forming the product.

A formula of Aitken[10] [3] provides an alternative way to solve $Ax = f$. We can write formally that $A^{-1} = (A^\star A)^{-1} A^\star$. Thus $x = (R^\star R)^{-1} A^\star f$, or equivalently,

$$(R^\star R)x = A^\star f, \tag{5.56}$$

where $A = QR$. Solving (5.56) just requires forward and backward solution with the triangular factors $R^\star$ and R.

5.5 MORE READING

The convexity argument used to prove Young's inequality (5.14) is just a glimpse of convex analysis [139]. Convex analysis plays a major role in optimization [17, 19]. The QR decomposition plays a central role in the computation of eigenvalues and eigenvectors [152].

---

[10]See page 27.

5.6 EXERCISES

Exercise 5.1 *Prove that p-norms (5.5) satisfy both (5.1) and (5.2).*

Exercise 5.2 *Prove that the 1-norm, that is, (5.5) for $p = 1$, satisfies the triangle inequality. (Hint: use the fact that $|a + b| \le |a| + |b|$ and apply induction.)*

Exercise 5.3 *Prove that (5.8) holds for any fixed x. (Hint: pick i such that $|x_i| = \|x\|_\infty$ and show that $\|x\|_p \ge |x_i|$ for all p.)*

Exercise 5.4 *Prove that $x^a \to 1$ as $a \to 0$ for all $x > 0$.*

Exercise 5.5 *Show that the kernel K defined in (5.10) is a linear subspace of V provided that the seminorm satisfies (5.2) and (5.3).*

Exercise 5.6 *Suppose that K is a linear subspace of V. Show that the quotient space V/K consisting of equivalence classes of elements of V modulo K is a vector space in a natural way. Show that a seminorm on V becomes a norm on V/K in a natural way if K is the kernel of the seminorm defined in (5.10).*

Exercise 5.7 *Prove that the p-norms are continuous with respect to p in the sense that*

$$\|x\|_q = \lim_{p \to q} \|x\|_p \tag{5.57}$$

for any $1 \le q < \infty$. Note that the case $q = \infty$ is (5.7).

Exercise 5.8 *Let $a(x, y)$ be any nonnegative, symmetric bilinear form (i.e., a real-valued function) defined on a vector space V, that is, $a(x, x) \ge 0$ for all $x \in V$, $a(x, y) = a(y, x)$ for all $x, y \in V$, and*

$$a(x + sy, w) = a(x, w) + sa(y, w) \tag{5.58}$$

for all $w, x, y \in V$ and scalar s. Prove that

$$a(x, y) \le \sqrt{a(x, x)} \sqrt{a(y, y)} \quad \forall x, y \in V \tag{5.59}$$

holds even if $a(\cdot, \cdot)$ is degencrate, that is, $u(x, x) = 0$ for some $x \ne 0$.

Exercise 5.9 *Use (5.59) to prove (5.12) for $p = q = 2$ in $\mathbb{R}^n$. (Hint: define the bilinear form*

$$a(x, y) = \sum_{i=1}^{n} x_i y_i, \tag{5.60}$$

a.k.a. the Euclidean inner product, for all $x, y \in \mathbb{R}^n$, and apply exercise 5.8.)

Exercise 5.10 *Let $a(x, y)$ be any nonnegative, symmetric bilinear form on a vector space V (see exercise 5.8). Prove the triangle inequality for $\|x\| = \sqrt{a(x, x)}$. Note that this may be only a seminorm. (Hint: expand the expression*

$$\|x + y\|^2 = a(x + y, x + y) = a(x, x) + 2a(x, y) + a(y, y) \tag{5.61}$$

and apply exercise 5.8.)

Exercise 5.11 *Prove (5.12) holds for $p = 1$ and $q = \infty$.*

Exercise 5.12 *Prove (5.12) for general x and y given that it holds under the condition (5.13). (Hint: scale x and y to have norm 1.)*

Exercise 5.13 *Prove that the exponential function is convex, i.e., verify (5.15) provided that (5.11) holds.*

Exercise 5.14 *Prove (5.21) and prove that the constants cannot be improved.*

Exercise 5.15 *Prove theorem 5.3 given that it is known for $V = \mathbb{F}^n$. (Hint: choose a basis for V and construct an isomorphism of V with $\mathbb{F}^n$, where n is the dimension of V. Show that a norm on V induces a norm on $\mathbb{F}^n$ in a natural way.)*

Exercise 5.16 *Suppose that x and y are any vectors such that $\|x\| > \|y\|$. Prove that*

$$\|x - y\|^{-1} \leq \frac{1}{\|x\| - \|y\|}. \tag{5.62}$$

(Hint: use the triangle inequality: $\|x\| \leq \|x - y\| + \|y\|$.)

Exercise 5.17 *In the proof of theorem 5.3, prove directly that $\nu > 0$ without resorting to the fact that a continuous function takes on its minimum on a compact set. (Hint: recall how that result is proved.)*

Exercise 5.18 *Prove (5.33). (Hint: show that it suffices to assume that $y \neq 0$. Define $\alpha = (x, y)/(y, y)$ and set $w = x - \alpha y$ and expand $0 \leq (w, w) = (x - \alpha y, x - \alpha y)$.)*

Exercise 5.19 *Show that the orthogonal vectors P_i (cf. (5.35)) are linearly independent.*

Exercise 5.20 *Prove (5.38) by an alternative calculation from the proof given in the text (hint: let q be arbitrary and consider the quadratic function of t defined by $\phi(t) := \|f - L_n^S f + tq\|_2^2$; use (5.39)).*

Exercise 5.21 *Suppose that P is a (complex) polynomial of degree n. Prove that for any $C > 0$, there is an $R > 0$ such that $|P(x)| \geq C$ for all $|z| \geq R$. Use this to show that the minimum of $|P(z)|$ occurs for some z satisfying $R = |z| < \infty$.*

Exercise 5.22 *Suppose that P is a (complex) polynomial of degree n and that $|P(z)|$ has a minimum at some z_0 satisfying $R = |z_0| < \infty$. Prove that $P(z_0) = 0$. (Hint: if $|P(z_0)| > 0$, then write $P(z) = a(1 + Q(z))$, where Q is a polynomial of degree n such that $Q(z_0) = 0$ and $a = P(z_0) \in \mathbb{C}$. Using exercise 9.12, write $Q(z) = re^{i\theta}(z - z_0)^k + q(z)$, where $r > 0$ and q is a polynomial of degree $n - 1$ such that $|q(z)| \leq C|z - z_0|^{k+1}$ for z near z_0. Show that $Q(te^{-i(\theta+\pi)/k} + z_0) < 0$ for $t > 0$ sufficiently small.)*

Exercise 5.23 *Suppose that P is a (complex) polynomial of degree n. Prove that there is a $z \in \mathbb{C}$ such that $P(z) = 0$. (Hint: use exercise 5.21 to pick z at the global minimum of $|P|$ and use exercise 5.22 to do the rest.)*

5.7 SOLUTIONS

Solution of Exercise 5.8. The proof of (5.59) is begun by expanding the quadratic function $q(t) = a(x + ty, x + ty)$ as a function of $t \in \mathbb{R}$:

$$0 \le a(x + ty, x + ty) = a(x, x) + t(a(x, y) + a(y, x)) + t^2 a(y, y)$$
$$= a(x, x) + 2ta(x, y) + t^2 a(y, y) = q(t). \tag{5.63}$$

If $a(y, y) = 0$, then q is linear. Since a (nontrivial) linear function has no minimum, it follows that $a(x, y) = 0$ as well, and (5.59) is satisfied trivially. So now suppose that $a(y, y) \neq 0$. Since the expression $q(t)$ has to be nonnegative, we can investigate what it means to have q nonnegative at its minimum. Since $q'(t) = 2(a(x, y) + ta(y, y))$, the minimum is at $t = -a(x, y)/a(y, y)$. But

$$0 \le q(-a(x, y)/a(y, y)) = a(x, x) - 2a(x, y)^2/a(y, y) + a(x, y)^2/a(y, y)$$
$$= a(x, x) - a(x, y)^2/a(y, y) \tag{5.64}$$

implies that $a(x, y)^2 \le a(x, x)a(y, y)$.

Solution of Exercise 5.13. A function f is convex if for $0 < t < 1$, we have

$$f(tX + (1 - t)Y) \le tf(X) + (1 - t)f(Y) \tag{5.65}$$

for all X and Y. Let M denote the point

$$M = tX + (1 - t)Y = Y + t(X - Y). \tag{5.66}$$

Then (5.65) is equivalent to

$$0 \le -t\left(f(M) - f(X)\right) + (1 - t)\left(f(Y) - f(M)\right) \tag{5.67}$$

(which is the statement that the second divided difference, cf. section 10.2.3, of f is positive). If f is C^1, we can write (5.67) as

$$0 \le -t \int_X^M f'(s)\, ds + (1 - t) \int_M^Y f'(s)\, ds. \tag{5.68}$$

Thus we will show that (5.68) holds under suitable conditions on f that we can verify for $f(x) = e^x$.

One simple criterion for convexity is based on the sign of the second derivative of f. If $f'' \ge 0$, then

$$f'(x) - f'(y) = \int_y^x f''(s)\, ds \ge 0, \tag{5.69}$$

so f' is nondecreasing. Applying this to the integrals in (5.68), we find

$$-t \int_X^M f'(s)\,ds + (1-t) \int_M^Y f'(s)\,ds \geq -t(M-X)f'(M)$$
$$+ (1-t)(Y-M)f'(M)$$
$$= (-t(M-X) + (1-t)(Y-M))\,f'(M) \qquad (5.70)$$
$$= ((Y-M) - t(Y-X))\,f'(M)$$
$$= 0,$$

by (5.66).

When $f(x) = e^x$, we have $f = f' = f'' > 0$, so e^x is (strictly) convex.

Solution of Exercise 5.17. If $\nu = 0$, there must be a sequence of points $x_j \in B$ such that $N(x_j) < 1/j$. Any infinite sequence of points in a bounded set in $\mathbb{R}^n$ must have an accumulation point x_∞, and thus there is a subsequence x_{j_k} such that $\lim_{k \to \infty} x_{j_k} = x_\infty$ [141]. In particular, we must have $x_\infty \in B$. But we also must have $N(x_\infty) = 0$ since $N(x_{j_k}) < 1/j_k$ and

$$N(x_\infty - x_{j_k}) \leq K \|x_\infty - x_{j_k}\|_\infty \to 0 \text{ as } k \to \infty,$$

by lemma 5.1. Thus $x_\infty = 0$. But this contradicts the fact that $x_\infty \in B$, so we must have $\nu > 0$.

Solution of Exercise 5.18. If $y = 0$, then $(x, y) = 0$, and the result is obvious. So we assume that $y \neq 0$. Define

$$s = (x, y)/(y, y) \qquad (5.71)$$

and $w = x - sy$. First, observe that

$$s(y, x) = s\overline{(x, y)} = |(x, y)|^2/(y, y) = \overline{s}(x, y). \qquad (5.72)$$

Now expand

$$0 \leq (w, w) = (x - sy, x - sy)$$
$$= (x, x) - s(y, x) - \overline{s}(x, y) + |s|^2(y, y)$$
$$= (x, x) - 2\frac{|(x, y)|^2}{(y, y)} + |s|^2(y, y) \qquad [\text{by } (5.72)] \qquad (5.73)$$
$$= (x, x) - \frac{|(x, y)|^2}{(y, y)} \qquad [\text{by } (5.71)].$$

Therefore,

$$|(x, y)|^2 \leq (x, x)(y, y), \qquad (5.74)$$

as claimed.

Chapter Six

Operators

Issai Schur (1875–1941) studied and worked in Berlin much of his life. He is known for his matrix decomposition and factorization, as well as many other results in mathematics. With the rise of Hitler and the acquiescence of colleagues, Schur was forced to resign his various academic posts. Schur emigrated to Palestine in 1939 [105].

We need to develop some further technology to measure the size of operators on vector spaces through a naturally associated "operator" norm. This is needed in several areas, in particular, in the study of iterative methods for approximation of the solution of both linear and nonlinear systems. However, many of the results are of interest just as abstract theorems.

One result in this chapter (theorem 6.8) is that we can *almost* think of the spectral radius of an operator (the size of the largest eigenvalue of the operator) as a norm. That is, we can always find a vector norm such that the corresponding operator norm is arbitrarily close to the spectral radius. This allows us to give a precise condition (theorem 6.11) that governs the convergence of many iterative processes, the result that can be viewed as the endpoint for the chapter. But the ingredients of the proof are of interest in their own right. In particular, we will show a rather surprising result, that *any* matrix A is similar to a matrix arbitrarily close to a diagonal matrix with the eigenvalues of A on the diagonal (theorem 6.7). That is, to any desired accuracy, a matrix may be viewed as diagonalizable. The proof of this result relies on the Schur decomposition (theorem 6.4) of a matrix. To keep track of all these results, we provide a roadmap in figure 6.1.

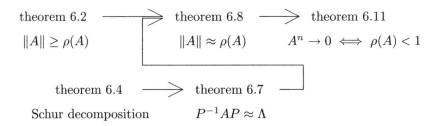

$$
\begin{array}{ccccc}
\text{theorem 6.2} & \longrightarrow & \text{theorem 6.8} & \longrightarrow & \text{theorem 6.11} \\
\|A\| \ge \rho(A) & & \|A\| \approx \rho(A) & & A^n \to 0 \iff \rho(A) < 1 \\
& & & & \\
& \text{theorem 6.4} & \longrightarrow & \text{theorem 6.7} & \\
& \text{Schur decomposition} & & P^{-1}AP \approx \Lambda &
\end{array}
$$

Figure 6.1 Roadmap of results in chapter 6.

6.1 OPERATORS

An operator is a mapping from one vector space to another vector space. These can be defined quite abstractly, as a machine, such as the mapping that takes a function, f, as input and produces its derivative, f', as output. To make this precise, we have to say what the linear space is. We leave as exercise 6.1 to show that the set $\mathcal{P}_n$ of polynomials of degree n in one variable can be viewed as a vector space; exercise 6.2 addresses the issue of showing that the derivative operator is well-defined on this space.

There is a special class of vector spaces that we want to distinguish: the linear space of operators on a vector space. Given any two operators A and B that map V to W, we define $A + B$ by setting $(A + B)v = Av + Bv$ for all $v \in V$, and we define scalar multiplication similarly: $(sA)v = s(Av)$ for $v \in V$ and scalars s. We denote the vector space of such operators by $\mathcal{O}(V, W)$. We can define norms on such vector spaces, but we will see that there is a special type of induced norm that reflects the product structure of operators.

6.1.1 Operator norms

There is a natural class of operator norms that come from duality. Given norms $\|\cdot\|_V$ and $\|\cdot\|_W$ on vector spaces V and W, and an operator $A : V \to W$, define

$$\|A\| = \|A\|_{V \to W} = \sup_{0 \neq v \in V} \frac{\|Av\|_W}{\|v\|_V}. \tag{6.1}$$

Then this forms a norm on the linear space of operators from V to W (see exercise 6.3). We drop the subscript $\|A\|_{V \to W}$ and write $\|A\|$ when there is no confusion about the spaces in question. However, in some cases, there are multiple spaces in the discussion.

What is significant about the operator norm is that it satisfies multiplicative properties that other norms do not. First, there is a natural multiplicative property relating the norms on V and W and the induced operator norm:

$$\boxed{\|Av\|_W \leq \|A\| \, \|v\|_V.} \tag{6.2}$$

This is essentially a tautology since the operator norm was defined as an infimum of quotients of the first and last terms in (6.2). But this has many important applications, including the second multiplicative property of operator norms. Suppose that B is a linear operator from a vector space U to V. Then

$$\boxed{\|A B\|_{U \to W} \leq \|A\|_{V \to W} \|B\|_{U \to V}.} \tag{6.3}$$

The proof requires just two applications of (6.2) and is left as exercise 6.4.

As cumbersome as it is, the subscript notation $\|A\|_{V \to V}$ does not necessarily provide complete information about the definition of the norm. For example, we will often be interested in the case $V = \mathbb{R}^n$, but this does not specify which norm on $\mathbb{R}^n$ we would be using. In section 5.1.1, we introduced the short-hand notation ℓ_p for the complete specification, such as $(\mathbb{R}^n, \|\cdot\|_p)$, to indicate the norm defined in (5.5). Since this special case occurs frequently, we define

$$\|A\|_p = \|A\|_{\ell_p \to \ell_p} = \sup_{0 \neq x \in \mathbb{F}^n} \frac{\|A\,x\|_p}{\|x\|_p} \tag{6.4}$$

for $A : \mathbb{F}^n \to \mathbb{F}^n$ with $\mathbb{F} = \mathbb{R}$ or $\mathbb{C}$. Note that this is quite different from the p-norm of the matrix associated with A represented as a vector in $\mathbb{F}^{n^2}$. To highlight this point, we define the Frobenius[1] norm by

$$\|A\|_F = \Big(\sum_{i,j=1}^{n} a_{ij}^2 \Big)^{1/2}, \tag{6.5}$$

which *is* the 2-norm of the matrix associated with A represented as a vector in $\mathbb{F}^{n^2}$. To avoid confusion, we will try to avoid using the symbol F as a vector space. In section 6.2.2, we will use the norm as a subscript, for a general norm, similar to the usage in definition (6.4).

6.1.2 Operator norms and eigenvalues

Suppose A is an operator that maps a vector space V to itself. An eigenvalue for A is a complex number λ such that $A\,x = \lambda x$ for some $x \neq 0$. The corresponding vector x is called the eigenvector associated with λ. We refer to λ, x as an eigenpair. There is a relationship between the eigenvalues of an operator and its norm:

$$\|A\,x\| = \|\lambda x\| = |\lambda|\,\|x\|, \tag{6.6}$$

which implies that

$$\|A\| \geq \frac{\|A\,x\|}{\|x\|} = |\lambda| \tag{6.7}$$

for any eigenvalue λ (and any norm).

Definition 6.1 *Suppose A is an operator that maps a vector space V to itself. The* spectral radius *of A, denoted $\rho(A)$, is the maximum modulus of all the eigenvalues:*

$$\rho(A) = \max \big\{ |\lambda| \mid \lambda \text{ is an eigenvalue of } A \big\}. \tag{6.8}$$

In (6.8), we wrote "max" instead of "sup" since the set of eigenvalues is finite (assuming V is finite-dimensional). Thus there is always an eigenvalue λ such that $|\lambda| = \rho(A)$. The inequality (6.7) yields the following theorem.

---

[1] Ferdinand Georg Frobenius (1849–1917) was a student of Weierstrass and an advisor of Issai Schur.

Theorem 6.2 *Suppose A is an operator that maps a vector space V to itself. For any norm on V, the associated operator norm on A satisfies*

$$\rho(A) \leq \|A\|, \tag{6.9}$$

where $\rho(A)$ is the spectral radius of A, defined in (6.8).

Since operators can be represented as matrices, one might think that it is sufficient just to have norms on Euclidean spaces. However, some operator norms cannot be written as a norm on a Euclidean space consisting of the coefficients of the corresponding matrix. For example, one can show (exercise 6.6) that for a Hermitian matrix A, the operator norm associated with the Euclidean norm satisfies

$$\boxed{\|A\|_2 = \rho(A) = \max\left\{ |\lambda| \mid \lambda \text{ is an eigenvalue of } A \right\}.} \tag{6.10}$$

We explain why there cannot be a formula for the eigenvalues of a matrix in section 14.4. Conversely, there are some norms on matrices that cannot be written as operator norms (exercises 6.8 and 6.9).

The identity (6.10) provides the guiding motivation for the chapter. Although it does not hold for general operators, we will see that it *almost* does provided we are willing to change to a different norm on $\mathbb{R}^n$. We will see in exercise 8.2 that other operator norms on matrices can also be identified quantitatively.

Note that operator norms have a special property not true for general norms. Let $\epsilon > 0$ be arbitrary. If $\| \cdot \|_V$ is a norm on V, then so is $\| \cdot \|_\epsilon$ a norm on V, where

$$\|x\|_\epsilon = \epsilon \|x\|_V \quad \forall x \in V. \tag{6.11}$$

Thus general norms can be scaled arbitrarily, whereas operator norms cannot (since the spectral radius is independent of the choice of norm).

We will return to the comparison of norms of an operator and its spectral radius in section 6.2.2, where we will provide a counterpoint to theorem 6.2.

6.2 SCHUR DECOMPOSITION

To understand norms of operators better, we need to develop some technology. Fortunately, this technology is interesting in its own right as it provides insight into fundamental properties of linear operators. The first result that we need is the *Schur decomposition*, which says that any matrix is unitarily equivalent to a triangular matrix.

Definition 6.3 *A matrix U is unitary if $U^\star U = I$.*

A unitary matrix corresponds to an operator that does not stretch coordinates in any direction.

Theorem 6.4 *For any square matrix A, there is a unitary matrix U such that $T = U^{-1}AU = U^\star AU$ is upper-triangular.*

There are several applications of the Schur decomposition, but one of them involves eigenvalues. In view of corollary 3.3, the diagonal entries of a triangular matrix are its eigenvalues. Since a similarity transformation does not change the eigenvalues (exercise 6.10), the eigenvalues of A are the diagonal entries of the triangular factor T in the Schur decomposition. For the special case of a Hermitian matrix A, we obtain the following well-known result.

Corollary 6.5 *For any Hermitian matrix A, there is a unitary matrix U such that $D = U^{-1}AU = U^{\star}AU$ is diagonal.*

The proof of this corollary is a simple application of the Schur decomposition and conjugation: $T^{\star} = (U^{\star}AU)^{\star} = U^{\star}A^{\star}U = U^{\star}AU = T$. A Hermitian triangular matrix must be diagonal. Similarly, we can prove the following.

Corollary 6.6 *For any matrix A, there is a unitary matrix U such that $D = U^{-1}AU = U^{\star}AU$ is diagonal if and only if A is normal, i.e., $A^{\star}A = AA^{\star}$.*

The "only if" is clear. The "if" is an interesting result itself: a triangular matrix T is normal $(T^{\star}T = TT^{\star})$ iff it is diagonal (exercise 6.12).

It might appear then that the Schur decomposition provides a means to compute eigenvalues. We will see that the converse is true: we use the existence of eigenvectors to establish the Schur decomposition. We postpone the proof of theorem 6.4 until section 6.2.3.

It is tempting to compare the QR factorization (5.54) with the Schur decomposition (theorem 6.4). There is a superficial similarity in that both involve unitary and triangular factors. But there are significant differences. First, the QR factorization applies more generally in that A need not be square. But more significantly, the Schur decomposition is a similarity transformation, whereas the QR factorization is unbalanced. However, most significant is the fact that we have a constructive algorithm that computes the QR factorization in a finite number of steps. We will see in section 14.4 why this is not possible in general for the Schur decomposition.

6.2.1 Nearly diagonal matrices

As a step toward the proof of our main theorem on the relationship between the spectral radius and operator norms, there is an interesting intermediate result, namely, that with a suitable similarity transformation, any matrix can be transformed so that it is essentially diagonal.

Theorem 6.7 *Suppose A is any square matrix and let $\epsilon > 0$ be arbitrary. Then there is an invertible matrix P such that*

$$P^{-1}AP = \Lambda + C, \tag{6.12}$$

where Λ is a diagonal matrix (having the eigenvalues of A on the diagonal) and C is a strictly upper-triangular matrix that satisfies $|C_{ij}| \leq \epsilon$ for all i, j (and $C_{ij} = 0$ for $i \geq j$).

This theorem says that we can make any matrix look as nearly diagonal as we want (but we will see that this comes at the expense of making P very large in general). What this means is that general similarity transformations can be somewhat misleading, unlike the unitary transformations in the Schur decomposition.

Proof. We begin with the Schur decomposition $U^\star AU = T$, where $T = \Lambda + B$ is an upper-triangular matrix and B is strictly upper-triangular (that is, B has zero diagonal entries and Λ is diagonal). Define

$$\mu = \max\left\{|b_{kl}| \mid k, l = 1, \ldots, n\right\}.$$

Let $\delta = \epsilon/\mu > 0$ and define D to be the diagonal matrix with $D_{ii} = \delta^{1-i}$ for $i = 1, \ldots, n$. Define $S = D^{-1}TD = \Lambda + C$, where $C = D^{-1}BD$. Then C is also strictly upper-triangular, and (see exercise 6.13)

$$c_{ij} = \delta^{i-1}b_{ij}\delta^{1-j} = \delta^{i-j}b_{ij}. \tag{6.13}$$

But since B is strictly upper-triangular, $b_{ij} \neq 0$ only if $j \geq i + 1$, so we have $|c_{ij}| \leq \delta\mu = \epsilon$ for all i and j. Define $P = UD$. Then $P^{-1}AP = D^{-1}U^\star AUD = D^{-1}TD = \Lambda + C$.   QED

6.2.2 The spectral radius is nearly a norm

We now consider one of the main results in the chapter, the gist of which is the title of this section.

Theorem 6.8 *Suppose A is any $n \times n$ matrix and let $\delta > 0$ be arbitrary. Then there is a norm N on $\mathbb{R}^n$ such that the corresponding operator norm $\|\cdot\|_N$ satisfies*

$$\rho(A) \leq \|A\|_N \leq \rho(A) + \delta, \tag{6.14}$$

where $\rho(A)$ is the spectral radius.

We emphasize here that that the operator norm is defined by

$$\|A\|_N = \sup_{x \in \mathbb{R}^n, \, x \neq 0} \frac{N(Ax)}{N(x)} \tag{6.15}$$

and depends on δ through the dependence of N on δ. In view of theorem 6.8, we are free to think of the spectral radius $\rho(A)$ as (essentially) a norm (recall the reverse inequality (6.7) which holds for all norms).

Proof. The first inequality in (6.14) is theorem 6.2, so we need to prove only the second. Let $\epsilon = \delta/n$ and choose P according to theorem 6.7. Note that

$$\rho(A) = \max\left\{|\Lambda_{ii}| \mid i = 1, \ldots, n\right\}. \tag{6.16}$$

Define $N(x) = \|P^{-1}x\|_\infty$ (see exercise 6.14). Then (making the substitution $x = Py$)

$$
\begin{aligned}
\|A\|_N &= \sup_{x \in \mathbb{R}^n,\, x \neq 0} \frac{\|P^{-1}Ax\|_\infty}{\|P^{-1}x\|_\infty} = \sup_{y \in \mathbb{R}^n,\, y \neq 0} \frac{\|P^{-1}APy\|_\infty}{\|y\|_\infty} \\
&= \sup_{y \in \mathbb{R}^n,\, y \neq 0} \frac{\|(\Lambda + C)y\|_\infty}{\|y\|_\infty} \\
&\leq \sup_{y \in \mathbb{R}^n,\, y \neq 0} \frac{\|\Lambda y\|_\infty}{\|y\|_\infty} + \sup_{y \in \mathbb{R}^n,\, y \neq 0} \frac{\|Cy\|_\infty}{\|y\|_\infty} \\
&\leq \sup_{y \in \mathbb{R}^n,\, y \neq 0} \frac{\|\Lambda y\|_\infty}{\|y\|_\infty} + n\epsilon = \rho(A) + \delta,
\end{aligned}
\tag{6.17}
$$

where we used exercise 6.15 in the penultimate step and exercise 6.16 in the last.   QED

6.2.3 Derivation of the Schur decomposition

The proof of the Schur decomposition is by induction on the matrix dimension. For $n = 1$, the theorem is trivial. So suppose it is true for $n - 1$, and let us show that it holds for n.

To begin with, we simply pick an eigenpair for A: $Ax = \lambda x$, normalized so that $x^\star x = 1$. This is the only nontrivial fact that we use in the proof. The property that every matrix has at least one eigenvector is worth reviewing; it stems from the fundamental theorem of algebra [35], which says that a polynomial always has a root (cf. exercise 5.23). In this case, the polynomial in question is the characteristic polynomial $p_n(\lambda) = \det(A - \lambda I)$. The second ingredient in this fact is that if the determinant of a matrix is zero, then the matrix ($A - \lambda I$ in our case) has a nontrivial null vector.

Given the eigenvector x, we construct

$$
U = [x \ B] \tag{6.18}
$$

by taking x to be the first column of U and filling in with the $n \times (n - 1)$ matrix B as needed. We can see that U is unitary if we are willing to change coordinates so that $x = [1 \ 0 \ \cdots \ 0]$; in this case, $U = I$. If we want to see this more concretely, we can study what it means for $U = [x \ B]$ to be unitary:

$$
U^\star U = \begin{pmatrix} x^\star x & x^\star B \\ B^\star x & B^\star B \end{pmatrix} = \begin{pmatrix} 1 & x^\star B \\ B^\star x & B^\star B \end{pmatrix}, \tag{6.19}
$$

where we have used a block-matrix multiplication formula (exercise 6.17) similar to (3.31). Let $b_1, \ldots, b_{n-1}$ be an orthonormal basis for the $(n - 1)$-dimensional space

$$
Y = \{ y \in \mathbb{C}^n \mid x^\star y = 0 \}. \tag{6.20}
$$

Such a basis could be generated by the Gram-Schmidt process as described in section 5.4.1. Then $B = [b_1, \ldots, b_{n-1}]$ has the required properties.

Now let us see what U does to A. By another block-matrix multiplication formula (exercise 6.18), we find

$$U^\star AU = \begin{pmatrix} x^\star A\,x & x^\star A\,B \\ B^\star A\,x & B^\star A\,B \end{pmatrix} = \begin{pmatrix} \lambda & x^\star A\,B \\ 0 & B^\star A\,B \end{pmatrix} = \begin{pmatrix} \lambda & z^\star \\ 0 & A^{(n-1)} \end{pmatrix}, \quad (6.21)$$

where $z = B^\star A^\star x$ is some (column) vector of length $n-1$ and $A^{(n-1)} = B^\star A\,B$ is an $(n-1)\times(n-1)$ matrix. We now invoke the induction hypothesis and let $\widetilde{V}$ be an $(n-1)\times(n-1)$ unitary matrix such that $\widetilde{V}^\star A^{(n-1)}\widetilde{V} = T^{(n-1)}$ is upper-triangular. Define

$$V = \begin{pmatrix} 1 & 0 \\ 0 & \widetilde{V} \end{pmatrix}. \quad (6.22)$$

Then the block-matrix multiplication formula (3.31) implies

$$\begin{aligned}
V^\star U^\star AUV &= V^\star \begin{pmatrix} \lambda & z^{\mathrm{T}} \\ 0 & A^{(n-1)} \end{pmatrix} V = \begin{pmatrix} \lambda & z^{\mathrm{T}} \\ 0 & \widetilde{V}^\star A^{(n-1)} \end{pmatrix} V \\
&= \begin{pmatrix} \lambda & z^{\mathrm{T}}\widetilde{V} \\ 0 & \widetilde{V}^\star A^{(n-1)}\widetilde{V} \end{pmatrix} = \begin{pmatrix} \lambda & z^{\mathrm{T}}\widetilde{V} \\ 0 & T^{(n-1)} \end{pmatrix}.
\end{aligned} \quad (6.23)$$

Thus UV is the required unitary matrix.   QED

The Schur decomposition is not an algorithm for determining eigenvalues. Unfortunately, its derivation is not constructive, in that it requires the provision of an eigenvector by some unspecified mechanism. Instead, we should think of the Schur decomposition as a way to catalog the eigenvectors (and eigenvalues) in a useful way.

6.2.4 Schur decomposition and flags

A *flag* is a nested sequence of subspaces V^k of a vector space V [9]. More precisely, a flag has the property that

$$\{0\} = V^0 \subset V^1 \subset V^2 \subset \cdots \subset V^k = V, \quad (6.24)$$

where the dimensions $d_i = \dim V^i$ satisfy $d_i > d_{i-1}$ for all $i \geq 1$. A *complete flag* is one in which $d_i = i$ for all i.

An operator $O : V \to V$ *supports a flag* if $O(V^j) \subset V^j$ for all $j = 1,\ldots,k$. For example,

$$V^i = \{(x_1,\ldots,x_n) \in \mathbb{R}^n \mid x_j = 0 \quad \forall j > i\} \quad (6.25)$$

is a complete flag, and any upper-triangular matrix T supports the flag (6.25) (exercise 6.19).

The Schur decomposition has an abstract representation in terms of flags and operators as follows.

Theorem 6.9 *Any operator on a finite-dimensional vector space V supports a complete flag.*

In other words, for any operator O there exists some complete flag (6.24) that it supports. Thus the Schur decomposition of an operator has a natural expression independent of any basis chosen to represent it as a matrix. Other matrix decompositions, such as the LU factorization (section 3.2), do not enjoy this property.

The proof of theorem 6.9 is left as exercise 6.20.

6.3 CONVERGENT MATRICES

There are many situations in which the result of an algorithm can be written as multiplication of a vector X by a fixed matrix A. Thus repeating the algorithm n times is equivalent (by induction) to applying the matrix A^k. Frequently, this represents the error in some iterative process. Thus we are interested in precise conditions when A is a *convergent matrix*, that is, $A^k \to 0$ as $k \to \infty$. We start with the following simple criterion.

Lemma 6.10 *If $\|A\| < 1$ for some operator norm, then $\|A^k\| \to 0$ as $k \to \infty$.*

Proof. By induction, (6.3) implies that $\|A^k\| \leq \|A\|^k \to 0$ as $k \to \infty$. QED

Combining lemma 6.10 with theorem 6.8, we get the following precise characterization.

Theorem 6.11 *For any matrix A, $A^k \to 0$ as $k \to \infty$ if and only if $\rho(A) < 1$.*

Proof. If $\rho(A) < 1$, choose a norm so that $\|A\| < 1$ by theorem 6.8 and apply lemma 6.10.

Conversely, suppose there is an eigenvalue λ of A such that $|\lambda| \geq 1$. There must be a vector $x \neq 0$ such that $Ax = \lambda x$. Thus $A^k x = A^{k-1} A x = \lambda A^{k-1} x$, so by induction we must have $A^k x = \lambda^k x$. But if $A^k \to 0$, we conclude that $\lambda^k x \to 0$. Since $|\lambda^k| = |\lambda|^k \geq 1$, we have a contradiction. QED

theorem 6.11 provides the basis for the convergence theory for stationary iterative methods (section 8.1), for the stability theory of time-stepping schemes for approximating the solutions to differential equations that will be discussed in section 17.2), as well as for many other applications.

6.4 POWERS OF MATRICES

theorem 6.11 provides some information about powers of matrices. Here we develop the theme in a bit more detail. Our objective is to prove the following result.

Theorem 6.12 *For any $n \times n$ (real or complex) matrix A and any norm $\|\cdot\|$ (on $\mathbb{R}^n$ or $\mathbb{C}^n$, respectively),*

$$\lim_{k \to \infty} \|A^k\|^{1/k} = \rho(A), \tag{6.26}$$

where $\rho(A)$ is the spectral radius of A and $\|A^k\|$ denotes the corresponding operator norm.

This again shows the close connection between the spectral radius and the operator norm. To begin with, an analog of (6.9) is that

$$\boxed{\rho(A) \leq \|A^r\|^{1/r}} \tag{6.27}$$

for any norm and any positive integer r. This is proved in the same way: $\rho(A) = |\lambda|$ for some eigenvalue λ, with eigenvector x such that $Ax = \lambda x$. Multiplying by A, we have $A^2 x = A(\lambda x) = \lambda A x = \lambda^2 x$. By induction, we have $A^r x = \lambda^r x$ for any r. Thus $\rho(A)^r = |\lambda|^r = \|A^r x\|/\|x\| \leq \|A^r\|$, proving (6.27). We now use this to prove theorem 6.12.

It seems remarkable at first that this would hold for any operator norm, but the fact is that it is sufficient to prove (6.26) for just one norm. For example, suppose that we know that

$$\lim_{k \to \infty} \|A^k\|_\infty^{1/k} = \rho(A). \tag{6.28}$$

Then by the equivalence of norms (lemma 5.1 or theorem 5.3), we have

$$\lim_{k \to \infty} \|A^k\|^{1/k} \leq \lim_{k \to \infty} \left(K\|A^k\|_\infty\right)^{1/k} = \lim_{k \to \infty} K^{1/k}\rho(A) = \rho(A). \tag{6.29}$$

Thus the limiting process quashes any constant factor. In view of (6.27), the theorem follows. Now let us verify (6.28).

The Schur decomposition shows that powers of a matrix tend to a very simple form. Suppose that $T = U^\star A U$ is the Schur decomposition, where U is unitary and T is triangular. We can turn the decomposition around and write $A = U T U^\star$. Then $A^2 = U^\star T U U^\star T U = U^\star T^2 U$. By induction,

$$A^k = U^\star T^k U \tag{6.30}$$

for any k. Write

$$T = D + N, \tag{6.31}$$

where $D = \text{diag}(T)$ is the diagonal matrix whose diagonal is the same as that of T. Thus $N = T - D$. Since the entries of D are the eigenvalues of T (and hence of A),

$$\|D\|_\infty = \rho(T) = \rho(A) \tag{6.32}$$

(see exercise 6.16). We need to calculate the norm of T^k, and since this is a somewhat lengthy step, we separate it as the following lemma.

Lemma 6.13 *Suppose that T is an upper-triangular $n \times n$ matrix and $N = T - D$, where $D = \text{diag}(T)$. Then for any matrix norm $\|\cdot\|$ satisfying the multiplicative property (6.3), we have*

$$\|T^k\| = \|(D + N)^k\| \leq k^n \|D\|^{k-n} \left(\|N\| + \|D\|\right)^n . \tag{6.33}$$

Proof. Note that N is *nilpotent*, that is, $N^n = 0$. More precisely, observe that N is strictly upper-triangular, that is, it is 0 on and below the main diagonal. Not all nilpotent matrices are strictly upper-triangular, but all strictly upper-triangular matrices are nilpotent. To work with such matrices, let us introduce some notation. We say that an upper-triangular matrix M has *shift index* μ if $M_{ij} = 0$ for $i > j - \mu$. A diagonal matrix has shift index 0, and the matrix $N = T - D$ has shift index 1.

The product MN of two upper-triangular matrices M and N, with shift indices μ and ν, respectively, has shift index $\mu + \nu$ (see exercise 6.23). In particular, since $N = T - D$ has shift index 1, $(N^k)_{ij} = 0$ for $i > j - k$ (see exercise 6.24). Thus $N^n \equiv 0$. If D is a diagonal matrix, then the shift index of DN and ND is no less than the shift index of N (see exercise 6.25).

We now want to expand the expression $T^k = (D + N)^k$. Since N and D need not commute, such an expression can be quite complicated. In particular, $(D+N)^2 = D^2 + DN + ND + N^2$. For $(D+N)^k$, there are 2^k such expressions. Fortunately, there is a one-to-one relationship between each such expressions and the binary representation of some integer $j \in [0, 2^k - 1]$. Define

$$P(j) = \prod_{i=1}^{k} D^{b_i} N^{1-b_i}, \tag{6.34}$$

where $b_k b_{k-1} \cdots b_1$ denotes the binary expansion of j, that is,

$$j = \sum_{i=1}^{k} b_i 2^{i-1}. \tag{6.35}$$

Note that $D^{b_i} N^{1-b_i}$ is D if $b_i = 1$, and N if $b_i = 0$. Then

$$(D + N)^k = \sum_{j=0}^{2^k - 1} P(j). \tag{6.36}$$

We can compute the shift index of $P(j)$: it is at least $\nu(j)$, the number of zeros in the binary expansion of j. Thus when $\nu(j) \geq n$, $P(j) = 0$ (exercise 6.26). Thus

$$(D + N)^k = \sum_{0 \leq j < 2^k, \nu(j) \leq n} P(j). \tag{6.37}$$

Note that

$$\|P(j)\| \leq \|D\|^{k - \nu(j)} \|N\|^{\nu(j)}. \tag{6.38}$$

From (6.38), we conclude

$$\|(D + N)^k\| \leq \sum_{0 \leq j < 2^k, \nu(j) \leq n} \|D\|^{k - \nu(j)} \|N\|^{\nu(j)}$$

$$= \sum_{\ell=0}^{n} \sum_{0 \leq j < 2^k, \nu(j) = \ell} \|D\|^{k - \ell} \|N\|^{\ell} \tag{6.39}$$

$$= \sum_{\ell=0}^{n} \binom{k}{\ell} \|D\|^{k - \ell} \|N\|^{\ell}$$

because the number of $j \in [0, 2^k - 1]$ such that $\nu(j) = \ell$ is $\binom{k}{\ell}$. We use the elementary estimate

$$\binom{k}{\ell} = \binom{n}{\ell}\frac{k!(n-\ell)!}{n!(k-\ell)!} = \binom{n}{\ell}\frac{(k-\ell+1)\cdots k}{(n-\ell+1)\cdots n} \leq \binom{n}{\ell}k^{\ell}. \qquad (6.40)$$

Therefore,

$$\|(D+N)^k\| \leq k^n\|D\|^{k-n}\sum_{\ell=0}^{n}\binom{n}{\ell}\|D\|^{n-\ell}\|N\|^{\ell}$$
$$= k^n\|D\|^{k-n}\left(\|D\| + \|N\|\right)^n, \qquad (6.41)$$

which completes the proof of the lemma.   QED

From (6.30), we have

$$\|A^k\|_{\infty} = \|U^{\star}T^kU\|_{\infty} \leq \|U^{\star}\|_{\infty}\|T^k\|_{\infty}\|U\|_{\infty} \leq C\|T^k\|_{\infty}, \qquad (6.42)$$

where $C = \|U^{\star}\|_{\infty}\|U\|_{\infty}$. Therefore, (6.32) and lemma 6.13 imply

$$\|A^k\|_{\infty} \leq Cc_nk^n\rho(A)^{k-n}, \qquad (6.43)$$

where $c_n = (\|D\|_{\infty} + \|N\|_{\infty})^n$. But since both C and c_n are positive constants,

$$\lim_{k\to\infty}(Cc_nk^n)^{1/k} = 1. \qquad (6.44)$$

Therefore, (6.43) yields

$$\lim_{k\to\infty}\|A^k\|_{\infty}^{1/k} \leq \lim_{k\to\infty}(Cc_nk^n)^{1/k}\rho(A)^{1-n/k} = \rho(A). \qquad (6.45)$$

Combined with (6.27), this proves (6.28) and completes the theorem.

6.5 EXERCISES

Exercise 6.1 *Show that the set $\mathcal{P}_n$ of polynomials of degree n in one variable can be viewed as a vector space of dimension $n + 1$. (Hint: define $(f + g)(x) = f(x) + g(x)$ for all x and define $(\alpha f)(x) = \alpha f(x)$ for all x. Use the monomials as a basis to determine the dimension.)*

Exercise 6.2 *Show that the derivative operator is a well-defined mapping from the set $\mathcal{P}_n$ of polynomials of degree n in one variable to itself (see exercise 6.1). Compute its matrix representation in the basis given by the monomials.*

Exercise 6.3 *Prove that the operator norm defined by (6.1) is a norm on the linear space of operators from V to W.*

Exercise 6.4 *Prove the product expression (6.3). (Hint: consider $(AB)u = A(Bu)$ and apply (6.2) twice.)*

Exercise 6.5 *Prove that* $\|U\|_2 = 1$ *for any unitary matrix. (Hint: just compute* $\|Ux\|_2^2 = (Ux)^\star Ux = x^\star U^\star Ux = x^\star x.)$

Exercise 6.6 *Prove that*

$$\|A\|_2 = \sqrt{\rho(A^\star A)} \tag{6.46}$$

for any matrix A. Use this to verify (6.10) for a Hermitian operator. (Hint: apply corollary 6.5 and compute

$$\|Ax\|_2^2 = (Ax)^\star Ax = x^\star A^\star Ax = x^\star U^\star DUx. \tag{6.47}$$

Apply exercise 6.5, or at least its hint.)

Exercise 6.7 *Suppose that A is a Hermitian, positive definite matrix. Show that for the operator norm associated with the Euclidean norm*

$$\|A^{-1}\|_2^{-1} = \min \left\{ |\lambda| \mid \lambda \text{ is an eigenvalue of } A \right\}. \tag{6.48}$$

(Hint: if λ is an eigenvalue of A, then λ^{-1} is an eigenvalue of A^{-1}.)

Exercise 6.8 *Show by example that the norm*

$$\|A\|_{\max} := \max \left\{ |A_{ij}| \mid i,j = 1,\ldots,n \right\} \tag{6.49}$$

does not satisfy the inequality (6.3).

Exercise 6.9 *Prove that the Frobenius norm (6.5) satisfies the property (6.3) regarding the norm of the product of matrices, even though it is not an operator norm in the sense of (6.1) for the case where $V = W$ (demonstrate that by example). What happens if we allow $V \neq W$? (Hint: use Cauchy (5.33) for the product formula. To prove it is not an operator norm for $V = W$, evaluate the Frobenius norm of the identity matrix. For the general case, see [33, 50].)*

Exercise 6.10 *Recall that a similarity transformation is of the form $B = S^{-1}AS$, where S is assumed to be invertible. Show that A and B have the same eigenvalues.*

Exercise 6.11 *Suppose that A and B are $n \times n$ matrices, and that A is invertible. Prove that AB and BA are similar matrices. (Hint: we seek an invertible matrix S such that $AB = S^{-1}BAS$. What if we choose $S = A^{-1}$?)*

Exercise 6.12 *Prove that if an upper-triangular matrix T commutes with its conjugate transpose $(T^\star T = TT^\star)$, then it must be diagonal. (Hint: use the block-matrix multiplication formula (3.31), where*

$$T = \begin{pmatrix} \alpha & b^\star \\ 0 & S \end{pmatrix} \tag{6.50}$$

and S is also upper-triangular.)

Exercise 6.13 *Suppose that D is a diagonal matrix with diagonal entries $d_1, \ldots, d_n$. Show that the matrices MD and DM are just M scaled by the diagonal entries of D via $(MD)_{ij} = M_{ij}d_j$ and $(DM)_{ij} = d_iM_{ij}$.*

Exercise 6.14 *Let P be an invertible matrix. Show that $N(x) := \|Px\|$ is a norm for any vector norm $\|\cdot\|$.*

Exercise 6.15 *Prove that*
$$\|Cy\|_\infty \leq n\|y\|_\infty \max\left\{|C_{ij}| \mid i, j = 1, \ldots, n\right\} \tag{6.51}$$
for any $n \times n$ matrix C and $y \in \mathbb{R}^n$.

Exercise 6.16 *Suppose that D is a diagonal $n \times n$ matrix. Prove that $\|D\|_\infty = \max\left\{|D_{ii}| \mid i = 1, \ldots, n\right\}$.*

Exercise 6.17 *Verify the block-matrix multiplication formula expressed by the first equality in (6.19). (Compare exercise 3.5.)*

Exercise 6.18 *Verify the block-matrix multiplication formula expressed by the first equality in (6.21). (Compare exercise 3.5.)*

Exercise 6.19 *Show that (6.25) is a complete flag that supports any upper-triangular matrix.*

Exercise 6.20 *Prove theorem 6.9. (Hint: first represent the operator as a matrix using a basis of V and use the Schur decomposition to decompose this matrix. Show that the operator can be represented as a triangular matrix in some basis, and apply exercise 6.19.)*

Exercise 6.21 *Suppose that $1/p + 1/q = 1$ ($q = \infty$ if $p = 1$, and $p = \infty$ if $q = 1$). Prove that for any $x \in \mathbb{F}^n$,*
$$\|x\|_p = \sup_{\|y\|_q=1} |y^\star x|, \tag{6.52}$$
where the supremum is over $y \in \mathbb{F}^n$ and $\mathbb{F}$ is either $\mathbb{R}$ or $\mathbb{C}$. (Hint: first use Hölder's inequality (5.12). Then for $p < \infty$, choose
$$y_i = \operatorname{sign}(x_i)|x_i|^{p-1}. \tag{6.53}$$
For $p = \infty$, note that $\|x\|_\infty = |x_i|$ for some i; choose y accordingly.)

Exercise 6.22 *Suppose that $1/p + 1/q = 1$ ($q = \infty$ if $p = 1$, and $p = \infty$ if $q = 1$). Prove that $\|A\|_p = \|A^\star\|_q$. (Hint: use exercise 6.21.)*

Exercise 6.23 *Suppose that M and N are $n \times n$ upper-triangular matrices with shift indices μ and ν, respectively. Prove that MN has shift index $\mu + \nu$. (Hint: observe that*
$$(MN)_{ij} = \sum_{k=i+\mu}^{j-\nu} N_{ik}M_{kj} \tag{6.54}$$
for all i, j.)

Exercise 6.24 *Suppose that N is an $n \times n$ matrix such that $N_{ij} = 0$ for $i > j - 1$. Prove that $N_{ij}^k = 0$ for $i > j - k$. (Hint: use induction and exercise 6.23.)*

Exercise 6.25 *Suppose that N is an $n \times n$ matrix such that $N_{ij} = 0$ for $i > j - \nu$, that is, the shift index of N is μ. Suppose that D is an $n \times n$ diagonal matrix. Prove that DN and ND have a shift index of at least ν. (Hint: see the hint for exercise 6.23.)*

Exercise 6.26 *Suppose that the binary expansion of j has at least n zeros, that is, $\nu(j) \geq n$. Thus there are at least n factors N in $P(j)$ defined in (6.34). Prove that $P(j) = 0$. (Hint: use exercise 6.25 to prove that the shift index of $P(j)$ is at least n.)*

6.6 SOLUTIONS

Solution of Exercise 6.10. First, note that by multiplying $B = S^{-1}AS$ on the left by S, we get $SB = AS$. Multiplying this on the right by S^{-1}, we find $SBS^{-1} = A$. Thus the similarity relationship is transitive. Suppose that $AX = \lambda X$. Then

$$\lambda X = AX = SBS^{-1}X.$$

Multiplying this by S^{-1} on the left shows that

$$\lambda S^{-1}X = BS^{-1}X.$$

Thus $S^{-1}X$ is an eigenvector of B with eigenvalue λ. Therefore, we have proved that all the eigenvalues of A are eigenvalues of B. Since the similarity relationship is reflexive, all the eigenvalues of B are thus eigenvalues of A, and so they are the same. QED

Solution of Exercise 6.12. Using the block-matrix multiplication formula (3.31), we find that

$$TT^\star = \begin{pmatrix} \alpha & b^\star \\ 0 & S \end{pmatrix} \begin{pmatrix} \overline{\alpha} & 0 \\ b & S^\star \end{pmatrix} = \begin{pmatrix} |\alpha|^2 + b^\star b & b^\star S^\star \\ Sb & SS^\star \end{pmatrix} \qquad (6.55)$$

and

$$T^\star T = \begin{pmatrix} \overline{\alpha} & 0 \\ b & S^\star \end{pmatrix} \begin{pmatrix} \alpha & b^\star \\ 0 & S \end{pmatrix} = \begin{pmatrix} |\alpha|^2 & \overline{\alpha}b^\star \\ \alpha b & S^\star S \end{pmatrix}. \qquad (6.56)$$

Thus we see that $TT^\star = T^\star T$ implies $b = 0$ (compare the upper-left entries) and $SS^\star = S^\star S$. Thus the result is easily completed via induction on the size of the matrix T.

Let us assume that T is $n \times n$. If $n = 2$, then S is just a scalar, and thus $b = 0$ implies T is diagonal. Now suppose that the result is known for matrices of size $(n-1) \times (n-1)$, for some $n \geq 3$. Then the equality of (6.55) and (6.56) implies that $SS^\star = S^\star S$. Since S is an upper-triangular matrix of size $(n-1) \times (n-1)$, we conclude that S must be diagonal. Together with the fact that $b = 0$ implies that T is diagonal.

Solution of Exercise 6.13. Just compute:

$$(MD)_{ij} = \sum_{k=1}^{n} M_{ik} D_{kj} = M_{ij} d_j$$

and

$$(DM)_{ij} = \sum_{k=1}^{n} D_{ik} M_{kj} = d_i M_{ij}.$$

QED

Solution of Exercise 6.14. It is clear that $N(sx) = |s| N(x)$ for any scalar s since $P(sx) = sPx$. The triangle inequality is equally easy: $P(x+y) = Px + Py$, so

$$N(x+y) = \|Px + Py\| \le \|Px\| + \|Py\| = N(x) + N(y).$$

Now suppose that $N(x) = 0$. Then $Px = 0$, and since P is invertible, we have $x = 0$.   QED

Chapter Seven

Nonlinear Systems

> In 1740, Thomas Simpson published "Essays on several cu-
> rious and useful subjects in speculative and mix'd mathe-
> maticks, illustrated by a variety of examples," in which he
> presents Newton's method essentially in the form (2.30),
> together with a generalization to systems of two equations,
> and shows that Newton's method can solve optimization
> problems by setting the gradient to zero [174].

We now turn to finding solutions of nonlinear systems of equations. There
are many ways in which nonlinear systems of equations arise. One common
one is in the minimization of a smooth, scalar-valued function ϕ, as antici-
pated already by Simpson [174]. Minima are characterized by the equation
$\nabla\phi(x) = 0$, so this suggests applying the techniques of this chapter to the
function $f(x) = \nabla\phi(x)$.

We will take as an example a simple form of a problem in geodesy, the science of determining locations in space from distance data. This has occupied
many mathematicians, including Gauss [20, 93]; it is pursued on a very large
scale [98] and is the basis for location by global positioning systems (GPS)
[153]. In figure 7.1, we depict the problem of determining the position of a
boat near a shoreline based on the distance from the boat to two markers on
shore. The distance might be determined by measuring the time it takes for
sound to travel to the boat. A flash of light could indicate when the sound
was emitted. The mathematical problem can be described in two equations

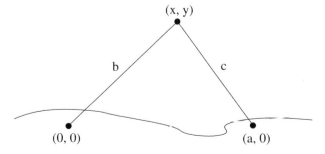

Figure 7.1 Determining position offshore. The unknown point (x, y) is to be determined from the known distances b and c to points on the shore.

for the distances:

$$b = \sqrt{x^2 + y^2} \qquad c = \sqrt{(x-a)^2 + y^2}. \tag{7.1}$$

We can turn this into a problem similar to what we studied in chapter 2 by defining the function

$$f(x_1, x_2) = \begin{pmatrix} x_1^2 + x_2^2 \\ (x_1 - a)^2 + x_2^2 \end{pmatrix}. \tag{7.2}$$

Our problem is thus reduced to finding a solution to $f(x_1, x_2) = (b^2, c^2)$. Now we consider such problems in general.

Thus we suppose that we have a function $f : \mathbb{R}^n \to \mathbb{R}^n$ and we want to find points $x \in \mathbb{R}^n$ where $f(x) = y$ for some $y \in \mathbb{R}^n$. Without loss of generality, we usually assume that $y = 0$ by simply subtracting y from f to create a new function. Correspondingly, we may also cast this as a fixed-point problem: $g(x) = x$. As we have seen in the one-dimensional case (chapter 2), there may be several different g's whose fixed points correspond to the solutions of $f(x) = 0$. However, the setting of fixed-point iteration is still, as in the one-dimensional case, the place to start.

7.1 FUNCTIONAL ITERATION FOR SYSTEMS

We just need to interpret the notation: g now maps R^n to itself, and fixed-point iteration seeks to find a fixed point

$$\xi = g(\xi), \tag{7.3}$$

where now $\xi \in R^n$. Fixed-point iteration

$$x^\nu = g(x^{\nu-1}) \tag{7.4}$$

still has the property that, if it converges, it converges to a fixed point (7.3), assuming only that g is continuous (exercise 7.1).

We emphasize now that g represents n functions $g_1, \ldots, g_n$ that each map $\mathbb{R}^n$ to $\mathbb{R}$ (we will limit our discussions to real-valued functions for simplicity). The basic behavior is the same as in the one-dimensional case, provided we use norms to measure vectors where we used absolute values before. If g is Lipschitz-continuous with constant $\lambda < 1$, that is,

$$\|g(x) - g(y)\| \le \lambda \|x - y\| \tag{7.5}$$

for some norm $\| \cdot \|$ on R^n, then convergence will happen for all starting points. More precisely, if we define $e^\nu = x^\nu - \xi$, then by induction

$$\|e^{\nu+1}\| = \|g(x^\nu) - g(\xi)\| \le \lambda \|e^\nu\| \le \lambda^\nu \|e^0\|. \tag{7.6}$$

In fact, we have used (7.5) with only one x, namely, $x = \xi$.

The rest of the story is similar to the one-dimensional case, except that we need some higher-dimensional calculus to figure out when and how fast it will converge. We can discover the local behavior by using a Taylor expansion:

$$g(y) = g(x) + J_g(x)(x - y) + R_g(x, y), \tag{7.7}$$

where the remainder $R_g(x, y)$ satisfies

$$\|R_g(x, y)\| \leq C\|x - y\|^2 \tag{7.8}$$

and J_g denotes the Jacobian of g, that is, the matrix with entries

$$(J_g(x))_{ij} = \frac{\partial g_i}{\partial x_j}(x). \tag{7.9}$$

We can present this visually as

$$J_g(x) = \begin{pmatrix} \frac{\partial g_1}{\partial x_1}(x) & \frac{\partial g_1}{\partial x_2}(x) & \cdots & \frac{\partial g_1}{\partial x_n}(x) \\ \frac{\partial g_2}{\partial x_1}(x) & \frac{\partial g_2}{\partial x_2}(x) & \cdots & \frac{\partial g_2}{\partial x_n}(x) \\ \vdots & \vdots & \vdots & \vdots \\ \frac{\partial g_n}{\partial x_1}(x) & \frac{\partial g_1}{\partial x_2}(x) & \cdots & \frac{\partial g_n}{\partial x_n}(x) \end{pmatrix}. \tag{7.10}$$

For example, for f defined in (7.2), then

$$J_f(x) = \begin{pmatrix} 2x_1 & 2x_2 \\ 2(x_1 - a) & 2x_2 \end{pmatrix}. \tag{7.11}$$

We recall that J_g may be thought of as a matrix-valued function, that is, a map $\mathbb{R}^n \to \mathbb{R}^{n^2}$. The expression $J_g(x)(x - y)$ in (7.7) is just a matrix-vector multiplication. Reviewing the expression (7.6), we have

$$\begin{aligned} \|e^{\nu+1}\| &= \|g(x^\nu) - g(\xi)\| \\ &= \|J_g(\xi)e^\nu + R_g(\xi, x^\nu)\| \\ &\leq \|J_g(\xi)e^\nu\| + \|R_g(\xi, x^\nu)\| \\ &\leq \|J_g(\xi)\| \, \|e^\nu\| + C\|e^\nu\|^2, \end{aligned} \tag{7.12}$$

where $\|J_g(\xi)\|$ is the operator norm of $J_g(\xi)$. Thus we see that the limiting behavior of fixed-point iteration will be determined by the value of $\|J_g(\xi)\|$.

There are several "theorems" that one could present based on the analysis above, but the following local result is the most important.

Theorem 7.1 *Suppose that the spectral radius $\rho(J_g(\xi)) < 1$ at a fixed point $\xi = g(\xi)$ and that the Taylor expansion (7.7) holds with the constant C in (7.8) (in some norm) fixed for all y in a neighborhood of $x = \xi$. Then fixed-point iteration (7.3) converges provided that x^0 is close enough to ξ.*

The main point of this result is that there is no reference to any particular norm in the estimation of the size of $J_g(\xi)$. We leave the proof of this result as exercise 7.2. The trick is to pick a norm sufficiently close to the spectral radius for the matrix $J_g(\xi)$ using theorem 6.8. Note that if the Taylor expansion (7.7) holds with the constant C in (7.8) fixed for all y in a neighborhood of $x = \xi$, then it holds for any norm by the equivalence of norms on $\mathbb{R}^n$ (section 5.3.2).

We need to justify the Taylor expansion (7.7) with the remainder term in (7.8). But to do so, we need to develop some notation, which we will do in section 7.1.2.

7.1.1 Limiting behavior of fixed-point iteration

Fixed-point iteration in higher dimensions can have a more complicated set of behaviors than in the one-dimensional case. Using the Taylor expansion (7.7), we can write

$$
\begin{aligned}
e^{\nu+1} = x^{\nu+1} - \xi &= g(x^\nu) - g(\xi) \\
&= J_g(\xi)(x^\nu - \xi) + R_g(\xi, x^\nu) \\
&\approx J_g(\xi)e^\nu + \mathcal{O}\left(\|e^\nu\|^2\right) \\
&\approx J_g(\xi)e^\nu.
\end{aligned}
\tag{7.13}
$$

Thus the errors evolve iteratively by multiplying the Jacobian matrix $J_g(\xi)$. These vectors are very similar to those generated by the power method for solving eigenproblems in section 15.1. Thus we will see that the generic behavior is that e^ν will tend to an eigenvector of $J_g(\xi)$ corresponding to an eigenvalue λ, where $|\lambda| = \rho(J_g(\xi))$, exactly in line with the statement of theorem 7.1. However, a simple example shows that other behaviors are possible. Define

$$
g(x) = A\,x + \xi \,\|x\|_2^2,
\tag{7.14}
$$

where A is a matrix such that $A\xi = 0$. The fixed point of interest is $x = 0$, and it is easy to see that $J_g(0) = A$. Suppose that we take $x^0 = \epsilon\xi$. Then by induction,

$$
x^\nu = \epsilon^{2^\nu}\xi.
\tag{7.15}
$$

Thus we find quadratic convergence in this special case.

7.1.2 Multi-index notation

For the time being, we will focus on scalar-valued (real- or complex-valued) functions. We will apply these ideas later to vector-valued functions. A multi-index, α, is an n-tuple of nonnegative integers, α_i. The length of α is given by

$$
|\alpha| := \sum_{i=1}^n \alpha_i.
\tag{7.16}
$$

For a smooth function $\phi : \mathbb{R}^n \to \mathbb{R}$, the notations

$$
D^\alpha\phi, \quad \left(\frac{\partial}{\partial x}\right)^\alpha \phi, \quad \phi^{(\alpha)}, \quad \partial_x^\alpha\phi, \text{ and } \phi_{,\alpha_1\alpha_2\cdots\alpha_n}
\tag{7.17}
$$

are used interchangeably to denote the partial derivative

$$
\left(\frac{\partial}{\partial x_1}\right)^{\alpha_1} \cdots \left(\frac{\partial}{\partial x_n}\right)^{\alpha_n} \phi.
\tag{7.18}
$$

Given a vector $x = (x_1, \ldots, x_n) \in \mathbb{R}^n$, we define

$$
x^\alpha := x_1^{\alpha_1} \cdot x_2^{\alpha_2} \cdots x_n^{\alpha_n}.
\tag{7.19}
$$

Note that if x is replaced formally by the symbol $\frac{\partial}{\partial x} := \left(\frac{\partial}{\partial x_1}, \ldots, \frac{\partial}{\partial x_n} \right)$, then this definition of x^α is consistent with the previous definition of $\left(\frac{\partial}{\partial x} \right)^\alpha$. The *order* of this derivative is given by $|\alpha|$.

For the moment, let us focus on scalar-valued functions, e.g., $u : \mathbb{R}^n \to \mathbb{R}$. The *Taylor polynomial* of order m expanded at y is given by

$$T_y^m u(x) = \sum_{|\alpha| < m} \frac{1}{\alpha!} D^\alpha u(y)(x - y)^\alpha, \tag{7.20}$$

where

$$\alpha! = \prod_{i=1}^{n} \alpha_i!. \tag{7.21}$$

For $\varphi \in C^m\big([0, 1]\big)$, we have (exercise 7.4)

$$\varphi(1) = \sum_{k=0}^{m-1} \frac{1}{k!} \varphi^{(k)}(0) + \int_0^1 \frac{1}{(m - 1)!} s^{m-1} \varphi^{(m)}(1 - s) \, ds. \tag{7.22}$$

Let u be a C^m function on $\mathbb{R}^n$. For $x \in \mathbb{R}^n$ and $y \in \mathbb{R}^n$, define

$$\varphi(s) = u\big(y + s(x - y)\big). \tag{7.23}$$

Then, by using the chain rule, we obtain

$$\boxed{\frac{1}{k!} \varphi^{(k)}(s) = \sum_{|\alpha| = k} \frac{1}{\alpha!} D^\alpha u\big(y + s(x - y)\big)(x - y)^\alpha.} \tag{7.24}$$

We will prove (7.24) shortly. Combining (7.22) and (7.24), we obtain

$$u(x) = \sum_{|\alpha| < m} \frac{1}{\alpha!} D^\alpha u(y)(x - y)^\alpha$$

$$+ \sum_{|\alpha| = m} (x - y)^\alpha \int_0^1 \frac{m}{\alpha!} s^{m-1} D^\alpha u\big(x + s(y - x)\big) \, ds \tag{7.25}$$

$$= T_y^m u(x) + m \sum_{|\alpha| = m} (x - y)^\alpha \int_0^1 \frac{1}{\alpha!} s^{m-1} D^\alpha u\big(x + s(y - x)\big) \, ds.$$

Applying (7.25) to each component of a function $g : \mathbb{R}^n \to \mathbb{R}^n$ for $m = 2$, we obtain the following expression for the ith component of the error term $R_g(x, y)$ in (7.7):

$$\boxed{R_g(x, y)_i = 2 \sum_{|\alpha| = 2} (x - y)^\alpha \int_0^1 \frac{1}{\alpha!} s D^\alpha g_i\big(x + s(y - x)\big) \, ds.} \tag{7.26}$$

To prove (7.24), we start with the chain rule:

$$\varphi'(s) = (x - y) \cdot \nabla u\big(y + s(x - y)\big)$$

$$= \sum_{i=1}^{n} (x_i - y_i) u_{,i}\big(y + s(x - y)\big) \tag{7.27}$$

$$= \sum_{|\alpha| = 1} (x - y)^\alpha D^\alpha u(y).$$

This covers the case $k = 1$. We apply the chain rule again to (7.27) to find

$$
\begin{aligned}
\varphi''(s) &= \sum_{i=1}^{n} (x_i - y_i)(x - y) \cdot \nabla u_{,i}\big(y + s(x - y)\big) \\
&= \sum_{i,j=1}^{n} (x_i - y_i)(x_j - y_j) u_{,ij}\big(y + s(x - y)\big) \\
&= (x - y)^{\mathrm{T}} H_u \big(y + s(x - y)\big) (x - y),
\end{aligned}
\tag{7.28}
$$

where the matrix H_u is called the Hessian[1] of u. This allows a useful representation of the Taylor approximation of order 2:

$$
u(x) \approx u(y) + \nabla u(y) \cdot (x - y) + \tfrac{1}{2}(x - y)^{\mathrm{T}} H_u(y)(x - y).
\tag{7.29}
$$

To establish the relationship with (7.24), we just count the terms in (7.28):

$$
\begin{aligned}
\varphi''(s) &= \sum_{i,j=1}^{n} (x_i - y_i)(x_j - y_j) u_{,ij}\big(y + s(x - y)\big) \\
&= \sum_{i=1}^{n} (x_i - y_i)^2 u_{,ii}\big(y + s(x - y)\big) \\
&\quad + 2 \sum_{i>j=1}^{n} (x_i - y_i)(x_j - y_j) u_{,ij}\big(y + s(x - y)\big) \\
&= \sum_{|\alpha|=2} \frac{2}{\alpha!}(x - y)^{\alpha} D^{\alpha} u(y + s(x - y)).
\end{aligned}
\tag{7.30}
$$

This covers the case $k = 2$; since this is all we need to derive (7.26), we leave the general case to exercise 7.5.

7.1.3 Higher-order convergence

Suppose that $J_g(\xi) = 0$ at a fixed point $\xi = g(\xi)$. By analogy with the one-dimensional case, we expect higher-order convergence in this case. Let us examine this now formally. We have by (7.7),

$$
g(y) - g(\xi) = R_g(\xi, y),
\tag{7.31}
$$

where the remainder $R_g(\xi, y)$ satisfies

$$
\|R_g(\xi, y)\| \le C \|\xi - y\|^2
\tag{7.32}
$$

in view of (7.8), as can be verified by using (7.26). Therefore

$$
\begin{aligned}
\|e^{\nu+1}\| &= \|g(\xi) - g(x^{\nu})\| \\
&= \|R_g(\xi, x^{\nu})\| \\
&\le C \|e^{\nu}\|^2,
\end{aligned}
\tag{7.33}
$$

and the convergence is second-order.

---

[1] Ludwig Otto Hesse (1811–1874) was a student of Jacobi (page 118) and was the advisor of Lipschitz (page 17).

7.1.4 Particular methods

Not all of the one-dimensional methods generalize to n dimensions. The chord method becomes

$$x^{\nu+1} = x^\nu - Af(x^\nu), \qquad (7.34)$$

where A is a matrix. Thus $g(x) = x - Af(x)$ and

$$J_g(x) = I - AJ_f(x) \qquad (7.35)$$

(see exercise 7.7). Thus we have $J_g(\xi)$ small if A is close to $J_f(\xi)^{-1}$. Therefore, adaptive methods will attempt to approximate $A \approx J_f(\xi)^{-1}$.

7.2 NEWTON'S METHOD

Newton's method takes $A = J_f(x^\nu)^{-1}$ in the chord method (7.35); that is, we solve the linear system

$$\boxed{J_f(x^\nu)(x^{\nu+1} - x^\nu) = -f(x^\nu) .} \qquad (7.36)$$

We can write this as a fixed-point iteration with

$$g(x) = x - J_f(x)^{-1}f(x). \qquad (7.37)$$

To compute J_g in this case will require some work.

To begin with, let us rewrite the expression (7.37) as

$$J_f(x)g(x) = J_f(x)x - f(x). \qquad (7.38)$$

Thus we need to differentiate the product $h(x) = J_f(x)g(x)$ for two different functions g, so let us consider this separately. Formally, we can expect this to be of the form

$$J_h(x) = J_f(x)J_g(x) + H_f(x)g(x), \qquad (7.39)$$

where $H_f(x)$ (the Hessian of the vector function f) involves second-order derivatives of f. The reasoning is just that the derivative of a product satisfies the rule $(uv)' = uv' + u'v$. Note the similarity to (7.35). What is a bit unusual about the expression (7.39) is that, as an equation for matrices, the "type" of $H_f(x)$ is new. It is an algebraic object that maps a vector to a matrix.

7.2.1 Tensors

Let us consider a function of the form

$$u(x) = J_f(x)\xi, \qquad (7.40)$$

where $\xi \in \mathbb{R}^n$ is a fixed (constant) vector. Thus

$$u_j(x) = \sum_{k=1}^n \frac{\partial f_j}{\partial x_k}(x)\xi_k. \qquad (7.41)$$

Differentiating, we find

$$\frac{\partial u_j}{\partial x_\ell}(x) = \sum_{k=1}^{n} \frac{\partial^2 f_j}{\partial x_k x_\ell}(x)\xi_k. \tag{7.42}$$

Let us define $H_f(x)$ as a map of $\mathbb{R}^n \to \mathbb{R}^{n^2}$ via

$$\left(H_f(x)\xi\right)_{j,\ell} = \sum_{k=1}^{n} \frac{\partial^2 f_j}{\partial x_k x_\ell}(x)\xi_k \quad \forall \xi \in \mathbb{R}^n. \tag{7.43}$$

Thus (7.42) can be written as

$$J_u(x) = H_f(x)\xi. \tag{7.44}$$

This allows us to justify (7.39) (see exercise 7.8).

The object H_f is called a *tensor*. We will not explore the algebraic properties of tensors in detail, but suffice it to say they are things with indices. The number of indices is sometimes called the "rank" but we prefer to reserve that word for another property, so we call the number of indices the *arity* of the tensor. The arity of H_f is 3. Tensors of arity 2 are matrices, and tensors of arity 1 are vectors (tensors of arity 0 are scalars).

The operation of multiplication of one tensor by another is often called *contraction*, and it reduces the arity correspondingly. A contraction of a tensor of arity k by one of arity ℓ produces a tensor of arity $k - \ell$, as we saw in (7.44). Derivatives of tensor functions produce tensors of higher arity, as we have seen.

The Hessian of a vector-valued function can be used to provide a Taylor approximation analogous to (7.29):

$$u(x) \approx u(y) + J_u(y)(x - y) + \tfrac{1}{2}(x - y)^\mathsf{T} H_u(y)(x - y), \tag{7.45}$$

which we leave as exercise 7.11.

7.2.2 Quadratic convergence of Newton's method

Convergence of Newton's method is again quadratic, as we now show. Returning to the expression (7.38), we differentiate it to get

$$H_f(x)g(x) + J_f(x)J_g(x) = H_f(x)x + J_f(x) - J_f(x) = H_f(x)x. \tag{7.46}$$

Thus we can solve for $J_g(x)$ to get

$$\begin{aligned}
J_g(x) &= J_f(x)^{-1}\left(H_f(x)x - H_f(x)g(x)\right) \\
&= J_f(x)^{-1}\left(H_f(x)(x - g(x))\right) \\
&= J_f(x)^{-1}\left(H_f(x)(J_f(x)^{-1}f(x))\right).
\end{aligned} \tag{7.47}$$

In the second step, we used the fact that tensors are linear operators (exercise 7.13), and the last step is just the definition (7.37) of g. Thus we have proved the following.

Theorem 7.2 *Suppose that g is the iteration function for Newton's method defined in (7.37). Then at a point x where $f(x) = 0$, we have $J_g(x) = 0$ provided that $J_f(x)$ is invertible. In this case, Newton's method is quadratically convergent.*

Most convergence results for Newton's method are essentially local in nature. In general, one expects chaotic behavior from Newton's method globally [91]. The following result is just one example of the type of local result that can be proved. See exercise 7.12 for a more sophisticated result of this type in which the Jacobian of f is allowed to be singular at the root $f(x) = 0$.

Theorem 7.3 *Suppose that $y \in \mathbb{R}^n$ is a root $f(y) = 0$. Suppose that $R > 0$ is chosen so that for the set*

$$\Omega = \left\{ x \in \mathbb{R}^n \mid |x - y| \leq R \right\}, \tag{7.48}$$

the following conditions hold for some constants α, β:

$$\sup_{x \in \Omega} \| J_f^{-1}(x) f(x) \| \leq \alpha,$$

$$\sup_{x \in \Omega} \| J_f^{-1}(x) H_f(x) z \| \leq \beta \| z \| \quad \forall z \in \mathbb{R}^n \ \ such \ that \ \ \| z \| \leq \alpha \tag{7.49}$$

for some norm $\| \cdot \|$ on $\mathbb{R}^n$. Then if $\lambda = \alpha\beta < 1$, Newton's method converges in Ω. That is, for all starting points $x^0 \in \Omega$, all subsequent iterates remain in Ω, and

$$\| x^k - y \|_\infty \leq \lambda^k R. \tag{7.50}$$

The gist of the theorem is that we can make α as small as we want by choosing R small enough (exercise 7.17), provided the Jacobian J_f of f is not too badly behaved (cf. exercise 7.12). Concrete bounds on β can be made provided, e.g., we also assume that

$$\sup_{x \in \Omega} \max_{j,k,\ell=1,\ldots,n} \left| \frac{\partial^2 f_j}{\partial x_k x_\ell}(x) \right| \leq c, \tag{7.51}$$

together with the assumption that $J_f(x)^{-1}$ is bounded for $x \in \Omega$. Note that we have written the assumptions (7.49) in an invariant way. Newton's method is invariant with respect to multiplication on the left by a nonsingular matrix A. That is, the iterates are the same for solving $f(x) = 0$ and $Af(x) = 0$ for fixed A. Thus assumptions regarding convergence should likewise be invariant [46].

The proof is just an application of (7.47): for $x \in \Omega$, we have

$$\| J_g(x) \| \leq \| J_f(x)^{-1} H_f(x) (J_f(x)^{-1} f(x)) \|$$
$$\leq \beta \| J_f(x)^{-1} f(x) \| \tag{7.52}$$
$$\leq \beta\alpha = \lambda < 1.$$

Observe that the norm of the expression $J_f(x)^{-1} H_f(x) z$ is the operator norm associated with $\| \cdot \|$. Applying (7.6) proves (7.50), which in turn guarantees that all iterates remain in Ω.

Note that if $J_f(x^\nu)$ is nearly singular at any point, then the change $x^{\nu+1} - x^\nu$ can be huge. This occurs even in one dimension: consider $f(x) = \cos x$ and start Newton's method near $x = 0$. However, the behavior of Newton's method in multiple dimensions with a singular Jacobian is much more complex than in the one-dimensional case (exercise 2.5). We give an example in section 7.2.4.

7.2.3 No other methods

Unfortunately, the other methods we studied in the one-dimensional case do not generalize to multidimensions. For example, Steffensen's method fails for two reasons. At the simplest level, it is not clear how to "divide" by the difference quotient in the vector case. But more fundamentally, the difference approximation $f(x + f(x)) - f(x)$ would provide differential information in only one direction.

7.2.4 Eigen problems

The eigenvalue problem for an $n \times n$ matrix is really a system of nonlinear equations in $n + 1$ variables, which we can write as

$$
\begin{aligned}
A\,x &= \lambda x \\
\|x\|_2 &= 1.
\end{aligned}
\tag{7.53}
$$

This is a good example to study as it is almost linear, being nonlinear only in the last equation and in the simple product λx. Since we are talking about eigenvalues, we allow everything to be complex. We have seen that Newton's method is generally effective at solving nonlinear problems, so it is reasonable to ask how it would apply to this problem. The extension to complex variables does not make a substantial change.

Let us write (7.53) formally as solving $F(x, \lambda) = 0$, where we can take F to be defined by

$$
F(x, \lambda) = \begin{pmatrix} A\,x - \lambda x \\ \frac{1}{2}\left(1 - \|x\|_2^2\right) \end{pmatrix} = \begin{pmatrix} A\,x - \lambda x \\ \frac{1}{2}\left(1 - x^\star x\right) \end{pmatrix}.
\tag{7.54}
$$

The Jacobian of F is given by

$$
J_F(x, \lambda) = \begin{pmatrix} A - \lambda I & -x \\ -x^\star & 0 \end{pmatrix}.
\tag{7.55}
$$

Note that if A is Hermitian, so is J_F for $\lambda \in \mathbb{R}$.

Newton's method for F is then

$$
J_F(x^k, \lambda^k) \begin{pmatrix} x^{k+1} - x^k \\ \lambda^{k+1} - \lambda^k \end{pmatrix} = -F(x^k, \lambda^k),
\tag{7.56}
$$

which translates componentwise to

$$
\begin{aligned}
(A - \lambda^k I)(x^{k+1} - x^k) - (\lambda^{k+1} - \lambda^k)x^k &= -(A - \lambda^k I)x^k \\
-(x^k)^\star(x^{k+1} - x^k) &= -\tfrac{1}{2}(1 - \|x^k\|_2^2).
\end{aligned}
\tag{7.57}
$$

Simplifying, we find

$$(A - \lambda^k I)x^{k+1} = (\lambda^{k+1} - \lambda^k)x^k$$
$$-(x^k)^\star x^{k+1} = -\tfrac{1}{2} - \tfrac{1}{2}\|x^k\|_2^2. \tag{7.58}$$

We can unravel this system by observing that $x^{k+1} = (\lambda^{k+1} - \lambda^k)y^k$, where y^k solves

$$(A - \lambda^k I)y^k = x^k. \tag{7.59}$$

Using the second equation in (7.58), we find that

$$\lambda^{k+1} = \lambda^k + \frac{\tfrac{1}{2} + \tfrac{1}{2}\|x^k\|_2^2}{(x^k)^\star y^k}. \tag{7.60}$$

We will see that Newton's method for the eigenvalue problem is essentially a version of what is known as inverse iteration (section 15.2).

Suppose that x, λ is an eigenpair for A. Then $J_F(x, \lambda)$ is singular if and only if there is a nontrivial solution to

$$0 = J_F(x, \lambda)\begin{pmatrix} y \\ \mu \end{pmatrix} = \begin{pmatrix} Ay - \lambda y - \mu x \\ -x^\star y \end{pmatrix}. \tag{7.61}$$

There are two types of solutions. We could have $\mu = 0$, in which case y must be another eigenvector corresponding to λ, orthogonal to x in view of the last component of (7.61). Or we could have $\mu \neq 0$, in which case $(A - \lambda I)^2 y = 0$ and y is a generalized eigenvalue of A corresponding to λ [11]. In either case, this implies that λ is not a simple eigenvalue. Thus for simple eigenvalues, $J_F(x, \lambda)$ is nonsingular.

It is somewhat surprising that for simple eigenvalues, the system (7.59) and (7.60), which is equivalent to (7.56), is not singular. We return to this question at more length in section 15.2.

7.2.5 An example

The range of behaviors of Newton's method for a singular Jacobian can best be seen by an example [44]. Define

$$A = \begin{pmatrix} 0 & 1 \\ 0 & 0 \end{pmatrix}, \tag{7.62}$$

for which $\lambda = 0$ is an eigenvalue of multiplicity 2. There is one eigenvector $(1, 0)^\mathsf{T}$. Then the Jacobian (7.55) satisfies

$$J_f(\nu, \mu, \lambda) = \begin{pmatrix} -\lambda & 1 & -\nu \\ 0 & -\lambda & -\mu \\ -\nu & -\mu & 0 \end{pmatrix}. \tag{7.63}$$

If we start Newton's method with $x^0 = (r, 0, \lambda)^\mathsf{T}$, then the Newton step can be computed explicitly to show that $x^1 = (\hat{r}, 0, \hat{\lambda})^\mathsf{T}$, where

$$\hat{r} = 1 + \frac{(r-1)^2}{2r} \quad \text{and} \quad \hat{\lambda} = \lambda \frac{r^2 - 1}{2r^2}. \tag{7.64}$$

Thus Newton's method is essentially a two-dimensional iteration in this case, with the second (middle) coordinate of the iterates always remaining zero. Moreover, the iteration $(r, \lambda) \to (\hat{r}, \hat{\lambda})$ converges to $(1, 0)$ quadratically, whereas for general starting values, $x^0 = (\nu, \mu, \lambda)$ with $\mu \neq 0$, the convergence is only linear (exercise 7.20).

7.3 LIMITING BEHAVIOR OF NEWTON'S METHOD

We are primarily interested in the rapid convergence of Newton's method and perhaps do not care how it gets there. But it is an interesting question in multiple dimensions as to whether the iterates wander around as they approach the limit or perhaps instead approach the limit in a systematic way. We can already get guidance from (2.36) in the one-dimensional case. The sign of $f''(\alpha)/f'(\alpha)$ at a root $f(\alpha) = 0$ determines the sign of the error. Thus the iterates approach systematically from one side or the other at the end of the iteration process. Moreover, there is a precise asymptotic relationship between the errors that expresses the quadratic convergence. But in multiple dimensions, a wider range of behaviors might be possible, so we consider this here.

To simplify the notation, let us assume that the root of interest for the function $f : \mathbb{R}^n \to \mathbb{R}^n$ is at the origin: $f(0) = 0$. Thus the errors $e^k = x^k$. Denote the Newton iterates by

$$x^{k+1} = x^k - J_f(x^k)^{-1} f(x^k). \tag{7.65}$$

We do not want to assume that the Jacobian J_f is always invertible, so we interpret the notation in (7.65) to mean that there is a solution $y^k \in \mathbb{R}^n$ to the equation

$$J_f(x^k) y^k = f(x^k), \tag{7.66}$$

in which case $x^{k+1} = x^k - y^k$. If $f(x^k)$ is not in the range of the linear operator $J_f(x^k)$, then no such solution is possible, and Newton's method fails. But we will simply ignore this situation.

Using the Taylor approximation (7.45), we have

$$\begin{aligned} f(x^{k+1}) &\approx f(x^k) + J_f(x^k)(x^{k+1} - x^k) \\ &\quad + \tfrac{1}{2}(x^{k+1} - x^k)^{\mathrm{T}} H_f(x^k)(x^{k+1} - x^k) \\ &= \tfrac{1}{2}(x^{k+1} - x^k)^{\mathrm{T}} H_f(x^k)(x^{k+1} - x^k), \end{aligned} \tag{7.67}$$

where we have used the definition of the Newton step to cancel terms. On the other hand, we can Taylor-expand around zero to find

$$f(x^{k+1}) \approx f(0) + J_f(0) x^{k+1} = J_f(0) x^{k+1}. \tag{7.68}$$

Combining (7.67) and (7.68) we find

$$\begin{aligned} J_f(0) x^{k+1} &\approx \tfrac{1}{2}(x^{k+1} - x^k)^{\mathrm{T}} H_f(x^k)(x^{k+1} - x^k) \\ &\approx \tfrac{1}{2}(x^k)^{\mathrm{T}} H_f(0) x^k \end{aligned} \tag{7.69}$$

since $x^{k+1} = \mathcal{O}\left((x^k)^2\right)$, cf. (2.36).

Suppose we are interested in tracking the direction of the approach to the solution; since we have assumed that $0 = f(0)$, we can do this by defining

$$\xi^k = \|x^k\|_2^{-1} x^k. \tag{7.70}$$

Similarly, let

$$t_k = \|x^k\|_2. \tag{7.71}$$

What we seek to understand is: do the vectors ξ^k on the unit sphere in $\mathbb{R}^n$ tend to a limit, cycle systematically, or wander chaotically?

Let us write (7.69) in terms of ξ's and t's:

$$\begin{aligned} t^{k+1} J_f(0)\xi^{k+1} = J_f(0)x^{k+1} &\approx \tfrac{1}{2}(x^k)^{\mathrm{T}} H_f(0)x^k \\ &= \tfrac{1}{2} t_k^2 (\xi^k)^{\mathrm{T}} H_f(0)\xi^k. \end{aligned} \tag{7.72}$$

Suppose there is a solution to the equation

$$\lambda J_f(0)\xi = \xi^{\mathrm{T}} H_f(0)\xi \tag{7.73}$$

and that the vectors $\xi^k \to \xi$ as $k \to \infty$. Then $t^{k+1} \approx (1/2\lambda)t_k^2$ and Newton's method behaves asymptotically like a one-dimensional iteration in the direction ξ. The problem (7.73) is a *tensor eigenproblem*, and the iteration implicit in (7.72) is similar to the power method (section 15.1) [133, 132] for ordinary eigenproblems.

Let us consider a simple example in two dimensions where $f(x) = x + Q(x)$ and

$$Q(x) = \begin{pmatrix} x^{\mathrm{T}} A x \\ x^{\mathrm{T}} B x \end{pmatrix}, \tag{7.74}$$

where A and B are 2×2 matrices. The limiting behavior of Newton's method is governed by the behavior of the iteration

$$x \leftarrow \|Q(x)\|_2^{-1} Q(x). \tag{7.75}$$

Let us consider the following examples. First, define

$$A = \begin{pmatrix} 1 & 0 \\ 0 & 1 \end{pmatrix}, \quad B = \begin{pmatrix} 0 & 1 \\ 1 & 0 \end{pmatrix}. \tag{7.76}$$

Then there are four eigenpairs given by

$$Q\begin{pmatrix} \pm 1 \\ 0 \end{pmatrix} = \pm 1 \begin{pmatrix} \pm 1 \\ 0 \end{pmatrix} \quad \text{and} \quad Q\begin{pmatrix} \pm 1 \\ 1 \end{pmatrix} = \pm 2 \begin{pmatrix} \pm 1 \\ 1 \end{pmatrix}. \tag{7.77}$$

Moreover, starting vectors x in (7.74) for which $\pm x_1 x_2 > 0$ tend rapidly to the eigenvector with eigenvalue ± 2. The four cases where the initial vector x satisfies $x_1 x_2 = 0$ go immediately to one of the eigenvectors with eigenvalue ± 1 (see exercise 7.21). However, if we instead define Q in (7.74) using

$$A = \begin{pmatrix} 1 & 0 \\ 0 & -1 \end{pmatrix}, \quad B = \begin{pmatrix} 0 & 1 \\ 1 & 0 \end{pmatrix}, \tag{7.78}$$

then the iterations (7.74) are chaotic (see exercise 7.22). Thus we see that there can be quite complex ways in which Newton's method can converge in terms of the directions of approach, even though the rate of convergence is quadratic. For more information, see [55].

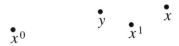

Figure 7.2 One step of Newton's method goes from a starting point x^0 to the next
iterate x^1 as part of the process of converging to x. Instead of finding
x^1 exactly, it may be useful to settle for the approximation y to x^1.

7.4 MIXING SOLVERS

So far, we have viewed the problem of solving a nonlinear system $f(x) = 0$
as a two-level process. At the first level, we define a sequence of linear
problems to be solved, e.g., (7.36) for Newton's method. The second level
is the solution of each of these linear systems, e.g., by one of the methods
studied earlier. But can we relax the separation barrier between these two
phases and potentially achieve the same result ($f(x) \approx 0$) more efficiently?
Given the generality of this question, there can be many answers. Here we
will focus on only two to make it clear what some of the possibilities are.

7.4.1 Approximate linear solves

Since the steps in the nonlinear process are themselves only approximate, it is
not necessary to solve the linear systems exactly. As long as the approximate
solution is closer to the exact solution for the next nonlinear iteration than
the old one, we have made progress. We depict in figure 7.2 how this might
work.

As we progress toward the solution, it may be necessary to solve the intermediate linear systems more and more accurately, and there may be other
restrictions along the way. But this suggests that we should be interested in
techniques that can approximate the solution of linear equations using less
work than the methods studied earlier. This will be the major subject of
chapters 8 and 9.

7.4.2 Approximate Jacobian

In section 2.2.4, we saw that the secant method can be more efficient than
Newton's method depending on the relative cost of evaluating the function
f and its derivative f'. In multidimensions, the dichotomy is even more
compelling. In the case of Newton's method, the linear system to be solved
at each step involves the Jacobian. Evaluating all the entries in the Jacobian
matrix $J_f(x)$ requires $\mathcal{O}\left(n^2\right)$ work, whereas evaluating $f(x)$ requires only
$\mathcal{O}(n)$ work. A more subtle approximation is thus to evaluate the Jacobian
operator only approximately, even if the resulting approximate linear system
is solved exactly. The class of quasi-Newton methods are of this form [120].
The mathematics surrounding these methods is very interesting, but lack of
time and space forces us to leave this for section 7.5.

We will see that many iterative methods require access only to the operator action corresponding to the matrix in the linear system and not to the matrix itself. In the case of Newton's method, the operator in question is the Jacobian. There are iterative methods that utilize this fact and are called *matrix free methods*. By Taylor's theorem,

$$f(x^0 + \epsilon x) - f(x^0) \approx J_f(x^0)\epsilon x + \mathcal{O}\left(\epsilon^2 \|x\|^2\right). \qquad (7.79)$$

Thus we can use the approximation

$$J_f(x^0)x \approx \epsilon^{-1}\left(f(x^0 + \epsilon x) - f(x^0)\right) \qquad (7.80)$$

with a suitably chosen ϵ. Thus the action of the Jacobian operator J_f can be approximated using only function evaluations of f [97].

7.5 MORE READING

The field of optimization is large and diverse, but the book [120] provides a good place to start. Also see [94, 95]. Optimization on manifolds is the subject of [1]. For more on how Newton's method behaves for a singular Jacobian, see [44].

7.6 EXERCISES

Exercise 7.1 *Suppose that g is a continuous function. Prove that, if fixed-point iteration (7.4) converges to some ξ, then ξ is a fixed point, i.e., it satisfies (7.3).*

Exercise 7.2 *Prove theorem 7.1. (Hint: pick a norm $\|\cdot\|$ on $\mathbb{R}^n$ such that the corresponding operator norm satisfies $\|J_g(\xi)\| < 1$. Now apply Taylor's theorem in this norm.)*

Exercise 7.3 *Suppose that $g(x) = x + f(x) - y$, where $y \in \mathbb{R}^n$ is fixed. Show that $J_g(x) = I + J_f(x)$, where I denotes the $n \times n$ identity matrix.*

Exercise 7.4 *Prove Taylor's theorem with an integral remainder in one dimension:*

$$f(x) - \sum_{k=0}^{m-1} \frac{f^{(k)}(y)}{k!}(x-y)^k$$

$$= \frac{(x-y)^m}{(m-1)!} \int_0^1 (1-s)^{m-1} f^{(m)}(y + s(x-y))\, ds \qquad (7.81)$$

$$= \frac{1}{(m-1)!} \int_y^x (t-y)^{m-1} f^{(m)}(t)\, dt.$$

(Hint: integrate by parts successively in the expression on the right-hand side of (7.81) and keep track of the terms that appear. Show that they are the same as on the left-hand side.)

Exercise 7.5 *Prove (7.24) for $k \geq 3$. (Hint: use induction. Apply the chain rule to the case $k - 1$ as in (7.28).)*

Exercise 7.6 *The Hessian H_ϕ of a scalar function ϕ is the matrix of second derivatives*

$$(H_\phi(x))_{ij} := \frac{\partial^2 \phi}{\partial x_i x_j}(x). \tag{7.82}$$

Prove that the Hessian of a scalar function is the Jacobian of its gradient.

Exercise 7.7 *Verify (7.35). (Hint: write it out coordinatewise. That is, consider each $g_j(x) = x_j - \sum_{k=1}^n a_{jk} f_k(x)$ and just differentiate.)*

Exercise 7.8 *Verify (7.39). (Hint: use (7.44), or at least follow its derivation.)*

Exercise 7.9 *Apply Newton's method to solve $f(x) = (b^2, c^2)$, where f is defined in (7.2).*

Exercise 7.10 *Determine the set of points x at which the Jacobian $J_f(x)$ is singular, where f is defined in (7.2).*

Exercise 7.11 *Prove (7.45). (Hint: apply (7.29) for each coordinate u_i and interpret the corresponding terms.)*

Exercise 7.12 *Weaken the α-condition in theorem 7.3 by the assumption that*

$$\|J_f(x^0)^{-1} f(x^0)\|_\infty = \|x^1 - x^0\|_\infty \leq a, \tag{7.83}$$

where x^1 is the Newton iterate obtained starting with x^0, and weaken the β-condition in theorem 7.3 by the assumption that only $\|J_f^{-1}(x^0)\|_\infty \leq b$. Note that this allows $J_f(y)$ to be singular. Let c be the constant in (7.51). Assuming that $abc \leq \frac{1}{2}$, prove that Newton's method converges. (Hint: see Theorem 3 in Chapter 3, Section 3.2 of [86]. Show that the Jacobians $\|J_f^{-1}(x^k)\|_\infty \leq M^k$ for some $M < \infty$ as $k \to \infty$. Also see exercise 8.15.)

Exercise 7.13 *Consider the operator $\xi \to H_f(x)\xi$ defined in (7.43) which maps $\mathbb{R}^n$ to $\mathbb{R}^{n^2}$. Show that this is a linear operator. That is, show that*

$$H_f(x)(\xi^1 + s\xi^2) = (H_f(x)\xi^1) + s(H_f(x)\xi^2) \tag{7.84}$$

for all $\xi^1, \xi^2 \in \mathbb{R}^n$ and all $s \in \mathbb{R}$. (Hint: just apply the definition and use associativity.)

Exercise 7.14 *Let A be an $n \times n$ matrix. Prove that*

$$\|A\|_\infty \leq n \max_{i,j=1,\dots,n} |a_{ij}|. \tag{7.85}$$

Exercise 7.15 *Consider the Hessian operator defined in (7.43). Show that*

$$\max_{j,\ell=1,\ldots,n} |(H_f(x)\xi)_{j,\ell}| \leq \max_{j,k,\ell=1,\ldots,n} \left|\frac{\partial^2 f_j}{\partial x_k x_\ell}(x)\right| \|\xi\|_1. \qquad (7.86)$$

Exercise 7.16 *Consider the Hessian operator defined in (7.43). Show that*

$$\|H_f(x)\xi\|_\infty \leq n^2 \max_{j,k,\ell=1,\ldots,n} \left|\frac{\partial^2 f_j}{\partial x_k x_\ell}(x)\right| \|\xi\|_\infty. \qquad (7.87)$$

(Hint: apply exercises 7.14 and 7.15 and use (5.9).)

Exercise 7.17 *Consider the set Ω defined in (7.48). Prove that the constant α in (7.49) may be bounded by*

$$\alpha \leq R \sup_{x \in \Omega} \|J_f(x)^{-1}\|_2 \|J_f(x)\|_2. \qquad (7.88)$$

(Hint: apply (7.25) for $m = 1$.)

Exercise 7.18 *Suppose that $g : \mathbb{R}^n \to \mathbb{R}^n$ is a C^2 function such that $\alpha = g(\alpha)$ and with the property that $\rho(J_g(\alpha)) < 1$. Prove that fixed-point iteration (2.5) converges asymptotically according to*

$$\limsup_{n \to \infty} \frac{|x_n - \alpha|}{\rho(J_g(\alpha))^n} < \infty, \qquad (7.89)$$

at least for an initial guess $x_0 \in \mathbb{R}^n$ sufficiently close to α. (Hint: compare exercise 2.4.)

Exercise 7.19 *Construct a smooth $g : \mathbb{R}^n \to \mathbb{R}^n$ with a fixed point $\alpha = g(\alpha)$ and with the property that $\rho(J_g(\alpha)) > 0$ but such that*

$$\lim_{n \to \infty} \frac{|x_n - \alpha|}{\rho(J_g(\alpha))^n} = 0 \qquad (7.90)$$

for a suitable initial guess $x_0 \in \mathbb{R}^n$. (Hint: compare exercises 7.18 and 2.4. Pick the starting guess so that all iterates lie in a lower-dimensional subspace orthogonal to the eigenvector corresponding to the largest eigenvalue of $J_g(\alpha)$.)

Exercise 7.20 *Prove that the iteration $(r, \lambda) \to (\hat{r}, \hat{\lambda})$ defined in (7.64) converges quadratically for r near 1 and λ near zero. Verify computationally that for general starting values the convergence is only linear.*

Exercise 7.21 *Verify the tensor eigenrelations stated in (7.77) for Q defined by (7.74) using the matrices (7.76). Verify computationally the subsequent statements regarding the limiting behavior of the iteration (7.75) for various starting vectors.*

Exercise 7.22 *Consider the iteration (7.75) with Q defined by (7.74) using the matrices (7.78). Show that the iterates appear to be chaotic for typical starting vectors.*

7.7 SOLUTIONS

Solution of Exercise 7.4. Suppose that $m = 1$. Then the statement is

$$f(x) - f(y) = (x - y) \int_0^1 f^{(1)}(y + s(x - y))\, ds. \tag{7.91}$$

If we let $t = y + s(x - y)$, then $dt = (x - y)ds$. Then we see that

$$(x - y) \int_0^1 f^{(1)}(y + s(x - y))\, ds = \int_y^x f^{(1)}(t)\, dt = f(x) - f(y). \tag{7.92}$$

Note that $s = (t - y)/(x - y)$ and $1 - s = (x - t)/(x - y)$. This shows that the two forms of the remainder are the same after a change of variables.

We will think of $s = s(t)$ as a function of t. Observe that $s(x) = 1$ and $s(y) = 0$. Using the same approach as in (7.92) for general m, we find by integrating by parts that

$$\begin{aligned}
&\frac{(x - y)^m}{(m - 1)!} \int_0^1 (1 - s)^{m-1} f^{(m)}(y + s(x - y))\, ds \\
&= \frac{(x - y)^{m-1}}{(m - 1)!} \int_y^x (1 - s(t))^{m-1} f^{(m)}(t)\, dt \\
&= \frac{(x - y)^{m-1}}{(m - 1)!} \int_y^x (m - 1)(1 - s(t))^{m-2} s'(t) f^{(m-1)}(t)\, dt \\
&\quad - \frac{(x - y)^{m-1}}{(m - 1)!} f^{(m-1)}(y) \\
&= \frac{(x - y)^{m-1}}{(m - 2)!} \int_0^1 (1 - s)^{m-2} f^{(m-1)}(y + s(x - y))\, ds \\
&\quad - \frac{(x - y)^{m-1}}{(m - 1)!} f^{(m-1)}(y).
\end{aligned} \tag{7.93}$$

That is, if we define

$$R_n = \frac{(x - y)^m}{(m - 1)!} \int_0^1 (1 - s)^{(m-1)} f^{(m)}(y + s(x - y))\, ds, \tag{7.94}$$

then we have proved that

$$R_{m-1} = R_{m-2} - \frac{(x - y)^{m-1}}{(m - 1)!} f^{(m-1)}(y). \tag{7.95}$$

Moreover, we also showed that $R_1 = f(x) - f(y)$. Thus iterating (7.95) completes the proof.

Solution of Exercise 7.10. We have from (7.11) that

$$\det J_f(x) = 4 (x_1 x_2 - (x_1 - a)x_2) = 4a x_2. \tag{7.96}$$

Thus J_f is singular only along the line joining the two stations along the shore.

Chapter Eight

Iterative Methods

> Niels Henrik Abel (1802–1829) achieved much in his short
> life and is memorialized by the Abel prize, sometimes re-
> ferred to as the Nobel prize of mathematics. His name
> is used to denote commutativity in algebra, and he made
> seminal contributions to the question of formulas for roots
> of polynomials, which have implications for algorithms for
> finding eigenvalues (section 14.4).

This chapter considers approximate solution techniques with potentially
fewer operations than the direct methods of chapters 3 and 4. Those methods
have the property of producing the exact answer (in exact arithmetic) in a
predictable (finite) number of operations. However, the number of operations
can approach astronomical proportions, and there is only limited benefit with
direct methods for sparse matrices (cf. section 4.3). Consider the family of
$(2n - 1) \times (2n - 1)$ matrices A which have

$$A_{ii} = 2, \quad A_{i,i+1} = -1, \quad A_{i+n-1,i} = -1 \quad \forall i = 1, \ldots, 2n - 1, \quad (8.1)$$

with all other entries zero. Note that the bandwidth (section 4.3) of A is
$n - 1$. In the case $n = 5$, we have

$$A = \begin{pmatrix}
2 & -1 & 0 & 0 & 0 & 0 & 0 & 0 & 0 \\
0 & 2 & -1 & 0 & 0 & 0 & 0 & 0 & 0 \\
0 & 0 & 2 & -1 & 0 & 0 & 0 & 0 & 0 \\
0 & 0 & 0 & 2 & -1 & 0 & 0 & 0 & 0 \\
-1 & 0 & 0 & 0 & 2 & -1 & 0 & 0 & 0 \\
0 & -1 & 0 & 0 & 0 & 2 & -1 & 0 & 0 \\
0 & 0 & -1 & 0 & 0 & 0 & 2 & -1 & 0 \\
0 & 0 & 0 & -1 & 0 & 0 & 0 & 2 & -1 \\
0 & 0 & 0 & 0 & -1 & 0 & 0 & 0 & 2
\end{pmatrix}. \quad (8.2)$$

Since the bandwidth grows with n, even the banded direct methods in section 4.3 would require $\mathcal{O}(n^3)$ operations. We will see that iterative methods
can produce acceptable results in far fewer steps.

In many cases, we may be interested only in an approximate solution, say,
with a specified number of digits of accuracy. Moreover we saw in section 4.2,
and will see in section 18.2, that floating-point errors render direct methods
to be only approximate in practice. Thus it is reasonable to ask if there might
be other methods that would provide approximate answers in potentially
fewer operations. Iterative methods may be of interest because they allow

one to stop when sufficient accuracy is reached, and they are self-correcting in that round-off errors tend only to defer convergence not to deter it. In some cases, one might be interested in a much lower level of accuracy from an approximation, so iterative methods similar to those used to solve nonlinear equations might be of interest if there is the possibility of monitoring the progress of the approximation. In particular, if the linear equation solution is the inner loop in an algorithm for solving a nonlinear system, a perfect solution in the intervening linear problems may not be useful.

8.1 STATIONARY ITERATIVE METHODS

We begin with a simple example, although it is one that captures the essence of stationary methods. Suppose M is a given matrix. If we multiply the matrix $I - M$ times the sum $I + M + M^2 + M^3 + \cdots$, we get a telescoping series:

$$(I - M) \sum_{k=0}^{n} M^k = \sum_{k=0}^{n} M^k - \sum_{k=1}^{n+1} M^k = I - M^{n+1}. \tag{8.3}$$

If we are allowed to let $n \to \infty$, then this provides a formula for the inverse of $I - M$:

$$(I - M)^{-1} = \sum_{k=0}^{\infty} M^k. \tag{8.4}$$

Fortunately, we have a criterion for the validity of (8.4) as follows.

Lemma 8.1 *Suppose that the spectral radius $\rho(M) < 1$. Then $I - M$ is an invertible matrix and the series in (8.4) converges to $(I - M)^{-1}$.*

Proof. The invertibility of $I - M$ follows directly from $\rho(M) < 1$; if $I - M$ is singular, then 0 is an eigenvalue, and thus 1 is an eigenvalue of M. We know by theorem 6.11 that $M^n \to 0$ as $n \to \infty$ if and only if $\rho(M) < 1$. Using (8.3) allows us to say that the partial sums of matrices $B^{(n)} = \sum_{k=0}^{n} M^k$ satisfy

$$B^{(n)} = (I - M)^{-1}(I - M^{n+1}). \tag{8.5}$$

Therefore, the partial sums $B^{(n)}$ of the series in (8.4) tend to $(I - M)^{-1}$ as $n \to \infty$.   QED

8.1.1 An algorithm

We can use the formula (8.4) as the basis for an iterative algorithm to solve $(I - M)x = f$. We can write $x = (I - M)^{-1}f$ and approximate the result by truncating the series (8.4). We will see that it is possible to compute efficiently the result for $n = 1, 2, \ldots$ iteratively, and we can stop the process once the approximation is sufficiently accurate. Thus suppose that we want to compute successively, for increasing n,

$$x^n = \left(\sum_{k=0}^{n} M^k \right) f = \sum_{k=0}^{n} M^k f. \tag{8.6}$$

To start with, we have $x^0 = f$, and

$$\begin{aligned} x^{n+1} &= \sum_{k=0}^{n+1} M^k f = f + \sum_{k=1}^{n+1} M^k f \\ &= f + M \left(\sum_{k=0}^{n} M^k f \right) = f + M x^n. \end{aligned} \tag{8.7}$$

Thus the sequence x^n defined in (8.6) can be computed by the simple iteration

$$x \leftarrow f + M x, \tag{8.8}$$

starting with $x \leftarrow f$. We leave the proof of the following as exercise 8.1.

Lemma 8.2 *Suppose that the spectral radius $\rho(M) < 1$. Then the iteration (8.8) converges to the solution of $(I - M)x = f$.*

Note that (8.8) is fixed-point iteration for solving the equation

$$x = f + M x, \tag{8.9}$$

i.e., $(I - M)x = f$. The error $e^n = x - x^n$ satisfies

$$e^n = M^n e^0, \tag{8.10}$$

so the convergence again follows from $\rho(M) < 1$.

8.1.2 General matrices

The algorithm (8.8) provides a way to solve a very special system, namely, $(I - M)x = f$. But what if we have a general matrix problem of the form $A x = f$ to solve? Simply define $M = I - A$. Then $A = I - M$, and we are in a position to apply the previous results. That is, if $\rho(I - A) < 1$, then A is invertible, and the iteration

$$x \leftarrow f + (I - A)x \tag{8.11}$$

converges to a solution of $A x = f$. We now consider more general algorithms of this type.

8.2 GENERAL SPLITTINGS

The idea is to use a general *splitting* of the matrix A into $A = N - P$, where N is an invertible matrix for which we are "willing" to solve systems directly.

The splitting in section 8.1.2 corresponds to the splitting $A = I - M = I - (I - A)$, so $N = I$. For example, N could be a diagonal matrix or a triangular matrix, as systems with such matrices are easily solved directly. The Jacobi[1] method has N diagonal, and the Gauss-Seidel[2] method has N a triangular matrix.

Using the splitting, we convert the equation $Ax = f$ to $Nx = Px + f$. Mathematically, the sequence that is generated using this splitting is defined by solving

$$Nx^{n+1} = f + Px^n. \tag{8.12}$$

We can cast this as a fixed-point problem (by formally inverting N) as $x = N^{-1}Px + g$, where $Ng = f$. Thus (8.12) takes the same form as (8.9), where $M = N^{-1}P$ (and $f = g$). The convergence theory for such methods is simple iff $M = N^{-1}P = I - N^{-1}A$ is convergent.

Theorem 8.3 *Suppose that $A = N - P$ with N invertible and that the spectral radius $\rho(N^{-1}P) = \rho(I - N^{-1}A) < 1$. Then the iteration (8.12) converges to the solution of $Ax = f$.*

8.2.1 Jacobi method

The Jacobi method [63] has N diagonal: $N = \text{diag}(A)$, where $\text{diag}(A)$ denotes the diagonal matrix whose diagonal is the same as the diagonal of A. Therefore, the iteration takes the special form

$$x \leftarrow \text{diag}(A)^{-1}f + (I - \text{diag}(A)^{-1}A)x. \tag{8.13}$$

The matrix $M_J = (I - \text{diag}(A)^{-1}A)$ for the Jacobi method is zero on the diagonal and elsewhere involves just a simple scaling of A. We will see that M_J also plays a role in the Gauss-Seidel method (section 8.2.2).

We show in figure 8.1 the spectral radius of the Jacobi iteration matrices for the the family of matrices (8.1). Thus we see that in this example, the Jacobi interaction can be very effective. We can show in certain cases that the diagonal scaling used to produce M_J is a good idea, namely, when A is *diagonally dominant*.

Definition 8.4 *An $n \times n$ matrix A is said to be diagonally dominant if*

$$|a_{ii}| > \sum_{j \neq i} |a_{ij}| \tag{8.14}$$

for all $i = 1, \ldots, n$.

---

[1] Carl Gustav Jacob Jacobi (1804–1851) (a.k.a. Jacques Simon) was applauded by Legendre as being "in the ranks of the best analysts of our era" together with Abel (page 115).

[2] Philipp Ludwig von Seidel (1821–1896) was closely associated with Jacobi and worked on analysis as well as in areas outside mathematics, including astronomy. He is credited with establishing, concurrently with Stokes, the notion of uniform convergence (cf. [103, pp. 131–141] and also exercise 16.2).

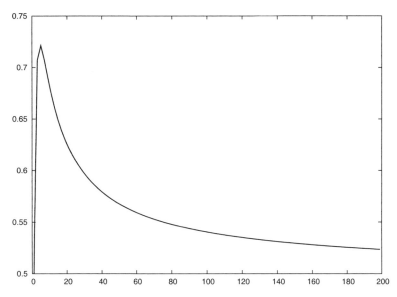

Figure 8.1 The spectral radius of the Jacobi iteration matrix for the family of matrices (8.1). The horizontal axis is the size $(2n - 1)$ of the nth matrix.

The convergence estimate of this section says that when A is diagonally dominant, then Jacobi converges. If A is diagonally dominant, then the Jacobi iteration matrix $M = M_J$ has the property that

$$\max_{i=1}^{n} \sum_{j=1}^{n} |m_{ij}| = \max_{i=1}^{n} \sum_{j \neq i}^{n} |a_{ij}/a_{ii}| = \mu < 1. \tag{8.15}$$

The maximal absolute row (respectively, column) sum of M can be identified as a particular norm, $\|M\|_\infty$ (respectively, $\|M\|_1$), as we describe in exercise 8.2. Thus we have the following result.

Lemma 8.5 *The matrix A is diagonally dominant if and only if*

$$\|M_J\|_\infty < 1, \tag{8.16}$$

where $M_J = I - \mathrm{diag}(A)^{-1}A$ denotes the Jacobi iteration matrix for A.

Corollary 8.6 *If A is diagonally dominant, the Jacobi iteration (8.13) converges to the solution of $A\,x = f$.*

Not only do we have a convergence condition, we also have a convergence estimate based on the computable quantity $\mu = \|M_J\|_\infty$ (exercise 8.2). From (8.10), we have

$$\|e^n\|_\infty \leq \mu^n \|e^0\|_\infty. \tag{8.17}$$

Note that if we start with $x^0 = 0$, then $e^0 = x$, so that (8.17) can be viewed as a relative-error estimate.

Unfortunately, corollary 8.6 is far from sharp. Diagonal dominance is not necessary for convergence of the Jacobi iteration. The matrices (8.1) represented in figure 8.1 are not diagonally dominant, yet their spectral radii are well below 1. We return to this issue in section 8.3.

8.2.2 Gauss-Seidel method

The easiest way to motivate the Gauss-Seidel algorithm[3] is by considering the Jacobi algorithm at the element-by-element level:

$$x_i^{n+1} = \frac{1}{a_{ii}}\left(f_j - \sum_{j\neq i} a_{ij}x_j^n\right) \tag{8.18}$$

for $i = 1, \ldots, n$. Note that precomputations could be done to avoid the repeated divisions, but for n large, these do not add appreciably to the overall work.

If we write this using the assignment notation, it reads

$$y_i \leftarrow \frac{1}{a_{ii}}\left(f_j - \sum_{j\neq i} a_{ij}x_j\right) \quad \forall i = 1, \ldots, n, \tag{8.19}$$

and once y has been computed, we update x: $x \leftarrow y$. Consider the modified Jacobi algorithm defined by

$$x_i \leftarrow \frac{1}{a_{ii}}\left(f_j - \sum_{j\neq i} a_{ij}x_j\right) \quad \forall i = 1, \ldots, n. \tag{8.20}$$

For one thing, (8.20) avoids the need to keep a separate temporary vector y as needed in (8.19). Moreover, it has the heuristic benefit of using the most recent values of x_j as soon as they are available. But to understand how it performs, we need to write it in different notation as

$$x_i^{n+1} = \frac{1}{a_{ii}}\left(f_j - \sum_{j=1}^{i-1} a_{ij}x_j^{n+1} - \sum_{j=i+1}^{n} a_{ij}x_j^n\right) \tag{8.21}$$

for $i = 1, \ldots, n$. We can write this as a system of equations for x^{n+1} as

$$\sum_{j=1}^{i} a_{ij}x_j^{n+1} = f_j - \sum_{j=i+1}^{n} a_{ij}x_j^n \tag{8.22}$$

for $i = 1, \ldots, n$. Now we can explain the algorithm (8.20) in terms of splitting. The system (8.22) corresponds to solving $Nx^{n+1} = f + Px^n$, where N is the lower-triangular part of A and $-P$ is the strictly upper-triangular part of A:

$$N = \begin{pmatrix} a_{11} & 0 & \cdots & 0 \\ a_{21} & a_{22} & \cdots & 0 \\ \cdots & & & \\ a_{n1} & a_{n2} & \cdots & a_{nn} \end{pmatrix} \text{ and } -P = \begin{pmatrix} 0 & a_{12} & \cdots & a_{1n} \\ 0 & 0 & \cdots & a_{2n} \\ \cdots & & & \\ 0 & 0 & \cdots & 0 \end{pmatrix}. \tag{8.23}$$

---

[3]The relationship between the method described here and those considered by Gauss and Seidel is tenuous [63], but Gauss definitely advocated iterative methods of this type in a letter to a colleague in which he said the "procedure can be done while half asleep, or while thinking about other things" [62].

The performance of Gauss-Seidel for the family of matrices in (8.1) is very similar to that of Jacobi (exercise 8.5). On the other hand, for the family of matrices indicated in (4.20), the spectral radii of Jacobi and of Gauss-Seidel differ more substantially. In this case, $\rho(M_J) \approx 1 - c_J n^{-2}$ and $\rho(M_{GS}) \approx 1 - c_{GS} n^{-2}$, where $C_{GS} = 2C_J$ and $C_J \approx 1.234$. Since the error behaves like ρ^k after k iterations, as indicated in (8.17), we can estimate the performance by considering the asymptotic behavior of

$$(1 - \epsilon)^k = e^{k \log(1-\epsilon)} \approx e^{-\epsilon k}. \tag{8.24}$$

Thus we see that the error reduction with Jacobi for these matrices is about $e^{-c_J k/n^2}$, and for Gauss-Seidel it is about $e^{-2c_J k/n^2} = (e^{-c_J k/n^2})^2$. This means that it takes twice as many iterations for Jacobi to reduce the error as much as Gauss-Seidel does. Thus Gauss-Seidel shows a substantial performance improvement over Jacobi in this case, but both methods require $k = \mathcal{O}(n^2)$ to have substantial error reduction. In this case, the banded direct methods are vastly superior.

A result analogous to corollary 8.6 can be proved (exercise 8.6). But we now examine the convergence properties of general splittings.

8.2.3 Convergence of general splittings

There are simple calculations that give a general criterion for the convergence of general splitting methods. This is useful in particular for establishing the convergence of Gauss-Seidel and of Jacobi for an important class of matrices. Although we are generally interested in only real matrices, we will consider complex matrices for the moment to get a complete characterization.

To begin with, we establish a simple estimate that reveals a key matrix that governs the success of general splittings.

Lemma 8.7 *Suppose the matrix A is Hermitian and that the splitting matrix N in (8.12) is nonsingular. Define the matrix*

$$Q = N + N^\star - A. \tag{8.25}$$

Then the subsequent iterations $y = Mx = x - N^{-1}Ax$ satisfy

$$y^\star Ay = x^\star Ax - (y - x)^\star Q(y - x). \tag{8.26}$$

The expression $\sqrt{v^\star Bv}$ forms a norm on $v \in \mathbb{C}^n$ under suitable conditions (exercise 8.7). The interpretation of (8.26) is then that the norm of y is less than the norm of x (unless $y = x$, in which case a fixed point has been reached).

Proof (of Lemma 8.7). We have $Ny = Nx - Ax$, so that $N(y-x) = -Ax$. Define $e = y - x$. Then $e^\star Ne = -e^\star Ax$, and by conjugating this expression, we also have

$$e^\star N^\star e = \overline{e^\star Ne} = \overline{-e^\star Ax} = -x^\star Ae.$$

Therefore,

$$
\begin{aligned}
(y - x)^\star Q(y - x) &= e^\star(N + N^\star - A)e = -e^\star A x - x^\star A e - e^\star A e \\
&= -e^\star A x - (x + e)^\star A e = -e^\star A x - y^\star A e \\
&= -y^\star A x + x^\star A x - y^\star A y + y^\star A x \\
&= x^\star A x - y^\star A y. \qquad\qquad \text{QED}
\end{aligned}
\tag{8.27}
$$

Lemma 8.8 *Suppose that A is a Hermitian matrix and that the splitting matrix N in (8.12) is nonsingular. Suppose that the matrix Q defined in (8.25) is positive definite. Then the matrix $M = I - N^{-1}A$ satisfies $\rho(M) < 1$ if and only if A is positive definite.*

Before proving the result, let us see why it is useful. For Gauss-Seidel, $Q = \mathrm{diag}(A)$, where $\mathrm{diag}(A)$ denotes the diagonal matrix whose diagonal is the same as the diagonal of A. When A is positive definite, $\mathrm{diag}(A)$ is always positive. Thus we have the following [136].

Corollary 8.9 *Suppose that A is Hermitian with a positive diagonal. Then Gauss-Seidel is convergent if and only if A is positive definite.*

It appears that Gauss-Seidel provides a test for positive definiteness for a certain class of matrices (Hermitian with a positive diagonal). This is not surprising since iterations of this type are closely linked to the power method for determining eigenvalues (section 15.1).

Proof (of Lemma 8.8). Suppose that $Mx = \lambda x$. Then $\lambda x = x - N^{-1}A x$, and so $(1 - \lambda)Nx = A x$. Thus

$$
(1 - \lambda)x^\star N x = x^\star A x. \tag{8.28}
$$

We are mainly interested in the "if" part of the theorem, so we start with that. So suppose that A is positive definite. Then $x^\star A x > 0$ since $x \neq 0$. In particular, this shows that $x^\star N x \neq 0$ and $\lambda \neq 1$. Moreover,

$$
\frac{1}{1 - \lambda} = \frac{x^\star N x}{x^\star A x}. \tag{8.29}
$$

The complex conjugate of (8.29) is

$$
\frac{1}{1 - \bar{\lambda}} = \frac{x^\star N^\star x}{x^\star A x}. \tag{8.30}
$$

Adding (8.29) and (8.30), we find

$$
2\,\mathcal{R}e\,\frac{1}{1 - \lambda} = \frac{x^\star(N + N^\star)x}{x^\star A x} = \frac{x^\star(A + Q)x}{x^\star A x} = 1 + \frac{x^\star Q x}{x^\star A x}, \tag{8.31}
$$

where $\mathcal{R}e\,z$ is the real part of z. Because we have assumed Q is positive definite, the last term in (8.31) is positive, so (8.31) implies that

$$
2\,\mathcal{R}e\,\frac{1}{1 - \lambda} > 1. \tag{8.32}
$$

Using the relation $1/z = \bar{z}/|z|^2$ for any complex z, we reduce (8.32) to

$$2\,\mathcal{R}e\,(1 - \bar{\lambda}) > |1 - \lambda|^2. \tag{8.33}$$

Writing $\lambda = \mu + i\nu$, (8.33) expands to give

$$2(1 - \mu) > |1 - \mu|^2 + \nu^2 = 1 - 2\mu + \mu^2 + \nu^2. \tag{8.34}$$

But (8.34) is precisely the condition $1 > \mu^2 + \nu^2$, i.e., $|\lambda| < 1$.
For the "only if" case, we refer the reader to exercises 8.8 and 8.9. QED

For the Jacobi iteration, $Q = 2\,\mathrm{diag}(A) - A$, where $\mathrm{diag}(A)$ denotes the diagonal matrix whose diagonal is the same as the diagonal of A. Suppose that $A = I + B$ for some matrix B. Then $Q = I - B$. Let the eigenvalues of B be denoted by λ_i^B. Then the eigenvalues of A are $1 + \lambda_i^B$, and those of Q are $1 - \lambda_i^B$. Thus the condition for Jacobi to be convergent is that

$$\min\{\lambda_i^B\} > -1\ (A > 0) \quad \text{and} \quad \max\{\lambda_i^B\} < 1\ (Q > 0). \tag{8.35}$$

The Q-condition is clearly quite restrictive and indicates that diagonal dominance is quite important for the success of the Jacobi method.
To give a concrete example, consider the matrix

$$A = \begin{pmatrix} 6 & -4 & 1 \\ -4 & 6 & -4 \\ 1 & -4 & 6 \end{pmatrix}, \tag{8.36}$$

which has positive eigenvalues (and is thus positive definite). The eigenvalues of the corresponding Jacobi iteration matrix

$$M_J = \frac{1}{6} \begin{pmatrix} 0 & 4 & -1 \\ 4 & 0 & 4 \\ -1 & 4 & 0 \end{pmatrix} \tag{8.37}$$

are $1/6$ and $(-1 \pm \sqrt{129})/12$. The eigenvalue $(-1 - \sqrt{129})/12 \approx -1.03$, and thus the Jacobi iteration is not convergent for the matrix A defined in (8.36).

8.3 NECESSARY CONDITIONS FOR CONVERGENCE

In section 8.2.3, we proved some theorems establishing necessary and sufficient conditions for convergence of general splitting methods for a limited class of matrices. For more general matrices, only limited results are available. However, it is possible to establish a result that provides a converse to corollary 8.6. Moreover, it exposes some important structural features of the Jacobi iteration.
The main weakness in using the concept of diagonal dominance to characterize the Jacobi iteration is that it is not independent of scaling. Moreover, the Jacobi iteration itself is invariant with respect to certain types of scaling. We can write the iteration matrix for the Jacobi iteration for A as

$$M_J(A) = I - \mathrm{diag}(A)^{-1}A, \tag{8.38}$$

where diag(A) denotes the diagonal matrix that agrees with A on the diagonal. Suppose that B is a diagonal matrix. Then $Ax = f$ iff $BAx = Bf$, and

$$M_J(BA) = I - \text{diag}(BA)^{-1}BA = I - \text{diag}(A)^{-1}B^{-1}BA = M_J(A) \quad (8.39)$$

since diag(BA) = B diag(A) (cf. exercise 8.11). Thus scaling by multiplication on the left by a diagonal matrix does not change the convergence properties of Jacobi, and indeed the two methods generate the same sequence of iterates (exercise 8.13). And by lemma 8.5, A is diagonally dominant iff BA is diagonally dominant. However, diagonal scaling by right-multiplication changes the concept of diagonal dominance substantially.

8.3.1 Generalized diagonal dominance

The notion of *generalized diagonal dominance* [89] captures the effect of scaling by right-multiplication. For any $v \in \mathbb{R}^n$, define diag(v) to be the diagonal matrix such that diag(v)$_{ii} = v_i$. Note that

$$(A \text{ diag}(v))_{ij} = a_{ij}v_j \quad (8.40)$$

for all i, j. Let $\mathbb{R}^n_+$ denote the subset of $\mathbb{R}^n$ consisting of vectors with positive entries.

Definition 8.10 *An $n \times n$ matrix A is said to satisfy* generalized diagonal dominance *(by rows) if, for some positive scaling vector $v \in \mathbb{R}^n_+$, $\widetilde{A} = A \text{ diag}(v)$ is diagonally dominant, that is,*

$$v_i|a_{ii}| > \sum_{j \neq i} v_j|a_{ij}| \quad (8.41)$$

for all $i = 1, \ldots, n$.

Analogous to lemma 8.5, we have the following result.

Lemma 8.11 *The matrix A satisfies generalized diagonal dominance iff for some positive scaling vector $v \in \mathbb{R}^n_+$*

$$\|I - \text{diag}(\widetilde{A})^{-1}\widetilde{A}\|_\infty < 1, \quad (8.42)$$

where $\widetilde{A} = A \text{ diag}(v)$.

We saw that diagonal scaling of A on the left did not change the Jacobi iteration matrix. Scaling on the right changes it via a similarity transformation.

Lemma 8.12 *Suppose that $v \in \mathbb{R}^n_+$ and A is any $n \times n$ matrix. Define $\widetilde{A} = A \text{ diag}(v)$. Then*

$$M_J(\widetilde{A}) = \text{diag}(v)^{-1}M_J(A)\text{diag}(v). \quad (8.43)$$

Thus the Jacobi iteration (8.13) converges to the solution of $Ax = f$ iff the Jacobi iteration converges to the solution of $\widetilde{A}y = g$.

The proof of the latter statement could proceed by relating the iterates of the Jacobi method for A and $\widetilde{A}$. We leave this approach as exercise 8.14. Instead, we argue more abstractly by exploiting the relationship (8.43) between their respective Jacobi iterations matrices.

Proof. By (8.40), we see that

$$\operatorname{diag}(A \operatorname{diag}(v)) = \operatorname{diag}(A)\operatorname{diag}(v). \tag{8.44}$$

Therefore,

$$
\begin{aligned}
M_J(A \operatorname{diag}(v)) &= I - \operatorname{diag}(A \operatorname{diag}(v))^{-1} A \operatorname{diag}(v) \\
&= I - \operatorname{diag}(v)^{-1}\operatorname{diag}(A)^{-1} A \operatorname{diag}(v) \\
&= \operatorname{diag}(v)^{-1}\left(I - \operatorname{diag}(A)^{-1} A\right) \operatorname{diag}(v) \\
&= \operatorname{diag}(v)^{-1} M_J(A)\operatorname{diag}(v).
\end{aligned}
\tag{8.45}
$$

Since (8.45) represents a similarity transformation,

$$\rho(M_J(A \operatorname{diag}(v))) = \rho(M_J(A)) \tag{8.46}$$

(exercise 6.10). Thus the Jacobi iteration for A converges iff the Jacobi iteration for $\widetilde{A}$ converges.   QED

Analogous to corollary 8.6, we have the following result.

Corollary 8.13 *If A satisfies generalized diagonal dominance, the Jacobi iteration (8.13) converges to the solution of $A x = f$.*

Proof. By lemma 8.12, theorem 6.2, and lemma 8.11,

$$\rho(M_J(A)) = \rho(M_J(A \operatorname{diag}(v))) \leq \|M_J(A \operatorname{diag}(v))\|_\infty < 1. \tag{8.47}$$

QED

Under certain conditions [89], generalized diagonal dominance provides a necessary condition for the convergence of Jacobi (and Gauss-Seidel). Rather than deriving conditions on A that are necessary for convergence, we take the point of view that the issue is to estimate the spectral radius of the Jacobi iteration matrix more effectively. However, we will see that this leads to the same results.

8.3.2 Estimating the spectral radius

We know that the convergence of Jacobi depends precisely on the size of the spectral radius $\rho(M_J)$. Thus what is needed is a better way to estimate $\rho(M_J)$. We will do this in the case where M_J is nonnegative, as is the case with the families of matrices (4.20) and (8.1). But first we establish an identity for a general matrix A.

For $v \in \mathbb{R}^n_+$, define a new norm on $\mathbb{R}^n$ by

$$\|x\|_v = \|\operatorname{diag}(v)x\|_\infty \tag{8.48}$$

(cf. exercise 6.14). Note that

$$\left(\text{diag}(v)\,A\,\text{diag}(v)^{-1}\right)_{ij} = \frac{v_i}{v_j}a_{ij} \tag{8.49}$$

for any matrix A (exercise 8.16). Then the associated operator norm satisfies

$$
\begin{aligned}
\|A\|_v &= \max_{x\neq 0} \frac{\|\text{diag}(v)A\,x\|_\infty}{\|\text{diag}(v)x\|_\infty} \\
&= \max_{y\neq 0} \frac{\|\text{diag}(v)\,A\,\text{diag}(v)^{-1}y\|_\infty}{\|y\|_\infty} \qquad [y = \text{diag}(v)x] \\
&= \|\text{diag}(v)\,A\,\text{diag}(v)^{-1}\|_\infty \\
&= \max_{i=1,\ldots,n} \sum_{j=1}^{n} \frac{v_i}{v_j}|a_{ij}|. \qquad [\text{by (8.49) and exercise 8.2}]
\end{aligned}
\tag{8.50}
$$

Note that by theorem 6.2,

$$\rho(A) \leq \inf_{v\in\mathbb{R}^n_+} \|A\|_v. \tag{8.51}$$

The following result says that we can effectively approximate the spectral radius using the ∞-norm with the appropriate weight.

Lemma 8.14 *Suppose that M is a nonnegative $n \times n$ matrix with a positive eigenpair $\lambda > 0$ and $x > 0$, i.e., $Mx = \lambda x$ and $x = (x_1,\ldots,x_n)$ with $x_i > 0$ for all $i = 1,\ldots,n$. Then*

$$\rho(M) = \min_{v\in\mathbb{R}^n_+} \|M\|_v = \|M\|_w, \tag{8.52}$$

where $w_i = 1/x_i$ for all $i = 1,\ldots,n$.

Proof. In view of (8.51), we just need to verify that $\rho(M) = \|M\|_w$. But by (8.50),

$$
\begin{aligned}
\rho(M) \leq \|M\|_w &= \max_{i=1,\ldots,n} \sum_{j=1}^{n} \frac{x_j}{x_i}|m_{ij}| = \max_{i=1,\ldots,n} \frac{1}{x_i}\sum_{j=1}^{n} x_j m_{ij} \\
&= \max_{i=1,\ldots,n} \frac{1}{x_i}\lambda x_i = \lambda \leq \rho(M).
\end{aligned}
\tag{8.53}
$$

QED

For the matrix in (8.1) for $n = 50$, figure 8.2 depicts the eigenvector corresponding to the eigenvalue $\lambda = \rho(M)$ for the associated Jacobi iteration matrix. This depicts the weighting vector w in lemma 8.14 for this matrix.

The proof of lemma 8.14, and more precisely (8.53), leads to the following result, which says that the only positive eigenvalue associated with a positive eigenvector for a nonnegative matrix M is $\lambda = \rho(M)$.

Corollary 8.15 *Suppose that M is a nonnegative $n \times n$ matrix with a positive eigenpair $\lambda > 0$ and $x > 0$. Then $\lambda = \rho(M)$.*

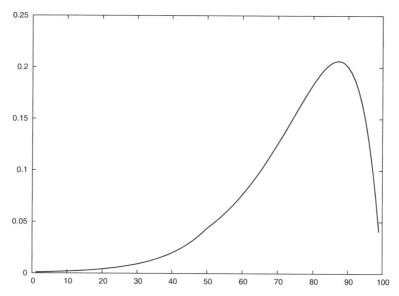

Figure 8.2 The eigenvector of the Jacobi iteration matrix for the family of matrices (8.1) for $n = 50$. The horizontal axis is the index of the eigenvector.

8.3.3 Convergence conditions

If A is a matrix with positive diagonal and negative off-diagonal entries, then the Jacobi iteration matrix $M_J(A) = I - \mathrm{diag}(A)^{-1}A$ is nonnegative. Thus if $M_J(A)$ has a positive eigenpair $\lambda > 0$ and $x > 0$, we conclude from lemma 8.14 that Jacobi converges only if $\|M_J(A)\|_w < 1$, where $w_i = 1/x_i$ for $i = 1, \ldots, n$. We now show that this implies that A satisfies generalized diagonal dominance. By (8.50), for all $i = 1, \ldots, n$,

$$a_{ii}x_i > a_{ii}x_i \sum_{j=1}^{n} \frac{x_j}{x_i} m_{ij} = a_{ii}x_i \sum_{i \neq j=1}^{n} \frac{x_j}{x_i} \frac{|a_{ij}|}{a_{ii}} = \sum_{i \neq j=1}^{n} x_j |a_{ij}|. \qquad (8.54)$$

Thus we have proved the following result.

Theorem 8.16 *Suppose A is a matrix with positive diagonal and negative off-diagonal entries and that the Jacobi iteration matrix*

$$M_J(A) = I - \mathrm{diag}(A)^{-1}A$$

has a positive eigenpair $\lambda > 0$ and $x > 0$. If the Jacobi method for A is convergent, then A satisfies generalized diagonal dominance (8.41) with $v = x$.

Fortunately, simple additional conditions are known that guarantee the existence of positive eigenpairs for positive matrices, as we describe in the section 8.3.4. Thus theorem 8.16 provides a broadly applicable converse to corollary 8.13.

8.3.4 Perron-Frobenius theorem

The Perron[4]-Frobenius[5] theorem provides a sufficient condition to guarantee the existence of the positive eigenpair appearing in lemma 8.14. The result holds for a nonnegative matrix provided that it cannot be decomposed in a particular way.

Definition 8.17 *An $n \times n$ matrix A is said to be* reducible *if the indices can be partitioned into two sets*

$$\{1, 2, \ldots, n\} = \{i_1, \ldots, i_k\} \cup \{j_1, \ldots, j_l\}, \tag{8.55}$$

where $n = k + l$, with $a_{i_\kappa, j_\nu} = 0$ for all $\kappa = 1, \ldots, k$ and $\nu = 1, \ldots l$. If no such partition exists, A is said to be irreducible.

Theorem 8.18 *Suppose that M is a nonnegative $n \times n$ irreducible matrix. Then it has a positive eigenpair $\lambda > 0$ and $x > 0$, i.e., $Mx = \lambda x$ and $x = (x_1, \ldots, x_n)$ with $x_i > 0$ for all $i = 1, \ldots, n$. Moreover, $\lambda = \rho(M)$.*

This theorem can be proved by considering the power method for computing eigenpairs, so we postpone it until chapter 14.

Note that the concept of reducibility relates only to off-diagonal elements of a matrix. Since indices of the nonzero off-diagonals of A and the Jacobi iteration matrix $M_j(A)$ are the same, we conclude that A is reducible iff $M_J(A)$ is reducible. Thus we have the following theorem.

Theorem 8.19 *Suppose that A is an $n \times n$ irreducible matrix that is positive on the diagonal and negative off-diagonal. Then the Jacobi iteration converges iff A satisfies generalized diagonal dominance.*

8.4 MORE READING

For further information, see [13, 73, 168]. There are many variants and generalizations of Jacobi and Gauss-Seidel iterations. In particular, parallel computation introduces new constraints and leads to novel algorithms [145]. Algorithms have been proposed [106] to determine whether a given matrix satisfies general diagonal dominance.

8.5 EXERCISES

Exercise 8.1 *Prove lemma 8.2. (Hint: see the proof of lemma 8.1; note that $x^n = B^{(n)}f$.)*

---

[4]Oskar Perron (1880–1975) was, like David Hilbert, Hermann Minkowski, Arnold J. W. Sommerfeld, and Martin Kutta (see page 275), a student of Lindemann, who was in turn a student of Felix Klein.

[5]See page 83.

Exercise 8.2 *Let B be an $n \times n$ matrix. Prove that the maximum absolute row sum can be identified as*

$$\max_{i=1}^{n} \sum_{j=1}^{n} |B_{ij}| = \|B\|_\infty \tag{8.56}$$

and that the maximum absolute column sum can be identified as

$$\max_{j=1}^{n} \sum_{i=1}^{n} |B_{ij}| = \|B\|_1. \tag{8.57}$$

Exercise 8.3 *Consider the iteration matrix for the general splitting method $M = I - N^{-1}A$, where N is any invertible matrix. Show that if $\lambda = 1$ is an eigenvalue of M, then A cannot be invertible. (Hint: see the derivation of (8.28).)*

Exercise 8.4 *Consider the iteration matrix for the general splitting method $M = I - N^{-1}A$, where N is any invertible matrix. Show that if $\lambda = 1$ is not an eigenvalue of M, then successive iterations $y = Mx$ must satisfy $y \neq x$.*

Exercise 8.5 *Compare the Jacobi iteration with the Gauss-Seidel iteration for the family of matrices in (8.1). Take for the right-hand side a vector F with all entries equal to 1, and take for starting vector $X^0 = 0$ in both cases. Compare the error for the two methods as a function of the number of iterations. How does this change with the size of the matrices in (8.1) (that is, for different n)?*

Exercise 8.6 *Suppose that the Jacobi iteration matrix $M_J = I - D^{-1}A$, where $D = \mathrm{diag}(A)$ is the diagonal matrix that agrees with A on the diagonal, satisfies $\|M_J\|_\infty < 1$. Prove that Gauss-Seidel converges, in particular, that $\|M_{GS}\|_\infty < 1$. (Hint: at each iteration, where $y = M_{GS}x$, proceed by induction on i to show that $|y_i| \leq \|M_J\|_\infty \|x\|_\infty$.)*

Exercise 8.7 *Prove that the expression $\|v\|_B := \sqrt{v^\star B v}$ defines a norm on $\mathbb{C}^n$ provided that B is Hermitian and positive definite. Compare this with exercise 6.14 and explain how the two results relate to each other.*

Exercise 8.8 *Suppose A, M, and Q are as in lemma 8.8 and that $\rho(M) < 1$. Show that $x^\star A x \geq 0$ for all eigenvectors of M. (Hint: reverse the proof of the "if" case in lemma 8.8.)*

Exercise 8.9 *Suppose A, M, and Q are as in lemma 8.8 and that $\rho(M) < 1$. Prove that A has to be positive definite. (Hint: Show that $x^\star A x \leq 0$ implies that $y^\star A y < 0$, where $y = Mx$, by using exercise 8.4. Show that all subsequent iterates $w = M^n x$ for $n > 1$ satisfy $w^\star A w < y^\star A y$ and hence that the w's cannot tend to zero, yielding a contradiction.)*

Exercise 8.10 *Show that the matrix in (8.36) is positive definite. Compute the corresponding Jacobi iteration matrix and determine its eigenvalues. Show that one of them exceeds 1 in absolute magnitude. Verify that the corresponding matrix Q, in the condition in lemma 8.8, is not positive definite.*

Exercise 8.11 *Suppose B is an $n \times n$ diagonal matrix. Show that for any $n \times n$ matrix A,*

$$\text{diag}(BA) = B\,\text{diag}(A). \tag{8.58}$$

Exercise 8.12 *Suppose that $A = I - M$, where M is nonnegative and irreducible and satisfies $\text{diag}(M) = 0$. Prove that if the Jacobi iteration for solving $Ax = f$ converges for any f and initial guess, then A has to satisfy generalized diagonal dominance. (Hint: use theorem 8.18 to guarantee a positive eigenpair and see lemma 8.14 to define an appropriate weight as needed in definition 8.10.)*

Exercise 8.13 *Suppose B is an $n \times n$ diagonal matrix. Show that the Jacobi iterations for solving $Ax = f$ and $BAx = Bf$ generate the same sequence of iterates x^n.*

Exercise 8.14 *Suppose A is an $n \times n$ matrix and $v \in \mathbb{R}^n_+$. Show that the Jacobi iterations for solving $Ax = f$ and $A\,\text{diag}(v)y = f$ generate sequences of iterates x^n and y^n related by $\text{diag}(A)y^n = x^n$.*

Exercise 8.15 *Suppose that $\|M\| < 1$ for some norm. Prove that $I - M$ is invertible and*

$$\|(I - M)^{-1}\| \leq \frac{1}{1 - \|M\|}. \tag{8.59}$$

(Hint: compare lemma 8.1.)

Exercise 8.16 *Prove (8.49).*

Exercise 8.17 *Show that the mapping $z \to 1/(1 - z)$ maps the unit circle (minus one point) $\{z = \cos\theta + i\sin\theta \mid 0 < \theta < 2\pi\}$ to the line*

$$\{z = \tfrac{1}{2} + it \mid t \in \mathbb{R}\}. \tag{8.60}$$

Exercise 8.18 *Suppose that A is an $n \times n$ Hermitian matrix $(A^\star = A)$ such that $\text{diag}(A)$ has all positive entries. Show that if A is diagonally dominant, then A is positive definite. (Hint: apply exercise 8.6 and corollary 8.9.)*

Exercise 8.19 *Prove a converse to lemma 8.2. That is, suppose that the iteration (8.8) converges for any f and any starting vector. Prove that the spectral radius $\rho(M) < 1$.*

8.6 SOLUTIONS

Solution of Exercise 8.2. Define

$$\mu = \max_{i=1}^{n} \sum_{j=1}^{n} |B_{ij}|. \tag{8.61}$$

Then for any $x \in \mathbb{R}^n$ and any $i = 1, \ldots, n$,

$$
\begin{aligned}
|(Bx)_i| &= \left| \sum_{j=1}^{n} B_{ij} x_j \right| \\
&\leq \sum_{j=1}^{n} |B_{ij}| \, |x_j| \\
&\leq \mu \|x\|_\infty.
\end{aligned}
\tag{8.62}
$$

Therefore, $\|Bx\|_\infty \leq \mu \|x\|_\infty$. Since $x \in \mathbb{R}^n$ was arbitrary, $\|B\|_\infty \leq \mu$.

Pick i so that the ith absolute row sum equals μ, which we can assume is positive (if $\mu = 0$, the previous estimate shows that $\|B\|_\infty = 0$). Define $x_j = \text{sign}(B_{ij})$ for $j = 1, \ldots, n$, where $\text{sign}(t)$ is $+1$ if $t > 0$, -1 if $t < 0$, and 0 for $t = 0$. Then

$$(Bx)_i = \sum_{j=1}^{n} B_{ij} x_j = \sum_{j=1}^{n} |B_{ij}| = \mu. \tag{8.63}$$

Therefore, $\|Bx\|_\infty \geq \mu$, which incidentally shows that $x \neq 0$. Since $\|x\|_\infty = 1$, we have $\|B\|_\infty \geq \mu$, as claimed.

Now define μ to be the maximum absolute column sum:

$$\mu = \max_{j=1}^{n} \sum_{i=1}^{n} |B_{ij}|. \tag{8.64}$$

Then for any $x \in \mathbb{R}^n$,

$$
\begin{aligned}
\|Bx\|_1 &= \sum_{i=1}^{n} |(Bx)_i| = \sum_{i=1}^{n} \left| \sum_{j=1}^{n} B_{ij} x_j \right| \\
&\leq \sum_{i,j=1}^{n} |B_{ij}| \, |x_j| \leq \mu \|x\|_1.
\end{aligned}
\tag{8.65}
$$

Since $x \in \mathbb{R}^n$ was arbitrary, $\|B\|_1 \leq \mu$.

Pick j so that the jth absolute column sum equals μ, which we can assume is positive (if $\mu = 0$, the previous estimate shows that $\|B\|_1 = 0$). Define $x_i = \delta_{ij}$ for $i = 1, \ldots, n$, where δ is the Kronecker symbol. Then

$$(Bx)_i = \sum_{k=1}^{n} B_{ik} x_k = B_{ij}. \tag{8.66}$$

Therefore, $\|Bx\|_1 = \mu$. Since $\|x\|_1 = 1$, we have $\|B\|_1 \geq \mu$. Therefore, $\|B\|_1 = \mu$, as claimed.

Solution of Exercise 8.8. Much of the "if" case argument is reversible. In particular, we showed that $|\lambda| < 1$ if and only if $2\,\mathcal{R}e\,(1-\lambda)^{-1} > 1$. Moreover, as long as $x^\star A\,x$ is not zero, (8.31) is still valid, and so $|\lambda| < 1$ implies that

$$\frac{x^\star Qx}{x^\star A\,x} = 2\,\mathcal{R}e\,\frac{1}{1-\lambda} - 1 > 0. \tag{8.67}$$

This implies that $x^\star A\,x > 0$.

Solution of Exercise 8.14. The sequence of iterates for the Jacobi method for solving $A\,x = f$ is

$$x^{n+1} = (I - \mathrm{diag}(A)^{-1}A)x^n + g, \tag{8.68}$$

where $g = \mathrm{diag}(A)^{-1}f$. The sequence of iterates for the Jacobi method for solving $\widetilde{A}y = f$ is

$$y^{n+1} = (I - \mathrm{diag}(\widetilde{A})^{-1}\widetilde{A})y^n + \tilde{g}, \tag{8.69}$$

where $\tilde{g} = \mathrm{diag}(\widetilde{A})^{-1}f$. Then by (8.44),

$$\mathrm{diag}(\widetilde{A})^{-1} = \mathrm{diag}(v)^{-1}\,\mathrm{diag}(A)^{-1},$$

so that

$$\begin{aligned}
\mathrm{diag}(v)y^{n+1} &= \big(\mathrm{diag}(v) - \mathrm{diag}(A)^{-1}A\,\mathrm{diag}(v)\big)\,y^n + \mathrm{diag}(v)\tilde{g} \\
&= \big(I - \mathrm{diag}(A)^{-1}A\big)\,\mathrm{diag}(v)y^n + g,
\end{aligned} \tag{8.70}$$

which agrees with (8.68) with $x^n = \mathrm{diag}(v)y^n$.

Chapter Nine

Conjugate Gradients

> In a letter to a colleague in 1824, Abel (see page 115) wrote "in analysis one is largely concerned with functions that can be represented by power-series. As soon as other functions enter—and this happens rarely—then [induction] does not work any more and an infinite number of incorrect theorems arise from false conclusions" [103].

The conjugate gradient (CG) method was perceived for some time as a direct method for solving systems of linear equations. In exact arithmetic, the method produces the exact solution in a finite number of steps. More precisely, for an $n \times n$ matrix, CG "converges" in at most n steps. The advantages of conjugate gradients as an iterative method were not widely appreciated until much later. Not only does CG provide good approximate solutions with fewer iterations than what is required to produce an exact result, it can also be seen to be a more adaptive procedure than the stationary iterative methods studied previously in section 8.1.

We will develop the method in a sequence of steps to put it in context. It can be applied to symmetric, positive definite matrices, so we will limit our discussion to linear systems with such matrices.

9.1 MINIMIZATION METHODS

If A is an $n \times n$ symmetric, positive definite matrix, then there is a naturally associated inner product

$$(u, v)_A = u^{\mathrm{T}} A v \tag{9.1}$$

defined for $u, v \in \mathbb{R}^n$. As with any inner product, there is an associated norm (cf. exercise 8.7)

$$\|u\|_A = \sqrt{(u, u)_A} . \tag{9.2}$$

We will see that minimization algorithms can be viewed as an interplay between the inner product $(u, v)_A$ and the natural Euclidean inner product, which in this notation can be written

$$(u, v)_I = u^{\mathrm{T}} v \tag{9.3}$$

for $u, v \in \mathbb{R}^n$. First, there is the obvious relationship (exercise 9.2)

$$(u, v)_A = (Au, v)_I = (u, Av)_I. \tag{9.4}$$

There is also a naturally associated quadratic function of $v \in \mathbb{R}^n$ defined by

$$Q_A(v)_A = \tfrac{1}{2}(v, v)_A - (f, v)_I \qquad (9.5)$$

defined for any $f \in \mathbb{R}^n$. The relationship between these inner products is expressed in the following lemma.

Lemma 9.1 *The minimum of Q_A occurs at $u \in \mathbb{R}^n$, which is the solution of the equation*

$$Au = f. \qquad (9.6)$$

Proof. The proof for $n = 1$ is elementary (exercise 9.3). In general, we can reduce it to a one-dimensional problem by expanding, for $u, v \in \mathbb{R}^n$ and $t \in \mathbb{R}$,

$$
\begin{aligned}
Q_A(u + tv)_A &= Q_A(u) + t(u, v)_A + \tfrac{1}{2}t^2(v, v)_A - t(f, v)_I \\
&= Q_A(u) + t(u^{\mathsf{T}}A - f^{\mathsf{T}})v + \tfrac{1}{2}t^2(v, v)_A \qquad (9.7) \\
&= Q_A(u) + t(Au - f, v)_I + \tfrac{1}{2}t^2(v, v)_A,
\end{aligned}
$$

where we used the symmetry of A in the last equality. Thus $Q_A(u + tv)_A = \alpha + \beta t + \gamma t^2$, where $\beta = (Au - f)^{\mathsf{T}}v = (Au - f, v)_I$. A quadratic $q(t) = \alpha + \beta t + \gamma t^2$ with $\gamma > 0$ has a minimum at $t = 0$ iff $\beta = 0$ (exercise 9.5). Therefore, the minimum u of $Q_A(u)$ is characterized by

$$(Au - f)^{\mathsf{T}}v = 0 \quad \forall v \in \mathbb{R}^n. \qquad (9.8)$$

Setting $v = Au - f$, we conclude that (9.6) must hold if (9.8) holds, and the converse is obvious. QED

Thus a natural strategy to approximate the solution of $Au = f$ is to minimize Q_A.

9.1.1 Descent methods

The geometry of the function (9.1) to be minimized is quite simple. When $f = 0$, the graph of Q_A is a simple elliptical bowl. When $f \neq 0$, the picture is tilted slightly. Thus a natural approach is to pick a *search direction* $s \neq 0$ in which Q_A is decreasing and proceed in that direction until you start to go up again. Suppose that u_0 is our starting point. We search for the optimum along a line including the point u_0 that is parallel to s; this part of the algorithm is called a *line search*. This corresponds to minimizing the problem

$$Q_A(u_0 + ts)_A = Q_A(u_0) + t(Au_0 - f, s)_I + \tfrac{1}{2}t^2(s, s)_A \qquad (9.9)$$

in view of (9.7), where t is a scalar that measures the distance we have gone.

Definition 9.2 *If u_0 is the current approximation of the solution u of the equation $Au - f$, then the* residual r *is defined by*

$$r = Au_0 - f. \qquad (9.10)$$

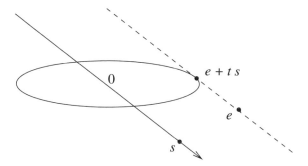

Figure 9.1 The ellipse indicates points in the plane determined by s and e that have the same norm. The solid line indicates the set of points ts for $t \in \mathbb{R}$, and the dashed line indicates the set of points $e + ts$ for $t \in \mathbb{R}$.

The residual provides a way to monitor the approximation process. For example, if $r = 0$ then $Au_0 = f$, i.e., $u_0 = u$. In general, we can define the error by

$$e = u_0 - u, \qquad (9.11)$$

and we find that the error satisfies an equation with the residual as the right-hand side:

$$Ae = r. \qquad (9.12)$$

Thus $e = A^{-1}r$ is small if r is small (and A^{-1} is not too large).

The minimum in (9.9) occurs where the derivative with respect to t is zero:

$$t_{\min} = -\frac{(Au_0 - f, s)_I}{(s, s)_A} = -\frac{(r, s)_I}{(s, s)_A} = -\frac{(Ae, s)_I}{(s, s)_A} = -\frac{(e, s)_A}{(s, s)_A} \qquad (9.13)$$

(cf. exercise 9.6 for the last step). Then the next position in our descent process will be $u_0 + t_{\min}s$.

There is a geometric interpretation of the value of t in (9.13) (see figure 9.1). The value of t is the length (in the norm $\|\cdot\|_A$) of e in the direction s, with the sign of t indicating whether e correlates with plus or minus s. Note that we do not know u, and thus we do not know e. Nevertheless, we can compute t (via the first or second equality) and find out the size of e in the direction s. Consider minimizing

$$\|e + ts\|_A^2 = \|e\|_A^2 + 2t(e, s)_A + t^2\|s\|_A^2 \qquad (9.14)$$

over all t. Differentiating with respect to t and setting the result to zero, we find that t is given again by (9.13). Therefore, we have proved the following.

Lemma 9.3 *Minimizing (9.9) is equivalent to minimizing (9.14). The minimum occurs at the value of t given in (9.13).*

This can be visualized in figure 9.1. Remarkably, we are able to cause the maximum decrease in the error (in the norm $\|\cdot\|_A$) in the descent direction s by minimizing (9.9), even though we do not know e. An alternative interpretation is found in exercise 9.7, which shows that

$$2Q_A(v) = (u - v, u - v)_A - (u, u)_A \tag{9.15}$$

for all $v \in \mathbb{R}^n$. That is, $Q_A(v)$ differs from $\frac{1}{2}\|u - v\|_A^2$ by a constant $(\frac{1}{2}\|u\|_A^2)$, so minimizing Q_A is equivalent to minimizing the error $u - v$ in the A-norm.

At the minimum value given by (9.13), we find that (9.14) reduces to

$$\|e + t_{\min}s\|_A^2 = \|e\|_A^2 - \frac{(e, s)_A^2}{\|s\|_A^2} = \|e\|_A^2 - t_{\min}^2\|s\|_A^2. \tag{9.16}$$

Thus unless $r^Ts = (r, s)_I = (e, s)_A = 0$, the error is decreased at each step. In fact, (9.13) implies that the largest step, and thus the largest decrease in error, will occur if $s = \alpha r$ for some scalar α.

9.1.2 Descent directions

Now we turn to the question of finding good descent directions. The best possible descent direction would be the error e, but we do not know how to evaluate this. Instead, we can minimize Q_A; to do so, we compute the gradient $\nabla Q_A(u_0)$ of the expression (9.5). This vector points in the direction of the maximum increase of $Q_A(u_0)$, so

$$s = -\nabla Q_A(u_0) \tag{9.17}$$

points in the direction of the greatest decrease.

Computing the gradient of the quadratic Q_A is possible by calculus, but we can also infer it from Taylor's theorem:

$$Q_A(u_0 + v) = Q_A(u_0) + \nabla Q_A(u_0) \cdot v + \mathcal{O}(v^2). \tag{9.18}$$

Comparing with the expansion (9.7), we find

$$\nabla Q_A(u_0) = Au_0 - f = r. \tag{9.19}$$

Thus we see that (minus) the residual provides the direction of maximum descent, in concert with the observation at the end of section 9.1.1 that $s = \alpha r$ provides the greatest error reduction.

9.1.3 The gradient descent method

Suppose we decide to descend in the direction of the gradient at each step (so $s = -r$) and to proceed as far as the minimum at each step (so $t = t_{\min}$). Since the notation gets denser at this point, we switch to denoting the scalar $t_{\min}$ by α. This corresponds to taking

$$r_k = Au_k - f$$
$$\alpha_k = \frac{(r_k, r_k)_I}{(r_k, r_k)_A} \tag{9.20}$$
$$u_{k+1} = u_k - \alpha_k r_k.$$

The error $e_k = u_k - u$ at each step satisfies

$$e_{k+1} = u_{k+1} - u = u_k - \alpha_k r_k - u = e_k - \alpha_k r_k. \tag{9.21}$$

Figure 9.2 The ellipse indicates points in the plane determined by r_k and u_k that have the same norm.

The error $e_k = u_k - u$ at each step is reduced, in view of (9.16), by

$$\|e_{k+1}\|_A^2 = \|e_k\|_A^2 - \frac{(r_k, r_k)_I^2}{\|r_k\|_A^2} = \|e_k\|_A^2 - \frac{(r_k, e_k)_A^2}{\|r_k\|_A^2}. \qquad (9.22)$$

Although the gradient descent method makes the maximal reduction in Q_A, it is not the direction that minimizes the error e_k; that direction would be $-e_k$. This sort of *greedy algorithm* can be improved by an algorithm with a more global view. The difficulty can be seen in figure 9.2. When the norm $\|\cdot\|_A$ is not very isotropic, the direction of greatest descent for Q_A may not be the best global strategy for reducing the error.

9.2 CONJUGATE GRADIENT ITERATION

The conjugate gradient method uses a more sophisticated choice for the search directions s_k. Instead of just using the current residual ($s_k = -r_k$), the direction is chosen to be orthogonal (conjugate) to previous residuals. This avoids repeating previous mistakes, or rather it avoids repeating directions already traversed.

9.2.1 The basic iteration

We switch notation temporarily and consider solving $Cy = g$, where C is an $n \times n$ positive definite matrix. We define sequences $\{y_k, r_k, s_k\} \subset \mathbb{R}^n \times \mathbb{R}^n \times \mathbb{R}^n$ and $\{\alpha_k, \beta_k\} \subset \mathbb{R} \times \mathbb{R}$ as follows. We assume that an initial approximation y_0 is given, and we define

$$s_0 = r_0 = Cy_0 - g. \qquad (9.23)$$

The main iteration steps proceed for $k = 0, 1, 2, \ldots$ by defining

$$\alpha_k = -\frac{(r_k, s_k)_I}{(s_k, s_k)_C} = -\frac{(r_k, s_k)_I}{(s_k, z_k)_I} \qquad (9.24)$$

$$y_{k+1} = y_k + \alpha_k s_k, \qquad (9.25)$$

$$z_k = Cs_k, \qquad (9.26)$$

$$r_{k+1} = Cy_{k+1} - g = r_k + \alpha_k Cs_k = r_k + \alpha_k z_k, \qquad (9.27)$$

$$\beta_k = -\frac{(r_{k+1}, s_k)_C}{(s_k, s_k)_C} = -\frac{(r_{k+1}, z_k)_I}{(s_k, z_k)_I} \tag{9.28}$$

$$s_{k+1} = r_{k+1} + \beta_k s_k. \tag{9.29}$$

We can interpret these definitions as follows.

Equations (9.24) and (9.25) represent the descent method with direction s_k.

Equation (9.26) can be viewed as just a definition, but it is also the only time in the iteration where the matrix C is applied to a vector. Thus it is the most computationally intense step.

The first equation in (9.27) defines the residual in the usual way, and the second equation follows because (9.25) implies that

$$r_{k+1} = Cy_{k+1} - g = C(y_k + \alpha_k s_k) - g = r_k + \alpha_k C s_k. \tag{9.30}$$

The third equation in (9.27) follows from (9.26).

Equations (9.28) and (9.29) define the new descent direction. The second equation in (9.28) follows from (9.26) and (9.4). Thus the conjugate gradient method is just a descent method in which the descent direction is derived from, but not equal to, the residual.

We have presented various forms for the coefficients α_k and β_k to indicate more efficient ways to evaluate them. Note that only one matrix multiplication is required in (9.24)-(9.29), the evaluation of z_k in (9.26). The different forms also simplify subsequent derivations. We will also derive other forms for these coefficients that can be used in alternate algorithms.

We will see that the convergence theory for the conjugate gradient method is quite sophisticated, but the derivation of these results takes some work. We begin with some basic orthogonality relations that are critical to the success of the CG method.

9.2.2 Orthogonality relations

There are three immediate orthogonality relations among the residuals and search directions for subsequent steps. Each of these has an important resulting equality and/or inequality. We develop all these in this section. We will ultimately see that other orthogonalities hold, but we will require a more complex induction to establish them.

Equations (9.28) and (9.29) say that the new search direction is based on the residual, modified so that it is orthogonal (with respect to the C-inner product) to the previous search direction:

$$(s_{k+1}, s_k)_C = (r_{k+1} + \beta_k s_k, s_k)_C = (r_{k+1}, s_k)_C + \beta_k (s_k, s_k)_C = 0. \tag{9.31}$$

Note that (9.29) and (9.31) imply that

$$\begin{aligned}
\|s_{k+1}\|_C^2 &= (s_{k+1}, s_{k+1})_C = (r_{k+1} + \beta_k s_k, s_{k+1})_C \\
&= (r_{k+1}, s_{k+1})_C + \beta_k (s_k, s_{k+1})_C \\
&= (r_{k+1}, s_{k+1})_C \leq \|r_{k+1}\|_C \|s_{k+1}\|_C,
\end{aligned} \tag{9.32}$$

so that dividing by $\|s_{k+1}\|_C$ yields

$$\boxed{\|s_{k+1}\|_C \le \|r_{k+1}\|_C} \tag{9.33}$$

(if $\|s_{k+1}\|_C = 0$, (9.33) holds trivially). We will see that a reverse inequality holds in a different norm.

The definition of α_k ensures that the current residual is orthogonal in the Euclidean inner product to the previous search direction:

$$
\begin{aligned}
(r_{k+1}, s_k)_I &= (r_k + \alpha_k C s_k, s_k)_I && \text{[by (9.27)]} \\
&= (r_k, s_k)_I + \alpha_k (s_k, s_k)_C = 0 && \text{[by (9.4) and (9.24)].}
\end{aligned}
\tag{9.34}
$$

There is an immediate consequence of this orthogonality:

$$
\begin{aligned}
(r_k, s_k)_I &= \|r_k\|_I^2 + (r_k, s_k - r_k)_I \\
&= \|r_k\|_I^2 + (r_k, \beta_{k-1} s_{k-1})_I && \text{[by (9.29)]} \\
&= \|r_k\|_I^2 && \text{[by (9.34)].}
\end{aligned}
\tag{9.35}
$$

We should note that when $k = 0$, (9.35) holds trivially since $r_0 = s_0$ by definition. Therefore, we get a reverse inequality to (9.33) (see its proof for details):

$$\boxed{\|s_k\|_I \ge \|r_k\|_I.} \tag{9.36}$$

The inequality (9.36) ensures that conjugate gradients can continue as long as $r_k \ne 0$ since this implies that $s_k \ne 0$. Once we have $r_k = 0$, we are done.

Using the definitions of α_k, r_k, s_k, and y_k, find that (for $k \ge 1$)

$$
\begin{aligned}
(r_{k+1}, r_k)_I &= (r_k + \alpha_k C s_k, r_k)_I && \text{[by (9.27)]} \\
&= \|r_k\|_I^2 + \alpha_k (s_k, r_k)_C && \text{[by (9.4)]} \\
&= \|r_k\|_I^2 + \alpha_k (s_k, s_k - \beta_{k-1} s_{k-1})_C && \text{[by (9.29)]} \\
&= \|r_k\|_I^2 + \alpha_k (s_k, s_k)_C && \text{[by (9.31)]} \\
&= \|r_k\|_I^2 - (r_k, s_k)_I && \text{[by (9.24)]} \\
&= \|r_k\|_I^2 - (r_k, r_k + \beta_{k-1} s_{k-1})_I && \text{[by (9.29)]} \\
&= \beta_{k-1}(r_k, s_{k-1})_I = 0 && \text{[by (9.34)].}
\end{aligned}
\tag{9.37}
$$

From (9.34), we have $0 = (r_1, s_0)_I = (r_1, r_0)_I$. Therefore,

$$(r_{k+1}, r_k)_I = 0 \quad \forall k \ge 0. \tag{9.38}$$

9.2.3 Further orthogonalities

Based on the three orthogonalities (9.31), (9.34), and (9.38) derived for subsequent iterations, we now derive some further orthogonalities among all iterates by induction.

Lemma 9.4 *The conjugate gradient process (9.23)-(9.29) terminates at the kth step only if $r_k = 0$; i.e., if $r_k \ne 0$, then $s_k \ne 0$. Further, if $r_l \ne 0$ for $l = 0, \ldots, k$, then*

$$0 = (r_{k+1}, s_q)_I = (r_{k+1}, r_q)_I = (s_{k+1}, s_q)_C \tag{9.39}$$

for $0 \le q \le k$. The conjugate gradient process will terminate in at most n steps for an $n \times n$ matrix C.

Proof. As we have noted, (9.36) ensures that the process will continue while $r_k \neq 0$. The fact that it will terminate after n steps is due to the orthogonality relations (9.39), which can hold for at most n nonzero vectors in $\mathbb{R}^n$.

We have already verified the first three orthogonalities for $k = 0$. So we assume that the lemma holds for some value of $k \geq 0$ and prove by induction that it also holds for $k + 1$.

We have already verified the three orthogonalities for $q = k$, so we may assume that $q < k$. By (9.27),

$$(r_{k+1}, s_q)_I = (r_k, s_q)_I + \alpha_k (Cs_k, s_q)_I = 0, \tag{9.40}$$

in view of (9.4) and the induction hypothesis. Also, by (9.27), (9.29), and the induction hypothesis,

$$(r_{k+1}, r_q)_I = (r_k, r_q)_I + \alpha_k (Cs_k, r_q)_I$$

$$\overbrace{}^{\text{omit if } q = 0}$$

$$= (r_k, r_q)_I + \alpha_k (Cs_k, s_q \overbrace{-\beta_{q-1} s_{q-1}}^{})_I \tag{9.41}$$

$$= 0 + \alpha_k (0 - 0) = 0.$$

Similarly, by (9.29), (9.25), and (9.27),

$$\overbrace{}^{= 0}$$

$$(s_{k+1}, Cs_q)_I = (r_{k+1}, Cs_q)_I + \beta_k \overbrace{(s_k, Cs_q)_I}^{}$$

$$= (r_{k+1}, \frac{1}{\alpha_q} (y_{q+1} - y_q))_I \tag{9.42}$$

$$= (r_{k+1}, \frac{1}{\alpha_q} (r_{q+1} - r_q))_I$$

$$= 0.$$

QED

Corollary 9.5 *The search directions and residuals for conjugate gradients satisfy*

$$0 = (s_{k+1}, r_q)_C \tag{9.43}$$

for $0 \leq q \leq k$.

Proof. Again, by definition we have $s_0 = r_0$, so that

$$(s_1, r_0)_C = (s_1, s_0)_C = 0 \tag{9.44}$$

since $s_0 = r_0$ by definition. So we assume that the lemma holds for some value of $k \geq 0$ and prove by induction that it also holds for $k + 1$. For $0 \leq q \leq k$, (9.29) implies that

$$\overbrace{}^{\text{omit if } q = 0}$$

$$(s_{k+1}, r_q)_C = (s_{k+1}, s_q \overbrace{-\beta_{q-1} s_{q-1}}^{})_C \tag{9.45}$$

$$= 0,$$

by (9.39).

QED

9.2.4 New formulas for α and β

It is of interest to have different forms for the coefficients in CG as they tend to have different behaviors in floating-point computations because cancellations related to the orthogonalities in CG are avoided. Equation (9.35) implies that we can write α_k as

$$\alpha_k = -\frac{\|r_k\|_I^2}{\|s_k\|_C^2} = -\frac{\|r_k\|_I^2}{(s_k, z_k)_I}. \tag{9.46}$$

Using the orthogonality (9.38), we can derive a new formula for β_k:

$$\begin{aligned}
\beta_k &= -\frac{(r_{k+1}, s_k)_C}{(s_k, s_k)_C} = -\frac{(r_{k+1}, z_k)_I}{(s_k, s_k)_C} && \text{[by (9.24)]} \\
&= -\frac{(r_{k+1}, \alpha_k^{-1}(r_{k+1} - r_k))_I}{(s_k, s_k)_C} && \text{[by (9.27)]} \\
&= -\frac{(r_{k+1}, r_{k+1})_I}{\alpha_k(s_k, s_k)_C} && \text{[by (9.38)]} && \tag{9.47} \\
&= \frac{(r_{k+1}, r_{k+1})_I}{(r_k, s_k)_I} && \text{[by (9.24)]} \\
&= \frac{\|r_{k+1}\|_I^2}{\|r_k\|_I^2} && \text{[by (9.35)]}.
\end{aligned}$$

9.3 OPTIMAL APPROXIMATION OF CG

Originally, we motivated conjugate gradients as an iterative method based on minimizing a quadratic form. We have seen that CG can also be viewed as a direct method, in that it is guaranteed to reduce the residual to zero in at most n steps for an $n \times n$ system. Now we change our point of view back to our original presentation of CG: as an approximation algorithm. To do so, we need to develop some technology.

9.3.1 Operator calculus

We have seen (exercise 6.1) that the set $\mathcal{P}_k$ of polynomials of degree k in one variable can be viewed as a vector space of dimension $k + 1$. We will now see that there is a way to map this space to the linear space $\mathcal{O}(\mathbb{R}^n, \mathbb{R}^n)$ of operators on $\mathbb{R}^n$ (cf. section 6.1). Suppose we have an $n \times n$ matrix C and consider the mapping $v \to Cv$ for all $v \in \mathbb{R}^n$. Then this defines such an operator in $\mathcal{O}(\mathbb{R}^n, \mathbb{R}^n)$, which we can also denote by C. Similarly, we can define C^2 by $v \to C(Cv)$ for all $v \in \mathbb{R}^n$. In fact, for any integer k, C^k is defined inductively by $v \to C(C^{k-1}v)$. Thus for any polynomial $P \in \mathcal{P}_k$, we can define $P(C)$ by

$$P(C) = \sum_{i=0}^{k} a_i C^i, \quad \text{where } P(x) = \sum_{i=0}^{k} a_i x^i. \tag{9.48}$$

By convention, we define $C^0 = I$, where I denotes the identity operator on $\mathbb{R}^n$, associated with the $n \times n$ identity matrix. Thus (exercise 9.8) (9.48) defines a linear operator $\mathcal{L}_C : \mathcal{P}_k \rightarrow \mathcal{O}(\mathbb{R}^n, \mathbb{R}^n)$. Note that if $P(x) \equiv 1$ is the constant polynomial, then $P(C) = I$ for any C.

9.3.2 CG error representation

Recall that the residual $r_k = Cy_k - g$ and the error $e_k = y_k - y$ are related by $r_k = Ce_k$. The next lemma relates the errors and search directions in CG to the initial error.

Lemma 9.6 *For $k \geq 0$, $e_k = P_k(C)e_0$ and $s_k = CQ_k(C)e_0$, where P_k and Q_k are polynomials of degree at most k and $P_k(0) = 1$.*

Proof. The proof is by induction. It is clear that $P_0 \equiv 1$ and $Q_0 \equiv 1$ work for $k = 0$ since $s_0 = r_0 = Ce_0$ by (9.12). For $k \geq 0$ we have

$$
\begin{aligned}
e_{k+1} &= e_k + y_{k+1} - y_k \\
&= e_k + \alpha_k s_k \qquad \text{[by (9.25)]} \\
&= (P_k(C) + \alpha_k C Q_k(C)) e_0 \\
&= P_{k+1}(C)e_0,
\end{aligned}
\tag{9.49}
$$

that is, $P_{k+1}(x) := P_k(x) + \alpha_k x Q_k(x)$ (cf. exercise 9.9). Note that $P_{k+1}(0) = P_k(0)$; hence $P_{k+1}(0) = 1$. Similarly,

$$
\begin{aligned}
s_{k+1} &= Ce_{k+1} + \beta_k s_k \qquad \text{[by (9.29) and (9.12)]} \\
&= (CP_{k+1}(C) + \beta_k C Q_k(C)) e_0 \\
&= CQ_{k+1}(C)e_0,
\end{aligned}
\tag{9.50}
$$

that is, $Q_{k+1}(x) := P_{k+1}(x) + \beta_k Q_k(x)$.   QED

Lemma 9.7 *Assuming that $r_k \neq 0$, the set $\{s_0, \ldots, s_k\}$ of vectors defined in (9.29) spans the Krylov[1] subspace*

$$
\mathcal{S}_k = \left\{ v \in \mathbb{R}^n \mid v = CT(C)e_0, T \in \mathcal{P}_k \right\}.
\tag{9.51}
$$

Proof. First, we need to show that $\mathcal{S}_k$ is a vector space and that its dimension is $k + 1$. By definition, $\mathcal{S}_k$ is a subset of $\mathbb{R}^n$. If $T_1, T_2 \in \mathcal{P}_k$, then set $v_i = CT_i(C)e_0$, $i = 1, 2$. Expanding, we find

$$
\begin{aligned}
v_1 + v_2 &= CT_1(C)e_0 + CT_2(C)e_0 = C(T_1(C) + T_2(C))e_0 \\
&= C(T_1 + T_2)(C)e_0 \in \mathcal{S}_k,
\end{aligned}
\tag{9.52}
$$

and similarly scalar multiples of $v \in \mathcal{S}_k$ are also in $\mathcal{S}_k$. Thus $\mathcal{S}_k$ is a linear subspace of $\mathbb{R}^n$ (cf. exercises 9.10 and 9.11).

---

[1] Alexei Nikolaevich Krylov (1863–1945) first used this subspace in 1931 to transform characteristic polynomials [58]. He was very active in the theory and practice of shipbuilding and is commemorated by the Krylov Shipbuilding Research Institute.

We can view $\mathcal{S}_k$ as the image of $\mathcal{P}_k$ in $\mathbb{R}^n$ via the mapping B defined by $BT = CT(C)e_0$. We have just shown that this mapping is linear. The image of a $(k+1)$-dimensional space can be no more than $k+1$, so we have shown that $\dim \mathcal{S}_k \leq k+1$.

By lemma 9.6, each s_l is in $\mathcal{S}_k$ for $l = 0, \ldots, k$, and by (9.31) they are orthogonal (in the C-inner product). Hence they are a basis.   QED

Lemma 9.8 *Define*

$$\mathcal{P}_k^0 = \{ P \in \mathcal{P}_k \mid P(0) = 1 \} . \tag{9.53}$$

The error e_k is optimal in the sense that

$$\|e_k\|_C = \min \{ \|P(C)e_0\|_C \mid P \in \mathcal{P}_k^0 \} . \tag{9.54}$$

Proof. Let P_k be the polynomial guaranteed by lemma 9.6. Since $P_k \in \mathcal{P}_k^0$, we have $\|e_k\|_C \geq \inf \{ \|P(C)e_0\|_C \mid P \in \mathcal{P}_k^0 \}$ because lemma 9.6 tells us that $e_k = P_k(C)e_0$. Now let us prove the reverse inequality.

Let $P \in \mathcal{P}_k^0$ and define $v = P(C)e_0$. Then

$$v = e_k + (P - P_k)(C)e_0. \tag{9.55}$$

But since $(P - P_k)(0) = 0$, $(P - P_k)(x) = xT_{k-1}(x)$, where T_{k-1} is a polynomial of degree $\leq k - 1$ (exercise 9.12). Hence

$$v = e_k + \delta_k, \tag{9.56}$$

$\delta_k \in \mathcal{S}_{k-1}$. Now $(e_k, \delta_k)_C = (r_k, \delta_k)_I = 0$ since $(r_k, s_q)_I = 0$ for all $q < k$. Thus

$$\begin{aligned} \|v\|_C^2 &= \|e_k\|_C^2 + 2(e_k, \delta_k)_C + \|\delta_k\|_C^2 \\ &= \|e_k\|_C^2 + \|\delta_k\|_C^2 \geq \|e_k\|_C^2, \end{aligned} \tag{9.57}$$

which proves that $\|e_k\|_C \leq \inf \{ \|P(C)e_0\|_C \mid P \in \mathcal{P}_k^0 \}$ and confirms the equality (9.54).   QED

9.3.3 Spectral theory

We now introduce a bit more technology from operator theory. Since we are working with a symmetric matrix C, we can expand in terms of its eigenvectors X_j, where $CX_j = \lambda_j X_j$. That is, for any $v \in \mathbb{R}^n$, we can write

$$v = \sum_{j=1}^{n} a_j X_j. \tag{9.58}$$

Then $C^k v - \sum_{j=1}^{n} a_j \lambda_j^k X_j$, and indeed (exercise 9.13)

$$\boxed{ P(C)v = \sum_{j=1}^{n} a_j P(\lambda_j) X_j } \tag{9.59}$$

for all $P \in \mathcal{P}_k$. Note that we can choose the eigenvectors to be orthonormal:

$$(X_j, X_k)_I = \delta_{jk}, \tag{9.60}$$

where δ_{jk} is the Kronecker δ. This means they are also orthogonal in the C-inner product:

$$(X_j, X_k)_C = (CX_j, X_k)_I = (\lambda_j X_j, X_k)_I = \lambda_j \delta_{jk}. \tag{9.61}$$

Also, it follows for v of the form (9.58) that (exercise 9.14)

$$\|v\|_I^2 = \sum_{j=1}^n a_j^2 \quad \text{and} \quad \|v\|_C^2 = \sum_{j=1}^n \lambda_j a_j^2. \tag{9.62}$$

Therefore, by (9.59),

$$\|P(C)v\|_I^2 = \sum_{j=1}^n P(\lambda_j)^2 a_j^2 \quad \text{and} \quad \|P(C)v\|_C^2 = \sum_{j=1}^n P(\lambda_j)^2 \lambda_j a_j^2. \tag{9.63}$$

Let us introduce the notation

$$\boxed{\|P(\lambda_{(\cdot)})\|_\infty = \max\left\{|P(\lambda_j)| \mid j = 1, \ldots, n\right\}.} \tag{9.64}$$

Then from (9.62), we see that

$$\|P(C)v\|_C^2 \le \|P(\lambda_{(\cdot)})\|_\infty^2 \sum_{j=1}^n \lambda_j a_j^2 = \|P(\lambda_{(\cdot)})\|_\infty^2 \|v\|_C^2. \tag{9.65}$$

The following result shows that error estimates for CG can be reduced to a polynomial approximation result on the spectrum of C.

Theorem 9.9 *Suppose that $\lambda_1, \ldots, \lambda_n$ are the eigenvalues of C. Then*

$$\|e_k\|_C \le \inf\left\{\|P(\lambda_{(\cdot)})\|_\infty \mid P \in \mathcal{P}_k^0\right\} \|e_0\|_C, \tag{9.66}$$

where $\|P(\lambda_{(\cdot)})\|_\infty$ is defined in (9.64) and $\mathcal{P}_k^0$ was defined in (9.53).

Proof. Apply (9.65) to lemma 9.8.   QED

9.3.4 CG error estimates

There are many results that can be derived from theorem 9.9. Here we are able to give only a sample.

Corollary 9.10 *Suppose that there are only k distinct eigenvalues for C. Then the CG iteration terminates (in the absence of rounding error) in k steps.*

Proof. The proof is an application of Lagrange interpolation (chapter 10). That is, we take P to be a polynomial of degree k that is 1 at 0 and 0 at all eigenvalues:

$$P(0) = 1, \ P(\lambda_j) = 0, \ j = 1, \ldots, k. \tag{9.67}$$

The existence of P will be proved in section 10.2.1. (For example, if $k = 1$, we take $P(x) = 1 - x/\lambda_1$.) Then $P \in \mathcal{P}_k^0$, and we conclude from theorem 9.9 that $e_k = 0$.   QED

Let us see what theorem 9.9 implies in a concrete case. Suppose there are only two eigenvalues, $\lambda_1 < \lambda_2$. Let $P(x) = 1 - x/\lambda$, where we will pick λ subsequently. Then $P \in \mathcal{P}_1^0$, and theorem 9.9 implies that

$$\|e_1\|_C \le \max\{|1 - \lambda_1/\lambda|, |1 - \lambda_2/\lambda|\}\|e_0\|_C \tag{9.68}$$

for all λ. Presumably, the minimum occurs when $\lambda_1 < \lambda < \lambda_2$, and

$$1 - \lambda_1/\lambda = -(1 - \lambda_2/\lambda), \tag{9.69}$$

which implies that $\lambda = \frac{1}{2}(\lambda_1 + \lambda_2)$. Thus (cf. exercise 9.15)

$$\|e_1\|_C \le \left(1 - \frac{2\lambda_1}{\lambda_1 + \lambda_2}\right)\|e_0\|_C = \left(\frac{\lambda_2 - \lambda_1}{\lambda_1 + \lambda_2}\right)\|e_0\|_C. \tag{9.70}$$

In general, we will prove the following result.

Lemma 9.11 *Suppose that $0 < \lambda_1 < \lambda_2$. Then there is a polynomial q_k of degree k such that $q_k(0) = 1$ and*

$$\|q_k\|_{\infty,[\lambda_1,\lambda_2]} \le \left(\Lambda + \sqrt{\Lambda^2 - 1}\right)^{-k}, \tag{9.71}$$

where $\Lambda = (\lambda_1 + \lambda_2)/(\lambda_2 - \lambda_1)$.

We postpone the proof of the lemma until section 11.1.1. We can write $\Lambda = (\kappa + 1)/(\kappa - 1)$, where $\kappa = \lambda_2/\lambda_1$. By exercise 9.16, we then have

$$\Lambda + \sqrt{\Lambda^2 - 1} = \frac{\sqrt{\kappa} + 1}{\sqrt{\kappa} - 1}. \tag{9.72}$$

Thus we have the following result.

Corollary 9.12 *Suppose that the eigenvalues of C lie in an interval $[\lambda_1, \lambda_2]$, where $0 < \lambda_1 < \lambda_2$. Define $\kappa = \lambda_2/\lambda_1$. Then*

$$\|e_k\|_C \le \left(\frac{\sqrt{\kappa} - 1}{\sqrt{\kappa} + 1}\right)^k = \left(1 - \frac{2}{\sqrt{\kappa} + 1}\right)^k \le e^{-2k/(\sqrt{\kappa}+1)}. \tag{9.73}$$

In the last inequality in (9.73), we used the fact that $1 - x \le e^{-x}$ for $x \ge 0$ (exercise 9.17). If we set

$$\lambda_{\{1,2\}} = \{\min, \max\}\left\{|\lambda| \mid \lambda \text{ is an eigenvalue of } A\right\},$$

then (6.10) and exercise 6.7 imply that (for A symmetric and positive definite)

$$\kappa = \lambda_2/\lambda_1 = \|A\|_2\|A^{-1}\|_2. \tag{9.74}$$

The quantity κ is called the *condition number* of A with respect to the Euclidean norm.

9.3.5 Preconditioned Conjugate Gradient bteration

Suppose that we want to solve
$$A x = b \tag{9.75}$$
and that M^{-1} is an approximate inverse for A, that is, M is a symmetric positive definite matrix such that $M^{-1}A$ is close to the identity in some sense. Suppose there is a symmetric positive definite matrix B such that $B^2 = M^{-1}$ (see exercise 9.18). We write B as $M^{-1/2}$. Suppose that we apply CG to
$$C = M^{-1/2} A M^{-1/2}$$
(at least in our heads). Then
$$Cy = g, \tag{9.76}$$
where $y = M^{1/2}x$ and $g = M^{-1/2}b$.

Let x_0 be given and take $y_0 = M^{1/2}x_0$. Let the sequences $\{y_k\}$, $\{r_k\}$, $\{\alpha_k\}$, and $\{\beta_k\}$ be defined by CG, as in (9.23)-(9.29). Let
$$x_k = M^{-1/2}y_k,$$
$$\sigma_k = M^{-1/2}s_k, \tag{9.77}$$
$$\rho_k = M^{1/2}r_k.$$
Then the iteration can be cast in terms of these variables:
$$\alpha_k = -(M^{-1}\rho_k, \rho_k)/(A\sigma_k, \sigma_k),$$
$$x_{k+1} = x_k + \alpha_k \sigma_k,$$
$$\rho_{k+1} = \rho_k + \alpha_k A\sigma_k, \tag{9.78}$$
$$\beta_k = (M^{-1}\rho_{k+1}, \rho_{k+1})/(M^{-1}\rho_k, \rho_k),$$
$$\sigma_{k+1} = M^{-1}\rho_{k+1} + \beta_k \sigma_k.$$
The verification of these formulas involves not only (9.23)-(9.29) but also lemma 9.4.

Note that the only additional work caused by the preconditioning is one application of M^{-1} per iteration.

If $e_k = y_k - C^{-1}g$ and $\varepsilon_k = x_k - A^{-1}b$, then
$$e_k = M^{1/2}x_k - M^{1/2}A^{-1}M^{1/2}b = M^{1/2}\varepsilon_k$$
and
$$(Ce_k, e_k) = \|e_k\|_C^2$$
$$= (M^{-1/2}A M^{-1/2}M^{1/2}\varepsilon_k, M^{1/2}\varepsilon_k)$$
$$= \|\varepsilon_k\|_A^2.$$
This observation and the previous results on CG give the following theorem.

Theorem 9.13 *Suppose that*
$$\mu_{\{1,2\}} = \{\min, \max\} \{|\lambda| \mid \lambda \text{ is an eigenvalue of } M^{-1}A\}$$
and define $\kappa = \mu_2/\mu_1$. *Then the error for the algorithm (9.77) and (9.78) satisfies*
$$\|\varepsilon_k\|_A \leq \left(\frac{\sqrt{\kappa}-1}{\sqrt{\kappa}+1}\right)^k = \left(1 - \frac{2}{\sqrt{\kappa}+1}\right)^k \leq e^{-2k/(\sqrt{\kappa}+1)}.$$

9.4 COMPARING ITERATIVE SOLVERS

It is possible to compare the behavior of stationary iterative methods with that of conjugate gradients. To simplify the analysis, let us assume that A is an $n \times n$ matrix with 1's on the the diagonal: $\text{diag}(A) = I$. Let us compare CG and Jacobi for A. First, note that the assumption $\text{diag}(A) = I$ is not an essential restriction. Jacobi is invariant with respect to scaling (on the left) by a diagonal matrix, and we can then compare with the diagonally scaled (preconditioned) CG.

We assume as well that A is symmetric and positive definite, so this means we can expand in eigenvectors as in (9.58). Let us write the initial error as

$$e^0 = \sum_{i=1}^{n} a_i X_i. \tag{9.79}$$

Combining (9.54) and (9.63), we see that the resulting CG error e_{CG}^k after k steps satisfies

$$\|e_{\text{CG}}^k\|_A^2 \le \sum_{i=1}^{n} a_i^2 P_k(\lambda_i)^2 \lambda_i. \tag{9.80}$$

On the other hand, the error in Jacobi satisfies $e_J^k = M_J(A)^k e^0$, where $M_J(A) = I - A$ is the Jacobi iteration matrix for A. The eigenvalues μ_i of $M_J(A)$ are related to the eigenvalues λ_i of A by $\mu_i = (1 - \lambda_i)$. Moreover, the eigenvectors of $M_J(A)$ and A are the same. In particular, we have

$$e_J^k = \sum_{i=1}^{n} (1 - \lambda_i)^k a_i X_i. \tag{9.81}$$

Thus (9.62) implies that

$$\|e_J^k\|_A^2 = \sum_{i=1}^{n} a_i^2 (1 - \lambda_i)^{2k} \lambda_i. \tag{9.82}$$

This corresponds to the choice $P(\lambda) = (1 - \lambda)^k$ in (9.80). Thus we see that CG adapts the choice of polynomial to the data, whereas for Jacobi it is fixed. Thus we see why the word "stationary" is appropriate for such iterative methods.

9.5 MORE READING

The conjugate gradient method is attributed primarily to Hestenes[2] and Stiefel[3] although there were additional influences [68], including Lanczos (see

---

[2]Magnus Rudolph Hestenes (1906–1991) obtained a Ph.D. at the University of Chicago with Gilbert Bliss in 1932. Bliss was an early graduate of Chicago, receiving a B.Sc. in 1897 and a Ph.D. in 1900 under the direction of Oskar Bolza, who had studied with Felix Klein in Göttingen. Hestenes was a professor at UCLA from 1947 to 1973.

[3]Eduard L. Stiefel (1909 -1978) was a student of Heinz Hopf, who was in turn a student of Erhard Schmidt. Stiefel was the advisor of Peter Henrici (page 288) as well as 63 other students over a period of 37 years. He is known for his work on the Stiefel-Whitney characteristic classes, and he was also an early user and developer of computers [154].

page 234) who was at UCLA at the time [157]. There are several monographs on conjugate gradients and related methods [71, 118, 165]. The theory of convergence rates for CG has an intriguing relation to potential theory [53]. The CG algorithm applies as well to infinite-dimensional operators, and in some cases CG for such operators can have surprising convergence rates [173].

9.6 EXERCISES

Exercise 9.1 *Suppose that A is an $n \times n$ symmetric, positive definite matrix. Prove that the expression (9.1) defines an inner product on $\mathbb{R}^n$. In particular, verify that $(u, v)_A = (v, u)_A$.*

Exercise 9.2 *Suppose that A is an $n \times n$ symmetric, positive definite matrix. Prove (9.4). (Hint: use the symmetry of A.)*

Exercise 9.3 *Use calculus to verify that the minimum of the expression (9.5) occurs at the solution of $Au = f$ in the case $n = 1$. (Hint: just write out $Q_A(u)$ and differentiate with respect to u and find where the derivative is zero.)*

Exercise 9.4 *Verify all the steps in the derivation of the expression (9.7). (Hint: expand the quadratic terms and use the symmetry of the inner product; cf. exercise 9.1 or 9.6).*

Exercise 9.5 *Suppose $q(t) = \alpha + \beta t + \gamma t^2$. Prove that $q(t)$ has a minimum at $t = 0$ iff $\beta = 0$.*

Exercise 9.6 *Suppose that A is an $n \times n$ symmetric, positive definite matrix. Prove that $(u, v)_A = (Au)^T v$ for all $u, v \in \mathbb{R}^n$. (Hint: use the symmetry of A.)*

Exercise 9.7 *Suppose that $Au = f$ and that $v \in \mathbb{R}^n$. Let $r = Av - f$ and $e = v - u$ (so that $Ae = r$). Prove that $2Q_A(v) = (e, e)_A - (u, u)_A$.*

Exercise 9.8 *Prove that the operator $\mathcal{L}_C : \mathcal{P}_k \to \mathcal{O}(\mathbb{R}^n, \mathbb{R}^n)$ defined by (9.48) is linear. That is, if $P(x) = \sum_{i=0}^k a_i x^i$, $Q(x) = \sum_{i=0}^k b_i x^i$, and $\alpha \in \mathbb{R}$, then*

$$\mathcal{L}_C P + \alpha \mathcal{L}_C Q = \mathcal{L}_C R, \tag{9.83}$$

where R is the polynomial given by $R(x) = \sum_{i=0}^k \alpha(a_i + b_i) x^i$.

Exercise 9.9 *Suppose that P is a polynomial given by $P(x) = \sum_{i=0}^k a_i x^i$ and that C is an $n \times n$ matrix. Prove that $CP(C) = Q(C)$, where $Q(x) = \sum_{i=0}^k a_i x^{i+1}$.*

Exercise 9.10 *Suppose that $M \subset \mathcal{O}(\mathbb{R}^n, \mathbb{R}^n)$ is a linear subspace. Show that the set $\{Av \mid A \in M\}$ is a linear subspace of $\mathbb{R}^n$ for any $v \in \mathbb{R}^n$.*

Exercise 9.11 *Suppose that $M \subset \mathcal{O}(\mathbb{R}^n, \mathbb{R}^n)$ is a linear subspace and that $B \in \mathcal{O}(\mathbb{R}^n, \mathbb{R}^n)$. Prove that the set $\{BA \mid A \in M\}$ is a linear subspace of $\mathcal{O}(\mathbb{R}^n, \mathbb{R}^n)$.*

Exercise 9.12 *Suppose that $p(x)$ is a polynomial that vanishes at 0: $p(0) = 0$. Prove that we can write $p(x) = xq(x)$ for a polynomial q whose degree is 1 less than the degree of p.*

Exercise 9.13 *Verify (9.59). (Hint: Expand.)*

Exercise 9.14 *Verify (9.62). (Hint: Expand.)*

Exercise 9.15 *Suppose that $\lambda_1 > 0$ and $\lambda_2 > 0$. Prove that*
$$\operatorname{argmin}\{\max\{|1 - \lambda_1/\lambda|, |1 - \lambda_2/\lambda|\} \mid \lambda > 0\} = \tfrac{1}{2}(\lambda_1 + \lambda_2). \qquad (9.84)$$
(Hint: show that $\max\{|a|, |b|\} = \tfrac{1}{2}|a + b| + \tfrac{1}{2}|a - b|$ for $a, b \in \mathbb{R}$.)

Exercise 9.16 *Prove that (9.72) holds. (Hint: just expand and note that $(\sqrt{\kappa} + 1)(\sqrt{\kappa} - 1) = \kappa - 1$.)*

Exercise 9.17 *Prove that $1 - x \le e^{-x}$ for $x \ge 0$. (Hint: expand e^{-x} as an alternating series and note that $1 - x$ represents the first two terms.)*

Exercise 9.18 *Prove that for any symmetric, positive definite matrix A, there is a symmetric, positive definite matrix B such that $B^2 = A$. (Hint: write $A = U^T D U$, where U is orthogonal and D is diagonal with positive entries; cf. corollary 6.5 and set $B = U^T \sqrt{D} U$.)*

Exercise 9.19 *Prove theorem 9.13.*

Exercise 9.20 *Prove (9.36).*

Exercise 9.21 *Prove that a polynomial of degree n can have at most n roots, counting multiplicity, unless it is identically zero. (Hint: use exercise 9.12 to represent the polynomial in terms of linear factors.)*

9.7 SOLUTIONS

Solution of Exercise 9.7. Recall that $Au = f$ and $e = v - u$. We expand and use the fact that $A^T = A$:

$$
\begin{aligned}
(e, e)_A = (v - u, v - u)_A &= (v - u)^T A(v - u) \\
&= v^T Av + u^T Au - u^T Av - v^T Au \\
&= v^T Av + u^T Au - u^T Av - v^T f \\
&= v^T Av + u^T Au - (A^T u)^T v - f^T v \qquad (9.85) \\
&= v^T Av + u^T Au - (Au)^T v - f^T v \\
&= v^T Av + u^T Au - 2f^T v \\
&= 2\, Q_A(v) + u^T Au.
\end{aligned}
$$

Thus $2Q_A(v) = (e, e)_A - u^{\mathsf{T}} A u$.

Solution of Exercise 9.15. First, we show that

$$\max\{|a|, |b|\} = \tfrac{1}{2}|a + b| + \tfrac{1}{2}|a - b| \qquad (9.86)$$

for $a, b \in \mathbb{R}$. We can assume that $|a| \geq |b|$ without loss of generality since we can just rename the variables if it is the other way around.

If $a \geq 0$, then $\max\{|a|, |b|\} = a$ and $a + b \geq 0$ (since $|b| \leq a$). Then

$$\max\{|a|, |b|\} - \tfrac{1}{2}|a + b| = a - \tfrac{1}{2}(a + b) = \tfrac{1}{2}(a - b) = \tfrac{1}{2}|a - b|. \qquad (9.87)$$

The last equality also follows from the fact that $|b| \leq a$.

If $a \leq 0$, then $\max\{|a|, |b|\} = -a$ and $a + b \leq 0$ and $a - b \leq 0$ (since $|b| \leq -a$). Then

$$\max\{|a|, |b|\} - \tfrac{1}{2}|a + b| = -a + \tfrac{1}{2}(a + b) = -\tfrac{1}{2}(a - b) = \tfrac{1}{2}|a - b| \qquad (9.88)$$

as well. Thus we have completed the proof of (9.86).

Now suppose that $\lambda_1 > 0$ and $\lambda_2 > 0$. Then

$$\max\{|1 - \lambda_1/\lambda|, |1 - \lambda_2/\lambda|\} = |1 - \tfrac{1}{2}(\lambda_1 + \lambda_2)/\lambda| + |\tfrac{1}{2}(\lambda_1 - \lambda_2)/\lambda|. \qquad (9.89)$$

Let us write $r = \tfrac{1}{2}(\lambda_1 + \lambda_2)/\lambda$, so that $\lambda = \tfrac{1}{2}(\lambda_1 + \lambda_2)/r$ and

$$\max\{|1 - \lambda_1/\lambda|, |1 - \lambda_2/\lambda|\} = |1 - r| + \frac{r|\lambda_1 - \lambda_2|}{\lambda_1 + \lambda_2}. \qquad (9.90)$$

Furthermore, our objective is now to show that

$$\operatorname{argmin}\left\{|1 - r| + \frac{r|\lambda_1 - \lambda_2|}{\lambda_1 + \lambda_2} \mid r > 0\right\} = 1. \qquad (9.91)$$

Define a function ϕ by

$$\phi(r) = |1 - r| + \frac{r|\lambda_1 - \lambda_2|}{\lambda_1 + \lambda_2}. \qquad (9.92)$$

For $r \geq 1$, $\phi(r) = r - 1 + r\frac{|\lambda_1 - \lambda_2|}{\lambda_1 + \lambda_2}$ is strictly increasing. For $r \leq 1$, $\phi(r) = 1 - r + r\frac{|\lambda_1 - \lambda_2|}{\lambda_1 + \lambda_2}$ is strictly decreasing since

$$\begin{aligned}
|\lambda_1 - \lambda_2| &= \max\{\lambda_1, \lambda_2\} - \min\{\lambda_1, \lambda_2\} \\
&< \max\{\lambda_1, \lambda_2\} \leq \lambda_1 + \lambda_2
\end{aligned} \qquad (9.93)$$

since both λ_i are positive. Thus

$$\operatorname{argmin} \phi(r) = 1. \qquad (9.94)$$

Chapter Ten

Polynomial Interpolation

> The web site http://www.blackphoto.com/glossary/i.asp
> describes interpolation as "a technique used by digital cam-
> eras, scanners and printers to increase the size of an image
> in pixels by averaging the colour and brightness values of
> surrounding pixels."

The approximation of general functions by simple classes of functions has
many applications as well as theoretical implications. The uniform approximation of a general continuous function on an interval by polynomials (a
theorem of Weierstrass[1]) is a fundamental result that casts light on the
nature of both polynomials and continuous functions. In the era of modern computers, approximation via interpolation has emerged as a general
paradigm for computing elementary functions as part of typical system software on current computers [107]. Probably one of the earliest applications
of interpolation was simply to link scattered data to provide some sense of
what a continuum representation might look like. The phrase "connecting
the dots" has become a common metaphor for problem solving, but this is
precisely what polynomial interpolation does.

One feature of the subject is that it introduces infinite-dimensional vector spaces in a natural way. Dealing with such spaces in a complete (pun
intended) way is beyond the scope of this book, but we hope that the ideas
stimulate interest in further study of functional analysis. We start by considering approximation by polynomials in one dimension. Some of the technology we develop applies to other classes of approximating spaces, as well
as multivariate approximation.

10.1 LOCAL APPROXIMATION: TAYLOR'S THEOREM

Taylor's theorem[2] in calculus provides a polynomial approximation to a sufficiently smooth function:

$$P_n(x) = \sum_{k=0}^{n} \frac{f^{(k)}(x_0)}{k!} (x - x_0)^k. \tag{10.1}$$

---

[1]Karl Theodor Wilhelm Weierstrass (1815–1897) was the only student of Christoph
Gudermann, who was, along with Friedrich Bessel, J. W. Richard Dedekind, Sophie Ger-
main, and Georg Riemann, a student of Gauss.

[2]This theorem appears to have been first discovered by Gregorie (see page 212) [162].

For x near x_0, this yields an accurate approximation to the function f provided we have the required data. Moreover, we have (exercise 7.4) a representation of the error:

$$f(x) - P_n(x) = \frac{(x - x_0)^{n+1}}{n!} \int_0^1 (1 - s)^n f^{(n+1)}(x_0 + s(x - x_0))\, ds. \quad (10.2)$$

Thus we can say that $f - P_n = \mathcal{O}((x - x_0)^{n+1})$ for x near x_0, provided that f is smooth enough.

This is a very powerful, but local, result. Moreover, it requires knowing the values of high-order derivatives of f to construct P_n. We now consider a more distributed approximation and one that does not require derivatives, only the values of f.

10.2 DISTRIBUTED APPROXIMATION: INTERPOLATION

Suppose that we gather data f_i associated with parameters x_i and that we want to depict these data as a function $f(x)$ with the property that $f(x_i) = f_i$. This is clearly not a well-defined problem since there are many functions with this property. (To simplify the discussion, we assume here that the x_i's are distinct, but also see exercise 12.12.) On the other hand, if we are restricted to the appropriate finite-dimensional space of functions, we can potentially make the process well-posed. One simple approach is to *interpolate* the data, e.g., with polynomials.

The name of Lagrange[3] is associated with the fact that a polynomial of degree n can be determined uniquely to match arbitrary values f_i at $n + 1$ distinct points x_i, e.g., with $i = 0, \ldots, n$. At the risk of further inflaming French-English relations, we point out that Newton (see page 15) had solved this problem earlier in an adaptive way (section 10.2.3).

We begin by considering arbitrary (but distinct) interpolation points x_i. In many data-fitting problems, there is little control on the spacing of the points, and in the application (9.67) the points are unknown. We will later consider specifying the points in a systematic way, but we will see that equally spaced points can be a bad choice. In chapter 11, we consider a better choice.

10.2.1 Existence of interpolant

The existence of the Lagrange interpolant can be proved by constructing polynomials ϕ_i such that

$$\phi_i(x_j) = \delta_{ij}. \quad (10.3)$$

---

[3] Joseph-Louis Lagrange (1736–1813), born in Turin and baptized Giuseppe Lodovico Lagrangia, became a dominant figure in his adopted country France. Best known for his work in mechanics, essentially establishing the variational calculus [101], his students included Poisson and Fourier. He is interred in the Pantheon in Paris.

Then the Lagrange interpolant is defined by

$$L_n f(x) = \sum_{i=0}^{n} f_i \phi_i. \tag{10.4}$$

The magic perhaps is in the fact that one can write a formula for each ϕ_i:

$$\phi_i(x) = \frac{\prod_{j \neq i}(x - x_j)}{\prod_{j \neq i}(x_i - x_j)} = \frac{\prod_{j \neq i}(x_j - x)}{\prod_{j \neq i}(x_j - x_i)}. \tag{10.5}$$

Here the indices j in the products range over the set $\{0, 1, \ldots, n\}$. Each ϕ_i is a polynomial of degree n since it is presented as a product of n monomials. Also $\phi_i(x_k) = 0$ for $k \neq i$ since one of the factors is $(x - x_k)$. Finally, $\phi_i(x_i) = 1$ because of the normalization. This completes the verification of (10.3).

The operator L_n defined in (10.4) maps the linear space $\mathbb{R}^{n+1}$ to the space $\mathcal{P}_n$ of polynomials of degree n. The operator L_n is clearly linear, by construction. That is, if $f \in \mathbb{R}^{n+1}$ and $g \in \mathbb{R}^{n+1}$ are two sets of data, then $L_n(f + g) = L_n f + L_n g$, and similarly $L_n(sf) = sL_n f$ for any scalar s.

The Lagrange interpolant can be extended to any continuous function f by

$$L_n f(x) = \sum_{i=0}^{n} f(x_i) \phi_i. \tag{10.6}$$

The operator L_n is a *projection*, i.e.,

$$L_n q = q \text{ for all } q \in \mathcal{P}_n. \tag{10.7}$$

To prove this, consider $p = L_n q - q$, which is also a polynomial of degree n such that $p(x_i) = 0$ for $n + 1$ points. By the fundamental theorem of algebra, a polynomial of degree n that vanishes at $n + 1$ distinct points must be identically zero (exercise 9.21). Therefore, $L_n q = q$. In fact, this approach can be used to show (exercise 12.19) the existence of the ϕ_i's.

10.2.2 Error expression

The error in Lagrange interpolation vanishes at all the interpolation points, so we can assert that the error is divisible by the function ω_k defined by

$$\omega_k(x) = \prod_{i=0}^{k-1}(x - x_i). \tag{10.8}$$

It is possible to prove (exercise 10.1) that

$$f(x) - L_n f(x) = \frac{\omega_{n+1}(x)}{(n+1)!} f^{(n+1)}(\xi(x)), \tag{10.9}$$

where $\xi(x)$ is just like the magic point in the pointwise remainder expression for Taylor's theorem. This expression has limitations as a way to understand the error in Lagrange interpolation in general, but it is useful theoretically and motivates certain things we will derive independently.

It is easy to see in one important case that (10.9) is valid, namely, when $p(x) = x^{n+1}$. Note that by expanding the expression (10.8), we find

$$\omega_{n+1}(x) = x^{n+1} + q(x) = p(x) + q(x), \tag{10.10}$$

where the degree of q is at most n. Thus $p = \omega_{n+1} - q$. Because $p - L_n p$ is a polynomial of degree $n + 1$ that vanishes at the roots of ω_{n+1}, we have $p - L_n p = \alpha \omega_{n+1}$ for some constant α. Thus

$$\alpha \omega_{n+1} = p - L_n p = \omega_{n+1} - q - L_n p, \tag{10.11}$$

so that $(1 - \alpha)\omega_{n+1} = q + L_n p$ is a polynomial of degree n. This can happen only if $\alpha = 1$ (and $q = -L_n p$). Thus we conclude that

$$p - L_n p = \omega_{n+1}, \tag{10.12}$$

consistent with (10.9).

10.2.3 Newton's divided differences

There is another approach to defining an interpolant that is associated with Newton (see page 15) and thus predates Lagrange (see page 152). It proceeds inductively based on the number of interpolation points, and it could be viewed as an adaptive procedure in which the interpolant with $n + 1$ points is derived from the one with n points. This allows one to derive (and use) the nth-order interpolation and assess whether further points need to be added.

Let f be fixed for the moment and let $p_n = L_n f$. If $n = 0$, there is only one point x_0 and p_0 is the constant function equal to $f(x_0)$. Now add another point $x_1 \neq x_0$. Then $p_1 = p_0 + q_1$, where q_1 is a linear polynomial. For p_1 to interpolate f at x_0, we must have $q_1(x_0) = 0$, and thus $q_1(x) = a_1(x - x_0)$. For p_1 to interpolate f at x_1, we must have

$$f(x_1) = f(x_0) + a_1(x_1 - x_0), \tag{10.13}$$

so that the coefficient a_1 must be

$$a_1 = \frac{f(x_1) - f(x_0)}{x_1 - x_0} = \frac{f(x_0) - f(x_1)}{x_0 - x_1}. \tag{10.14}$$

The coefficient a_1 is what is known as a *divided difference*, i.e., a difference quotient approximating the derivative of f near x_0, x_1. The standard notation for divided differences is

$$f[x_0, x_1] = \frac{f(x_1) - f(x_0)}{x_1 - x_0}. \tag{10.15}$$

With this notation, we have $p_1 = p_0 + f[x_0, x_1](x - x_0)$. If we define $\omega_0 \equiv 1$ and $f[x_0] = f(x_0)$, then we can write

$$p_1 = f[x_0]\omega_0 + f[x_0, x_1]\omega_1(x), \tag{10.16}$$

where ω_k for $k \geq 0$ is defined in (10.8).

Lagrange interpolation was defined by exhibiting particular basis functions for polynomials. The Newton approach can be viewed as one in which a different basis is chosen. It is not hard to see that the error polynomials ω_k defined in (10.8) form a basis for polynomials. The key observation is that they are linearly independent (exercise 10.3). Thus we can write any polynomial $p \in \mathcal{P}_n$ as

$$p(x) = \sum_{i=0}^{n} a_i \omega_i(x), \tag{10.17}$$

for suitable coefficients a_i. Thus for any continuous function f, we define coefficients a_i^f by

$$L_n f(x) = \sum_{i=0}^{n} a_i^f \omega_i(x). \tag{10.18}$$

Motivated by (10.16), we define

$$f[x_0, x_1, \ldots, x_i] = a_i^f \tag{10.19}$$

for $i \geq 0$. Thus by definition we have the Newton divided difference form of (Lagrange) interpolation:

$$\boxed{L_n f(x) = \sum_{k=0}^{n} f[x_0, x_1, \ldots, x_k] \omega_k(x).} \tag{10.20}$$

Divided differences are defined for any set of points $x_0, \ldots, x_n$, and they obey rules similar to those of calculus. First, it is evident from (10.15) that $f[x_0, x_1] = f[x_1, x_0]$. But from the definition (10.19), we see that the order of the x_i's does not matter in general, that is,

$$f[x_0, x_1, \ldots, x_n] = f[x_{\sigma(0)}, x_{\sigma(1)}, \ldots, x_{\sigma(n)}] \tag{10.21}$$

for any permutation σ of $\{0, 1, \ldots, n\}$, as we get the same interpolant independently of what order we use to introduce the points x_i into the interpolation process.

It is useful to derive explicitly the expression of p_2 in terms of second divided differences, but we leave this as exercise 10.4 and proceed to the general properties.

Theorem 10.1 *For distinct points $x_0, \ldots, x_n$, the nth coefficient of the Newton interpolation (10.20) defined by (10.19) satisfies*

$$f[x_0, x_1, \ldots, x_n] = \sum_{k=0}^{n} \frac{f(x_k)}{\prod_{i \neq k}(x_k - x_i)}, \tag{10.22}$$

where $f[x_0] = f(x_0)$ in the case of $n = 0$.

Note that (10.15) and (10.22) are consistent when $n = 1$.

Proof. We have two representations of the interpolant: (10.20) and (10.4). If we differentiate these n times, we find

$$f[x_0, x_1, \ldots, x_n]\omega_n^{(n)}(x) = \sum_{i=0}^{n} f(x_i)\phi_i^{(n)}(x) \tag{10.23}$$

since all the other terms in (10.20) vanish because ω_k is a polynomial of degree k. By inspection, we see that

$$\phi_i^{(n)}(x) = \frac{n!}{\prod_{j \neq i}(x_i - x_j)} \tag{10.24}$$

because the leading-order term in the numerator in (10.4) is x^n. Similarly, the leading-order term in $\omega_n(x)$ is also x^n, so that $\omega_n^{(n)} \equiv n!$. Plugging these values for the derivatives into (10.23) completes the proof of the theorem. QED

Lemma 10.2 *For distinct points* $x, x_0, \ldots, x_n$ *in an interval* I,

$$f(x) - L_n f(x) = f[x_0, x_1, \ldots, x_n, x]\omega_{n+1}(x). \tag{10.25}$$

Notice the similarity to Taylor's theorem with a remainder. Both approximations are written in terms of a sequence of basis functions $((x - x_0)^k$ versus $\omega_k(x))$, and the error terms are written in terms of the next basis functions in the sequence.

Proof. We think of x as an additional interpolation point and use (10.20) to write

$$L_{n+1} f(y) = L_n f(y) + f[x_0, x_1, \ldots, x_n, x]\omega_{n+1}(y). \tag{10.26}$$

But since x is an interpolation point, we have $L_{n+1} f(x) = f(x)$, so applying (10.26) with $y = x$ completes the proof.   QED

Corollary 10.3 *For distinct points* $x_0, \ldots, x_n$ *in an interval* I, *there is a point* $\xi \in I$ *satisfying*

$$f[x_0, x_1, \ldots, x_n] = \frac{f^{(n)}(\xi)}{n!}, \tag{10.27}$$

provided that $f \in C^{(n+1)}(I)$.

Proof. Apply (10.9) and lemma 10.2.   QED

Further manipulation (exercise 10.6) of the indices in (10.22) shows that the nth divided difference is a divided difference of divided differences of order $n - 1$, as described in the following result.

Corollary 10.4 *For distinct points* $x_0, \ldots, x_n$, *the* nth *divided difference satisfies*

$$f[x_0, x_1, \ldots, x_n]$$
$$= \frac{f[x_1, \ldots, x_{i-1}, x_{i+1}, \ldots, x_n] - f[x_0, \ldots, x_{j-1}, x_{j+1}, \ldots, x_n]}{x_j - x_i} \tag{10.28}$$

for any indices $i \neq j$.

10.3 NORMS IN INFINITE-DIMENSIONAL SPACES

We need to formalize a statement made about the definition of the Lagrange interpolation operator. Initially, in (10.4) it was defined for $f \in \mathbb{R}^{n+1}$, but then it was extended in (10.6) to be defined for $f \in C^0$, the space of continuous functions. This space is also a vector space. Addition in C^0 is defined by saying that

$$(f + g)(x) = f(x) + g(x) \text{ for all } x, \tag{10.29}$$

and scalar multiplication is defined by

$$(sf)(x) = sf(x) \text{ for all } x, \tag{10.30}$$

for any scalar s. But C^0 is an *infinite-dimensional* vector space. However, there is a natural norm defined on it, the *maximum norm*. To be more precise, for any closed interval I, we can define

$$\|f\|_\infty = \sup_{x \in I} |f(x)|. \tag{10.31}$$

We leave as exercise 10.7 the verification that this is indeed a norm.

10.3.1 Instability of Lagrange interpolation

Lagrange interpolation is more robust than Taylor approximation, but it has properties that make it sensitive to use in practice. In particular, the placement of the interpolation points is extremely influential regarding its performance. The main difficulty stems from a lack of stability of the interpolation operator in the maximum norm for many choices of interpolation points.

For the moment, let us think of Lagrange interpolation as defining an operator $L_n : C^0(I) \to \mathcal{P}_n$. If we take $\|\cdot\|_\infty$ as the norm for both of these spaces, we can define an operator norm in the same way as we did in (6.1) in the finite-dimensional case:

$$\|L_n\|_\infty = \sup_{f \in C^0(I)} \frac{\|L_n f\|_\infty}{\|f\|_\infty}. \tag{10.32}$$

In view of (6.2), we can think of $\|L_n\|_\infty$ as the smallest constant C such that $\|L_n f\|_\infty \leq C\|f\|_\infty$. (See exercise 10.8 for a finite-dimensional interpretation of the norm defined in (10.32).)

We will see that $\|L_n\|_\infty$ can be very large for typical choices of interpolation points. Suppose in this case that some small errors are made in the specification of the interpolation data f. That is, suppose we apply interpolation to $f + \delta$, where δ is small but somewhat arbitrary (e.g., due to round-off error). Since L_n is linear, the result is

$$L_n(f + \delta) = L_n f + L_n \delta. \tag{10.33}$$

We can then assert that the error, which is equal to $L_n\delta$, must be no larger than $\|L_n\|_\infty \|\delta\|_\infty$. But since we may have little control over δ, it could

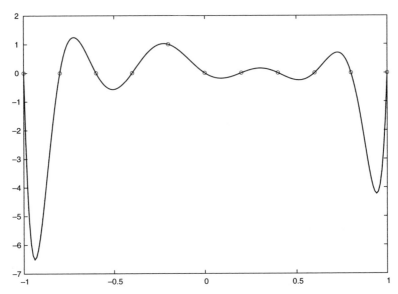

Figure 10.1 The Lagrange basis function ϕ_5 for 11 equally spaced interpolation
points on the interval $[-1, 1]$. The circles indicate the interpolation
data.

possibly be as large as $\|L_n\|_\infty \|\delta\|_\infty$. Thus $\|L_n\|_\infty$ represents the error am-
plification factor.

Let us estimate the size of $\|L_n\|_\infty$. Since we can choose f in (10.32)
arbitrarily, we can pick f to be 1 at the ith interpolation point and 0 at the
others. In this case, $L_n f = \phi_i$, the corresponding basis function. Thus we
conclude that

$$\|L_n\|_\infty \geq \max_{i=0,\ldots,n} \|\phi_i\|_\infty . \tag{10.34}$$

On the other hand, it is not hard (exercise 10.10) to show that

$$\|L_n\|_\infty \leq \sum_{i=0}^n \|\phi_i\|_\infty . \tag{10.35}$$

Thus we need to look carefully at the size of the Lagrange basis functions.

Since the Lagrange basis functions are chosen to be 1 at one point and
0 at the others, one might hope that the maximum of the basis functions
will not be too big. In particular, it might be the case that the maximum
occurs at the interpolation point where it is specified to be 1. But in general,
this does not happen. A typical Lagrange basis function for equally spaced
interpolation points

$$x_i = -1 + (i - 1)/k, \ i = 1, \ldots, 2k + 1, \tag{10.36}$$

is depicted in figure 10.1. For $n = 2k + 1$ equally spaced interpolation
points on $[-1, 1]$ as defined in (10.36), the largest norm appears to occur

γ	$n = 5$	7	9	11	13	15	17	19	21
$\sqrt{2}$	0.07	0.04	0.02	0.01	0.007	0.004	0.003	0.002	0.001
2	0.16	0.12	0.10	0.09	0.080	0.075	0.072	0.070	0.068
3	0.30	0.32	0.38	0.50	0.67	0.94	1.3	1.9	2.7

Table 10.1 Errors $\|r_\gamma - L_n r_\gamma\|_\infty$ in Lagrange interpolation L_n of the Runge function for various values of γ and polynomial order n.

for ϕ_k (which is associated with the interpolation point $x_k = -1/k$), and its maximum appears to occur in the first interval $[x_1, x_2]$ (see exercise 10.15). Computing the Lagrange basis functions for various values of k and the maximum value ϕ_k in this interval yields lower bounds via (10.34), which grow exponentially (see exercise 10.16). In particular, for $k = 31$, $\|L_{2k+1}\|_\infty$ exceeds 10^{16} (we explain in section 11.3 how this operator norm can be computed with some confidence). This means that round-off errors in the data f could lead to order one errors in computing $L_n f$ for large n no matter how well-behaved f may be.

10.3.2 Data-dependence of Lagrange interpolation

In addition to the inherent instability of Lagrange interpolation for large n, there are also classes of functions that are not suitable for approximation by certain types of interpolation. There is a celebrated example of Runge[4] [86] based on interpolating the function $1/(1 + x^2)$ on various intervals. For simplicity, here we instead look at interpolating the function

$$r_\gamma(x) = 1/(1 + (\gamma x)^2) \tag{10.37}$$

on the fixed interval $[-1, 1]$.

The special feature of the Runge function r_γ is that it has a singularity in the complex plane at $z = \pm i/\gamma$ despite the fact that it is infinitely differentiable on the real line (where the interpolation is being done). As γ gets large, the singularity encroaches upon the domain $[-1, 1]$ of approximation. For $\gamma = 2$, the maximum error

$$\|r_\gamma - L_n r_\gamma\|_\infty \tag{10.38}$$

slowly decreases as a function of n for uniformly spaced points (see table 10.1). However, for $\gamma = 3$, the error (10.38) increases; figure 10.2 depicts the error with $n = 11$. We again see that the maximum error occurs in the first and last segments.

---

[4]Carl David Tolmé Runge (1856–1927) was a student of Weierstrass and Kummer, and he married the niece of Paul Du Bois-Reymond (see page 69). He was a close friend of, and student with, Max Planck, and he was a teacher and mentor of Max Born.

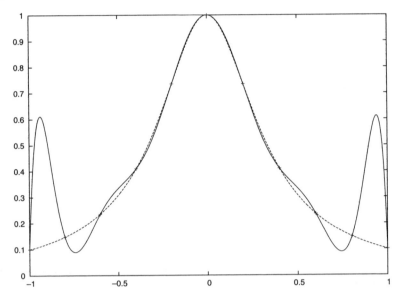

Figure 10.2 The Runge function r_3 (dashed line) and its Lagrange interpolant for 11 equally spaced interpolation points on the interval $[-1, 1]$. The "+" signs indicate the interpolation data.

10.4 MORE READING

Polynomial interpolation has been motivated by a variety of factors. With the advent of modern computing machines, the need to provide accurate approximations of various sorts [78, 113] includes interpolation [107] as a major technique.

The approach to the Newton form of interpolation was guided by [42], which should be consulted for information about other types of interpolation as well.

We have shown that Lagrange interpolation requires some thought if we want to compute stable approximations. See [83] regarding the influence of floating-point and stable ways of computing Lagrange interpolants.

10.5 EXERCISES

Exercise 10.1 *Prove (10.9). (Hint: explain why*

$$f(x) - L_n f(x) = C_x \omega_{n+1}(x), \tag{10.39}$$

where C_x depends on x but will be held fixed for the calculations. Use Rolle's theorem repeatedly to show that the function

$$\chi(y) = f(y) - L_n f(y) - C_x \omega_{n+1}(y) \tag{10.40}$$

has the property that $\chi^{(k)}$ vanishes at $n+2-k$ distinct points in an interval containing all the points $x_0, \ldots, x_n$ and x. Let ξ be the point where $\chi^{(n+1)}$ vanishes.)

Exercise 10.2 *Show that the Lagrange interpolation problem may be reduced to a linear system of equations with a Vandermonde matrix*

$$\begin{pmatrix} 1 & x_0 & x_0^2 & \cdots & x_0^n \\ 1 & x_1 & x_1^2 & \cdots & x_1^n \\ & & \cdots & & \\ 1 & x_n & x_n^2 & \cdots & x_n^n \end{pmatrix}. \tag{10.41}$$

Prove that a Vandermonde matrix is invertible if and only if the x_i's are distinct. (Hint: represent polynomials in terms of the standard monomial basis functions x^i.)

Exercise 10.3 *Prove that the error polynomials ω_n defined in (10.8) are linearly independent. (Hint: use induction. Note that you cannot have $\omega_n = \sum_{i=0}^{n-1} a_i \omega_i(x)$ because the right-hand side is a polynomial only of degree n.)*

Exercise 10.4 *Derive the expression*

$$p_2(x) = f[x_0] + f[x_0, x_1](x - x_0) + f[x_0, x_1, x_2](x - x_0)(x - x_1) \quad (10.42)$$

following the approach used to construct the representation for p_1.

Exercise 10.5 *Suppose that $f[x_0, x_1, x_2]$ is defined as a divided difference of two first-order divided differences, e.g.,*

$$f[x_0, x_1, x_2] = \frac{f[x_0, x_2] - f[x_0, x_1]}{x_2 - x_1}. \tag{10.43}$$

Show that

$$f[x_0, x_1, x_2] = f[x_{\sigma(0)}, x_{\sigma(1)}, x_{\sigma(2)}] \tag{10.44}$$

for any permutation σ of $\{0, 1, 2\}$.

Exercise 10.6 *Prove corollary 10.4. (Hint: use theorem 10.1 to show that*

$$f[x_0, x_1, \ldots, x_n] = \frac{f[x_1, \ldots, x_n] - f[x_0, \ldots, x_{n-1}]}{x_n - x_0} \tag{10.45}$$

and then apply (10.21).)

Exercise 10.7 *Show that $\| \cdot \|_\infty$ defined in (10.31) is a norm, that is, satisfies the properties (5.1)-(5.3).*

Exercise 10.8 *Show that the definition (10.32) yields the same number $\|L_n\|_\infty$ if we replace $C^0(I)$ by $\mathbb{R}^{n+1}$ as the domain of L_n:*

$$\|L_n\|_\infty = \sup_{f \in \mathbb{R}^{n+1}} \frac{\|L_n f\|_\infty}{\|f\|_\infty}, \tag{10.46}$$

where we use the alternative interpretation $L_n : \mathbb{R}^{n+1} \to \mathcal{P}_n$ defined in (10.4).

Exercise 10.9 *Give an example of a function f such that $L_n f = \phi_i$ and has the required properties that allow us to conclude the estimate (10.34).*

Exercise 10.10 *Prove (10.35).*

Exercise 10.11 *Prove that the determinant of a Vandermonde matrix as described in (10.41) satisfies*

$$\det \begin{pmatrix} 1 & x_0 & x_0^2 & \cdots & x_0^n \\ 1 & x_1 & x_1^2 & \cdots & x_1^n \\ & & \cdots & \\ 1 & x_n & x_n^2 & \cdots & x_n^n \end{pmatrix} = \prod_{0 \le i < j \le n} (x_j - x_i) \qquad (10.47)$$

and is thus invertible when all x_i's are distinct. (Hint: reduce this to a problem involving Lagrange interpolation by expanding the determinant along the last column.)

Exercise 10.12 *Consider Lagrange interpolation L_n based on distinct interpolation points x_i. Prove that, for any continuous function f,*

$$(L_n(xf) - xL_n f)(x) = (-1)^{n+1} f[x_0, x_1, \ldots, x_n] \omega_n(x). \qquad (10.48)$$

Here the commutator $L_n(xf) - xL_n f$ is defined as follows. We let $g(x) = xf(x)$ and $g_n(x) = x(L_n f(x))$. Then $(L_n(xf) - xL_n f)(x) = L_n g(x) - g_n$.

Exercise 10.13 *Determine a bound for the derivatives of the Runge function r_γ defined in (10.37) as a function of n and γ. Use this in (10.9) to give an estimate for the error in Lagrange interpolation.*

Exercise 10.14 *Suppose that f is a polynomial of degree $n + 1$. Prove that*

$$f[x_0, x_1, \ldots, x_n] = \frac{f^{(n+1)}}{(n+1)!}, \qquad (10.49)$$

where we note that $f^{(n+1)}$ is a constant. (Hint: see corollary 10.3.)

Exercise 10.15 *Let ϕ_i denote the ith basis function for Lagrange interpolation for $n = 2k + 1$ equally spaced interpolation points $x_i = -1 + (i-1)/k$ on $[-1, 1]$, $i = 1, \ldots, 2k + 1$. Verify computationally that the largest norm occurs for ϕ_k associated with the interpolation point $x_k = -1/k$ and that its maximum occurs in the first interval $[x_1, x_2]$.*

Exercise 10.16 *Let ϕ_i denote the ith basis function for Lagrange interpolation for $n = 2k + 1$ equally spaced interpolation points $x_i = -1 + (i-1)/k$ on $[-1, 1]$, $i = 1, \ldots, 2k + 1$. Verify computationally that*

$$M_k = \max_{i=1,\ldots,2k+1} \|\phi_i\|_\infty \qquad (10.50)$$

grows exponentially with k. In particular, verify that $M_{63} > 10^{16}$. (Hint: plot $(\log M_k)/k$.)

Exercise 10.17 *Verify table 10.1 computationally.*

Exercise 10.18 *Consider equally spaced points $x_i = i/n$ $(i = 0, \ldots, n)$ on $[0, 1]$ and consider $\omega_{n+1}(x) = \prod_{i=0}^{n}(x - x_i)$. Verify computationally that, for n odd, $|\omega_{n+1}(1/2n)| \approx e^{-an}$ and $|\omega_{n+1}(1/2)| \approx e^{-bn}$, where $a \approx 1.0$ and $b \approx 1.7$.*

Exercise 10.19 *Consider any distinct points $x_0, x_1, \ldots x_n$. Show that each Lagrange basis functions ϕ_i defined in (10.5) can be written as*

$$\phi_i(x) = \frac{c_i \omega_{n+1}(x)}{x - x_i} \qquad (10.51)$$

for some constant c_i, where $\omega_{n+1}(x) = (x - x_0)(x - x_1) \cdots (x - x_n)$. What is c_i?

10.6 SOLUTIONS

Solution of Exercise 10.4.

Suppose we add now x_2 and try to determine p_2 from p_1 in the way that we determined p_1 from p_0 and q_1. Then we have $p_2 = p_1 + q_2$, where q_2 is quadratic and must vanish at x_0, x_1 (to avoid spoiling the interpolation at those points). Thus

$$q_2(x) = a_2(x - x_0)(x - x_1) = a_2 \omega_2, \qquad (10.52)$$

and the value of a_2 will be determined by the requirement $p_2(x_2) = f(x_2)$:

$$f(x_2) = f(x_0) + f[x_0, x_1](x_2 - x_0) + a_2(x_2 - x_0)(x_2 - x_1). \qquad (10.53)$$

Rearranging terms in (10.53), we find

$$f(x_2) - f(x_0) = (x_2 - x_0)(f[x_0, x_1] + a_2(x_2 - x_1)), \qquad (10.54)$$

which says that

$$f[x_2, x_0] = f[x_0, x_1] + a_2(x_2 - x_1). \qquad (10.55)$$

Thus we have found that

$$a_2 = \frac{f[x_2, x_0] - f[x_0, x_1]}{x_2 - x_1}. \qquad (10.56)$$

The result then follows from theorem 10.1.

Solution of Exercise 10.7.
In the derivation of the properties of the norm, we have to calculate the supremum of the set $\{|f(x)| \mid x \in I\}$ for various functions f. By definition, $\|f\|_\infty \geq 0$ since it is the supremum of nonnegative numbers.

If $\|f\|_\infty = 0$, then f must be identically zero, since in this case we must have

$$\{|f(x)| \mid x \in I\} = \{0\}. \qquad (10.57)$$

Now let $f \in C^0(I)$ be arbitrary, let s be any scalar, and let $r = |s|$. Note that

$$|f(x)| \leq \|f\|_\infty \quad \forall x \in I. \tag{10.58}$$

Then

$$|sf(x)| = r|f(x)| \leq r\|f\|_\infty \tag{10.59}$$

for all $x \in I$. So $r\|f\|_\infty$ is an upper bound for $\{|sf(x)| \mid x \in I\}$. Thus

$$\|sf\|_\infty = \sup\{|sf(x)| \mid x \in I\} \leq r\|f\|_\infty. \tag{10.60}$$

If $r = 0$, then (10.60) implies that

$$\|sf\|_\infty = 0 = |s|\,\|f\|_\infty. \tag{10.61}$$

So now suppose $r > 0$. If b is an upper bound for $\{|sf(x)| \mid x \in I\}$, then

$$r|f(x)| = |sf(x)| \leq b \tag{10.62}$$

for all $x \in I$. Then $|f(x)| \leq b/r$ for all $x \in I$. Hence

$$\|f\|_\infty = \sup\{|f(x)| \mid x \in I\} \leq b/r. \tag{10.63}$$

Since this holds for all b, we find

$$r\|f\|_\infty \leq \sup\{|sf(x)| \mid x \in I\} = \|sf\|_\infty. \tag{10.64}$$

Combining (10.64) and (10.60), we conclude that $|s|\,\|f\|_\infty = \|sf\|_\infty$.

Now let us address the triangle inequality. By (10.58), we have

$$\begin{aligned}|(f+g)(x)| &= |f(x) + g(x)| \\ &\leq |f(x)| + |g(x)| \leq \|f\|_\infty + \|g\|_\infty \quad \forall x \in I.\end{aligned} \tag{10.65}$$

That is, $\|f\|_\infty + \|g\|_\infty$ is an upper bound for $\{|(f+g)(x)| \mid x \in I\}$. Thus

$$\|f+g\|_\infty = \sup\{|(f+g)(x)| \mid x \in I\} \leq \|f\|_\infty + \|g\|_\infty. \tag{10.66}$$

Solution of Exercise 10.11. For $n = 1$, we have

$$\det\begin{pmatrix} 1 & x_0 \\ 1 & x_1 \end{pmatrix} = x_1 - x_0 = \prod_{0 \leq i < j \leq 1}(x_j - x_i), \tag{10.67}$$

so we may proceed by induction. For $x \in \mathbb{R}^{n+1}$, define $d_n(x)$ via

$$d_n(x_0, x_1, \ldots, x_n) = \prod_{0 \leq i < j \leq n}(x_j - x_i). \tag{10.68}$$

Expanding the Vandermonde determinant around the last column, we have by induction that

$$(-1)^n \det\begin{pmatrix} 1 & x_0 & x_0^2 & \cdots & x_0^n \\ 1 & x_1 & x_1^2 & \cdots & x_1^n \\ \vdots & \vdots & \vdots & \vdots & \vdots \\ 1 & x_n & x_n^2 & \cdots & x_n^n \end{pmatrix} = \begin{cases} x_0^n d_{n-1}(x_1, x_2, \ldots, x_n) - \\ x_1^n d_{n-1}(x_0, x_2, \ldots, x_n) + \\ \qquad\qquad \vdots \\ (-1)^n x_n^n d_{n-1}(x_0, x_1, \ldots, x_{n-1}) \end{cases}$$

$$= \sum_{k=0}^n (-1)^k x_k^n d_{n-1}(\hat{x}^{(k)}), \tag{10.69}$$

where $\hat{x}^{(k)}$ denotes the point in $\mathbb{R}^n$ obtained by deleting the kth coordinate of x. Expanding the expression (10.68), we find

$$d_n(x_0, x_1, \ldots, x_n) = d_{n-1}(\hat{x}^{(k)}) \prod_{0 \leq j < k} (x_k - x_j) \prod_{k < j \leq n} (x_j - x_k). \quad (10.70)$$

Therefore,

$$d_{n-1}(\hat{x}^{(k)}) = \frac{d_n(x_0, x_1, \ldots, x_n)}{(-1)^k \prod_{0 \leq j \leq n, \, j \neq k}(x_j - x_k)}. \quad (10.71)$$

Thus (10.69) becomes

$$(-1)^n \det \begin{pmatrix} 1 & x_0 & x_0^2 & \cdots & x_0^n \\ 1 & x_1 & x_1^2 & \cdots & x_1^n \\ & & \cdots & & \\ 1 & x_n & x_n^2 & \cdots & x_n^n \end{pmatrix} \quad (10.72)$$

$$= d_n(x_0, x_1, \ldots, x_n) \sum_{k=0}^{n} \frac{x_k^n}{\prod_{0 \leq j \leq n, \, j \neq k}(x_j - x_k)}.$$

Thus we have reduced the problem to prove the (unlikely looking) expression

$$(-1)^n = \sum_{k=0}^{n} \frac{x_k^n}{\prod_{0 \leq j \leq n, \, j \neq k}(x_j - x_k)}. \quad (10.73)$$

The denominator of this expression is the same as in the Lagrange interpolation basis functions, so we can write

$$\sum_{k=0}^{n} \frac{x_k^n}{\prod_{0 \leq j \leq n, \, j \neq k}(x_j - x_k)} = \sum_{k=0}^{n} \frac{x_k^n \phi_k(x)}{\prod_{0 \leq j \leq n, \, j \neq k}(x_j - x)}$$

$$= (-1)^n \sum_{k=0}^{n} \frac{x_k^n \phi_k(x)}{\prod_{0 \leq j \leq n, \, j \neq k}(x - x_j)}$$

$$= (-1)^n \sum_{k=0}^{n} \frac{x_k^n \phi_k(x)(x - x_k)}{\prod_{0 \leq j \leq n}(x - x_j)} \quad (10.74)$$

$$= (-1)^n \omega_{n+1}(x)^{-1} \sum_{k=0}^{n} x_k^n \phi_k(x)(x - x_k)$$

for any $x \neq x_j$ for all $j = 0, \ldots, n$, by (10.8). Let $p(x) = x^{n+1}$ and $q(x) = x^n$. Then

$$\sum_{k=0}^{n} x_k^n \phi_k(x)(x - x_k) = x \left(\sum_{k=0}^{n} x_k^n \phi_k(x) \right) - \left(\sum_{k=0}^{n} x_k^{n+1} \phi_k(x) \right)$$

$$= x L_n q(x) - L_n p(x) = x q(x) - L_n p(x) \quad (10.75)$$

$$= p(x) - L_n p(x) = \omega_{n+1}(x),$$

as required. Note that (10.73) implies that

$$(-1)^n = \sum_{k=0}^{n} \frac{x_k^n}{\prod_{0 \leq j \leq n, \, j \neq k}(x_j - x_k)} = \sum_{k=0}^{n} \frac{1}{\prod_{0 \leq j \leq n, \, j \neq k}\left(\frac{x_j}{x_k} - 1\right)}$$

$$= \sum_{k=0}^{n} \prod_{0 \leq j \leq n, \, j \neq k} \left(\frac{x_j}{x_k} - 1\right)^{-1}. \quad (10.76)$$

Solution of Exercise 10.12. Using the definition of the Lagrange interpolation basis functions,

$$
\begin{aligned}
f[x_0, x_1, \ldots, x_n] &= \sum_{k=0}^{n} \frac{f(x_k)}{\prod_{j \neq k} (x_j - x_k)} \\
&= \sum_{k=0}^{n} \frac{f(x_k) \phi_k(x)(x_k - x)}{\prod_{j=0}^{n} (x_j - x)} \\
&= (-1)^{n+1} \omega_{n+1}(x)^{-1} \sum_{k=0}^{n} f(x_k) \phi_k(x)(x_k - x) \\
&= (-1)^{n+1} \omega_{n+1}(x)^{-1} \left(L_n(xf) - x L_n f \right)(x)
\end{aligned} \tag{10.77}
$$

for any $x \neq x_j$ for all j. This proves the result for $x \neq x_j$; for $x = x_j$, the result is true because

$$
x_j L_n f(x_j) = x_j f(x_j) = L_n(xf)(x_j). \tag{10.78}
$$

Chapter Eleven

Chebyshev and Hermite Interpolation

"This is a mathematical textbook rather than a compendium of computational rules. It is hoped that the material included will provide a useful background for those seeking to devise and evaluate routines for numerical computation." (Alston S. Householder in [84])

For equally spaced points, two examples have suggested that bad things happen at the ends of the interpolation interval: the basis functions are large there, and the approximation of the Runge function (10.37) can be as well. Also, the size of the error function ω_k defined in (10.8) is relatively larger (exercises 10.18 and 11.1) at the ends of the interval. The Chebyshev[1] points are a special set of Lagrange interpolation points that cluster at the ends of the interval:

$$x_j = x_{j,n} = -\cos\left(\frac{\pi(2j+1)}{2(n+1)}\right) = \cos\left(\frac{\pi(2(n-j)+1)}{2(n+1)}\right) \qquad (11.1)$$

for $j = 0, 1, \ldots, n$ (for interpolation on $[-1, 1]$). We include the extra subscript $x_{j,n}$ to be precise, as we will compare error terms for different values of n. The key feature is that these points are clustered around the ends of the interval: $x_1 - x_0 = \mathcal{O}(n^{-2})$, whereas in the middle of the interval, the spacing is $\mathcal{O}(n^{-1})$.

We will also consider more general forms of interpolation involving derivative and other information in addition to function values.

11.1 ERROR TERM ω

There is a simple interpretation of the Chebyshev points in terms of the error function ω_{n+1} defined in (10.8), as follows.

Theorem 11.1 *With the Chebyshev interpolation points (11.1), the error function ω_{n+1} defined in (10.8) for Lagrange interpolation satisfies*

$$\omega_{n+1}(x) = \begin{cases} 2^{-n}\cos((n+1)\cos^{-1} x) & \forall x \in [-1, 1] \\ 2^{-n-1}\left(\left(x + \sqrt{x^2 - 1}\right)^{n+1} + \left(x - \sqrt{x^2 - 1}\right)^{n+1}\right) & \forall x \in \mathbb{R}. \end{cases}$$
$$(11.2)$$

---

[1]Pafnuty Lvovich Chebyshev (1821–1894) "was probably the first mathematician to recognize the general concept of orthogonal polynomials" [140]. His students included Andrei Andreyevich Markov and Aleksandr Mikhailovich Lyapunov.

Moreover, the following three-term recursion relation holds:

$$\omega_{n+1}(x) = x\omega_n(x) - \tfrac{1}{4}\omega_{n-1}(x), \tag{11.3}$$

where $\omega_{n+1-i}(x) = \prod_{j=0}^{n-i}(x - x_{j,n-i})$ *for* $i = 0,1,2$ *and* $x_{j,n}$ *is defined in (11.1).*

Proof. The function $\cos((n+1)\cos^{-1}x)$ has the right roots:

$$\cos((n+1)\cos^{-1}x_j) = \cos((\pi/2)(2(n-j)+1)) = 0 \tag{11.4}$$

since $2(n-j)+1 = 1,3,5,\ldots$. It is somewhat remarkable that

$$\cos((n+1)\cos^{-1}x)$$

is a polynomial in x, of degree $n+1$, but let us suppose for the moment that it is true (cf. exercise 11.2). Since $\cos((n+1)\cos^{-1}x)$ and $\omega_{n+1}(x)$ have the same roots, they are a constant multiple of each other.

It is also surprising that square-roots appear in (11.2) in a representation of a polynomial. But for $n = 0$, the two square-root terms cancel, giving $\omega_1(x) = x$, as required. Also disconcerting is the fact that, for $|x| < 1$, the square-roots are complex numbers. However, expanding the terms in the second representation in (11.2) via the binomial theorem (exercise 11.3) shows that all the square-root terms cancel, leaving only polynomial terms:

$$\left(x + \sqrt{x^2 - 1}\right)^k + \left(x - \sqrt{x^2 - 1}\right)^k = 2 \sum_{0 \le j \le k/2} \binom{k}{2j}(x^2 - 1)^j x^{k-2j}. \tag{11.5}$$

Thus if we can identify the two expressions in (11.2), then we have shown that the cosine expression is a polynomial in x.

We use some facts about complex variables, namely, that

$$(\cos\theta + i\sin\theta)^k = \left(e^{i\theta}\right)^k = e^{ik\theta} = \cos k\theta + i\sin k\theta, \tag{11.6}$$

where $i = \sqrt{-1}$. The equality of the leftmost and rightmost terms in (11.6) is known as a theorem of De Moivre, which was used by Euler[2] to prove the equalities involving the complex exponential and trigonometric functions [54]. For $|x| \le 1$, write $x = \cos t$. Then

$$x \pm \sqrt{x^2 - 1} = \cos t \pm i\sqrt{1 - x^2} = \cos t \pm i\sin t = e^{\pm it}. \tag{11.7}$$

Combining (11.7) with (11.6), we find

$$\left(x + \sqrt{x^2 - 1}\right)^k + \left(x - \sqrt{x^2 - 1}\right)^k = e^{ikt} + e^{-ikt} = 2\cos kt, \tag{11.8}$$

which confirms that the two expressions in (11.2) are the same for $|x| \le 1$.

The expression (11.2) shows that the leading term of $\cos((n+1)t)$ is $2^n x^{n+1}$, and since the leading term of $\omega_{n+1}(x)$ is x^{n+1}, we must have $\cos((n+1)t) = 2^n \omega_{n+1}(x)$.

We leave the recursion relation (11.3) as exercise 11.4.   QED

---

[2]Leonhard Euler (1707–1783) lived in the same period as Benjamin Franklin (1706–1790) [125] and was said by Laplace to be "the master of us all" [54].

11.1.1 Chebyshev asymptotics

The formula (11.2) provides a complete description of the error polynomial $w_{n+1}(x)$ for all x. Between -1 and 1, the first representation shows that it just oscillates between $\pm 2^{-n}$ (cf. exercise 10.18). For x outside this interval, $w_{n+1}(x)$ increases in size dramatically. Fortunately, it is easy to describe its behavior precisely for large n based on the second representation in (11.2). Although the square-root terms cancel algebraically, they can be used to obtain a precise estimate of the growth.

To simplify notation, we switch subscripts and consider

$$w_n(x) = (x - x_{0,n-1})(x - x_{1,n-1}) \cdots (x - x_{n-1,n-1}) \qquad (11.9)$$

in the remainder of the section, where the Chebyshev points are given in (11.1).

Theorem 11.2 *Suppose that w_n is as in (11.9). For $\pm x > 1$,*

$$\lim_{n\to\infty} |w_n(x)|^{1/n} = \tfrac{1}{2}\left(\pm x + \sqrt{x^2 - 1}\right). \qquad (11.10)$$

Moreover,

$$|w_n(x)| > \left(\tfrac{1}{2}\left(\pm x + \sqrt{x^2 - 1}\right)\right)^n. \qquad (11.11)$$

We leave the proof as exercise 11.6. figure 11.1 depicts the ratio

$$\frac{w_n(x)}{\left(\tfrac{1}{2}\left(x + \sqrt{x^2 - 1}\right)\right)^n} \qquad (11.12)$$

for $n = 10, 20, 40, 80$ on the interval $[1, 1.001]$.

11.1.2 Application to CG

One way to describe the error term w_n is to say that it is the polynomial of degree n that deviates least from zero on the interval $[-1, 1]$ among all polynomials that are asymptotic to x^n at infinity. We can make this precise by turning (11.11) around. Suppose we want a polynomial p_n of degree n such that $p_n(\Lambda) = 1$, for some fixed $\Lambda > 1$, and p_n is small on $[-1, 1]$. Define

$$p_n(x) = w_n(x)/w_n(\Lambda). \qquad (11.13)$$

Then

$$\|p_n\|_{\infty,[-1,1]} = 2^{-n}/w_n(\Lambda) < \left(\Lambda + \sqrt{\Lambda^2 - 1}\right)^{-n}. \qquad (11.14)$$

Thus we see that even though $p_n(\Lambda) = 1$, p_n is exponentially small on the interval $[-1, 1]$. We now use this to prove lemma 9.11.

Recall that $0 < \lambda_1 < \lambda_2$ and define $b = \tfrac{1}{2}(\lambda_2 - \lambda_1)$ and

$$\Lambda = \frac{\lambda_1 + \lambda_2}{\lambda_2 - \lambda_1} = \frac{\lambda_1 + \lambda_2}{2b}. \qquad (11.15)$$

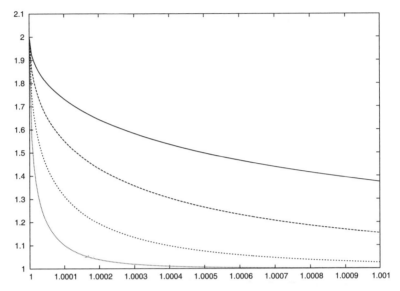

Figure 11.1 Plot of the ratio (11.12) for $x \in [1, 1.001]$ and $n = 10, 20, 40, 80$ (top to bottom).

Then we have

$$\lambda_1 = (\Lambda - 1)b \qquad \text{and} \qquad \lambda_2 = (\Lambda + 1)b. \qquad (11.16)$$

Define $q_n(x) = p_n(\Lambda - x/b)$. Then $q_n(0) = 1$, and

$$\|q_n\|_{\infty, [\lambda_1, \lambda_2]} = \|q_n\|_{\infty, [(\Lambda-1)b, (\Lambda+1)b]} = \|p_n\|_{\infty, [-1,1]}. \qquad (11.17)$$

Thus (11.14) implies

$$\|q_n\|_{\infty, [\lambda_1, \lambda_2]} \le \left(\Lambda + \sqrt{\Lambda^2 - 1}\right)^{-n}, \quad \text{where} \quad \Lambda = \frac{\lambda_1 + \lambda_2}{\lambda_2 - \lambda_1}. \qquad (11.18)$$

This completes the proof of lemma 9.11.

11.2 CHEBYSHEV BASIS FUNCTIONS

We can identify the jth basis function for the Chebyshev interpolation points (see exercise 10.19) as

$$\phi_j(x) = \phi_j(\cos t) = c'_j \frac{\omega_{n+1}(x)}{(x - x_j)} = \frac{c_j \cos((n+1)t)}{\cos t - \cos t_j}, \qquad (11.19)$$

where $t_j = \frac{\pi(2(n-j)+1)}{2(n+1)}$ and $c_j = c'_j 2^n$ is defined such that $\phi_j(x_j) = 1$. By l'Hopital's Rule, we have

$$c_j = \lim_{t \to t_j} \frac{\cos t - \cos t_j}{\cos((n+1)t)} = \frac{\sin t_j}{(n+1)\sin((n+1)t_j)} = \frac{(-1)^{n-j}\sin t_j}{(n+1)} \qquad (11.20)$$

because $(n+1)t_j = \frac{1}{2}\pi(2(n-j)+1) = (n+\frac{1}{2})\pi, (n-\frac{1}{2})\pi, \ldots, \frac{1}{2}\pi$.

Now let us develop some bounds for how big the Chebyshev basis functions can get. For $|\cos t - \cos t_j| \geq |c_j|$, we have $|\phi_j(t)| \leq 1$. Thus we need to estimate $|\phi_j(t)|$ for t only near t_j. For the numerator of $|\phi_j(t)|$, we see

$$|\cos((n+1)t)| = |\cos((n+1)t) - \cos((n+1)t_j)|$$
$$= \left| \int_{(n+1)t_j}^{(n+1)t} \sin x \, dx \right| \leq (n+1)|t_j - t| \tag{11.21}$$

for any t. Similarly, if $t_j < t \leq \frac{1}{2}\pi$, then the denominator of $|\phi_j(t)|$ can be estimated by

$$|\cos(t) - \cos(t_j)| = \int_{t_j}^{t} \sin x \, dx \geq |t_j - t| \sin t_j. \tag{11.22}$$

Combining (11.22) and (11.21), we see that $|\phi_i(x)| \leq 1$ for $x_j < x \leq 0$ by (11.20).

Unfortunately, the maximum of ϕ_j does not occur at x_j, so we have $|\phi_i(x)| > 1$ for $x < x_j$. Thus the norm of the Chebyshev interpolation operator is greater than 1. Rather than trying to estimate the basis functions in more detail, we approach the problem more generally.

11.3 LEBESGUE FUNCTION

We have seen by examples that the norm of the Lagrange interpolation operator can be challenging to compute and that its size is strongly dependent on the choice of interpolation points. We now consider the issue in more depth. The norm of the Lagrange interpolation operator is a double supremum, and it is useful to take the suprema one at a time. The *Lebesgue function* is defined to be

$$\lambda_n(x) = \sup \left\{ |(L_n f)(x)| \mid \|f\|_\infty = 1 \right\}. \tag{11.23}$$

In figure 11.2, a typical case with Chebyshev points is presented (cf. exercise 11.9). Note that the maximum value occurs at the ends of the interval [147].

The *Lebesgue constant* is the name for the norm of the Lagrange interpolation operator, and we see (exercise 11.8) that

$$\|L_n\|_\infty = \|\lambda_n\|_\infty. \tag{11.24}$$

Fortunately, the Lebesgue function is easy to compute:

Theorem 11.3 *For any set of interpolation points* $x_0, \ldots, x_n$ *and corresponding Lagrange basis functions* ϕ_j, *the Lebesgue function (11.23) satisfies*

$$\lambda_n(x) = \sum_{j=0}^{n} |\phi_j(x)| \tag{11.25}$$

for all x.

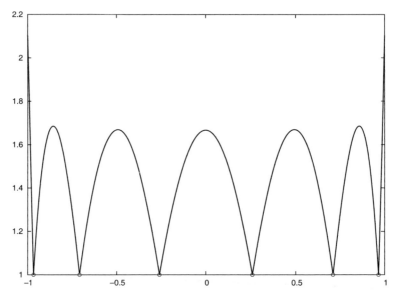

Figure 11.2 The Lebesgue function for Chebyshev interpolation points on the interval $[-1, 1]$. The circles indicate the interpolation points where the Lebesgue function has the value 1.

As a result of (11.25), we see that $\lambda_n(x_j) = 1$ for all interpolation points x_j; cf. figure 11.2.

Proof. For any $x \in I$, there is an $f \in C^0(I)$ such that

$$
\begin{aligned}
|L_n f(x)| &= \left| \sum_{j=0}^{n} f(x_j) \phi_j(x) \right| \\
&= \sum_{j=0}^{n} |\phi_j(x)|,
\end{aligned}
\tag{11.26}
$$

by taking $f(x_j) = \operatorname{sign}(\phi_j(x))$. For example, we can take f to be the piecewise linear function with these values at the interpolation points, and we thus have $\|f\|_\infty = 1$. This proves that $\lambda(x) \geq \sum_{j=0}^{n} |\phi_j(x)|$. The reverse inequality follows because

$$
\begin{aligned}
|L_n f(x)| &= \left| \sum_{j=0}^{n} f(x_j) \phi_j(x) \right| \\
&\leq \|f\|_\infty \sum_{j=0}^{n} |\phi_j(x)|
\end{aligned}
\tag{11.27}
$$

for any $f \in C^0(I)$.   QED

It is beyond our scope to compute analytical expressions for the Lebesgue constants, but for completeness we report on what is known. For equally spaced points, the behavior is exponential:

$$\|L_n\|_\infty \approx \frac{2^n}{en \log n},$$ (11.28)

and for the Chebyshev points,

$$\|L_n\|_\infty \approx \tfrac{1}{2}\pi \log n$$ (11.29)

(see [147] for references). We might think that the logarithmic growth in the Lebesgue constant for Chebyshev points could be improved by a better choice of interpolation points (cf. exercise 11.13). In fact, it was proved by Erdös[3] [57] that for all choices of n interpolation points on $[-1, 1]$,

$$\|L_n\|_\infty \geq (\tfrac{1}{2}\pi \log n) - E \quad \forall n$$ (11.30)

for some constant E (see exercise 11.14).

11.4 GENERALIZED INTERPOLATION

Suppose that we have data associated not just with function values but also with other quantities, such as derivatives, that we want to use in an interpolation scheme. An example of this is the Hermite[4] interpolation scheme which involves the value and derivative at each end of an interval $[a, b]$. These four parameters determine uniquely a cubic polynomial, but it is less simple to write down the basis functions than it was for Lagrange interpolation. Moreover, we might be interested in other interpolation data, such as the integral over the interval. We will show that quadratics are determined uniquely to interpolate a function at a and b and to match the integral over $[a, b]$.

Rather than approaching each of these interpolation problems in an ad hoc manner, we develop a systematic approach. The key concept that we use is that of a *linear functional* (or *linear form*) defined on a vector space V. Suppose that V is a vector space over the scalar field $\mathbb{F}$. Then a linear functional $\mathcal{L}$ is a function $\mathcal{L} : V \to \mathbb{F}$ such that

$$\mathcal{L}(v + \alpha w) - \mathcal{L}(v) + \alpha \mathcal{L}(w)$$ (11.31)

for all $v, w \in V$ and $\alpha \in \mathbb{F}$. For example, if $V = C^0(I)$, we can define $\mathcal{L}_x(v) = v(x)$ for a given $x \in I$. Note that we can write the conditions for the Lagrange interpolant P of a given function f as $\mathcal{L}_{x_i} P = \mathcal{L}_{x_i} f$ for all interpolation points x_i.

We assume in general that there are n linear functionals $\mathcal{L}_j$ for which we want to enforce interpolation, that is, $\mathcal{L}_j p = \mathcal{L}_j f$ to determine a polynomial

---

[3]Paul Erdös (1913–1996) was one of the most prolific mathematicians of all time and a proponent of the importance of beauty in proofs. He was a student of Leopold Fejér (see page 220).

[4]Charles Hermite (1822–1901) was, among many distinguished appointments, Maître de Conférence at École Polytechnique, despite having had a low score on the entrance exam as a student and ultimately leaving before graduating. His name is also commemorated in the name "Hermitian matrix."

p of degree $n-1$. For Lagrange interpolation, $\mathcal{L}_j f = f(x_j)$ for all j, but with Hermite $\mathcal{L}_j f = f'(x_j)$ for some j. And similarly we can define

$$\mathcal{L}f = \int_a^b f(x)\,dx \tag{11.32}$$

as well as an infinite variety of other functionals.

11.4.1 Existence of interpolant

The existence of the generalized interpolant can be proved by constructing polynomials $\phi_i \in \mathcal{P}_n$ such that

$$\mathcal{L}_j \phi_i = \delta_{ij}. \tag{11.33}$$

Then the generalized interpolant is defined by

$$G_n f(x) = \sum_{i=0}^n (\mathcal{L}_i f)\phi_i. \tag{11.34}$$

We can again also think of G_n being defined on data $f \in \mathbb{R}^n$ by

$$G_n f(x) = \sum_{i=0}^n f_i \phi_i. \tag{11.35}$$

As in the case of the Lagrange interpolant, G_n is a projection, and error estimates can be determined once a bound for the operator norm $\|G_n\|$ is determined.

The key issue is to have a simple condition that is equivalent to (11.33).

Definition 11.4 *A set of linear functionals* $\{\mathcal{L}_0, \mathcal{L}_1, \ldots, \mathcal{L}_n\}$ *uniquely determines* $\mathcal{P}_n$ *if* $\mathcal{L}_j p = 0$ *for all* $j = 0, \ldots, n$ *implies* $p \equiv 0$ *for any* $p \in \mathcal{P}_n$.

We recall that the kernel of a linear function is the set where it vanishes:

$$\ker \mathcal{L} = \{p \in \mathcal{P}_n \mid \mathcal{L}p = 0\} \tag{11.36}$$

in our case. Then we can say that $\mathcal{L}_0, \ldots, \mathcal{L}_n$ uniquely determines $\mathcal{P}_n$ iff

$$\cap_{i=0}^n \ker \mathcal{L}_j = \emptyset. \tag{11.37}$$

Another term often used is *unisolvent*; we say $\{\mathcal{L}_0, \ldots, \mathcal{L}_n\}$ is unisolvent on $\mathcal{P}_n$ iff (11.37) holds.

Lemma 11.5 *There is a basis of* $\mathcal{P}_n$ *satisfying (11.33) if and only if the set of linear functionals* $\{\mathcal{L}_0, \mathcal{L}_1, \ldots, \mathcal{L}_n\}$ *uniquely determines* $\mathcal{P}_n$.

Proof. Define a matrix A by $a_{ij} = \mathcal{L}_i(x^j)$. Then if $p(x) = \sum_{i=0}^n c_j x^j$, we have $(AC)_i = \mathcal{L}_i p$. Thus $\mathcal{L}_j p = 0$ for all $j = 0, \ldots, n$ if and only if $AC = 0$. Therefore, $\mathcal{L}_0, \ldots, \mathcal{L}_n$ uniquely determines $\mathcal{P}_n$ if and only if $AC = 0$ implies $C = 0$, which in turn is equivalent to A being invertible. The condition (11.33) means that we need, for each i, a vector C of coefficients such that

$AC = D$, where $D_j = \delta_{ij}$. But this is just another condition equivalent to the invertibility of A. Thus we have shown that both conditions are equivalent to the invertibility of A and hence are equivalent to each other. QED

We note that lemma 11.5 provides the basis for constructing multidimensional approximations of a very general type [21]. The space $\mathcal{P}_n$ can be essentially any space of functions of dimension $n + 1$ for which the linear functionals $\mathcal{L}_j$ are defined.

11.4.2 Applications

We now see how the (easy half of the) fundamental theorem of algebra guarantees the existence of the Lagrange basis functions. In view of lemma 11.5, all we need to show is that any polynomial of degree n that vanishes at $n+1$ distinct points is identically zero, and this is part of what the fundamental theorem of algebra states (cf. exercise 9.21).

But now we can use the same idea with more complicated forms of interpolation. In particular, we can again use the fundamental theorem of algebra to guarantee the existence of the Hermite basis functions simply by observing that we count zeros with multiplicity. That is, a cubic function that vanishes to second-order at two distinct points must be identically zero. In view of lemma 11.5, there are cubic polynomials ϕ_i such that $\phi_{ij}^{(k)}(\ell) = \delta_{ik}\delta_{j\ell}$. This notation is suitable for the Hermite cubic basis on $[0, 1]$, but the approach is the same for any interval.

Finally, consider quadratic polynomials p that vanish at a, b and have

$$\int_a^b p(x)\, dx = 0. \tag{11.38}$$

We must have

$$p(x) = c(a - x)(b - x) \tag{11.39}$$

because p vanishes at a and b. But since $\int_0^1 x(1 - x)\, dx \neq 0$, we must have $c = 0$ (cf. exercise 11.16). Thus we have an interpolation scheme for quadratics based on the values at the endpoints and the integral over the interval.

Lemma 11.5 does not always produce a positive result. It can also be used to show that a particular interpolation scheme will not work. For example, it is easy to see that we cannot use both the integral over an interval and the value at the midpoint to determine a linear interpolant. Both of these linear functionals vanish on the linear function that is zero at the midpoint.

To demonstrate the power of the abstract approach, we consider an interpolation problem related to the Euler-Maclaurin formula (see section 13.3.4). This is also a good example of the general Birkhoff[5] interpolation problem [15, 112].

---

[5]George David Birkhoff (1884–1944) was born in Overisel Township, Michigan, and was a graduate student at the University of Chicago where he worked with E. H. Moore.

Lemma 11.6 *Let k be a positive integer. Suppose that p is a polynomial of degree $2k + 1$ that vanishes at -1 and $+1$, together with its odd-order derivatives up through order $2k - 1$. That is, $p(\pm 1) = 0$ and*

$$p^{(2i-1)}(\pm 1) = 0 \qquad\qquad (11.40)$$

for $i = 1, \ldots, k$. Then $p \equiv 0$.

Proof. For $k = 1$, this is just Hermite interpolation. But for larger k we are missing even-order derivative information, so we cannot apply the fundamental theorem of algebra. However, a simple application of Rolle's theorem provides the missing information. Since $p(\pm 1) = 0$ as well, there is a point $-1 < \xi_1 < 1$ such that $p^{(1)}(\xi_1) = 0$. Since $p^{(1)}(\pm 1) = 0$, there must be two points $-1 < \mu_1^- < \mu_1^+ < 1$ such that $p^{(2)}(\mu_1^\pm) = 0$ (in particular, $\mu_1^- < \xi_1 < \mu_1^+$). Thus there is a point $-1 < \xi_2 < 1$ such that $p^{(3)}(\xi_2) = 0$. Since $p^{(3)}(\pm 1) = 0$, we conclude that there must be two points $-1 < \mu_2^- < \mu_2^+ < 1$ such that $p^{(4)}(\mu_2^\pm) = 0$. Continuing in this way, we find ultimately that there is a point $-1 < \xi_k < 1$ such that $p^{(2k-1)}(\xi_k) = 0$. Also since $p^{(2k-1)}(\pm 1) = 0$ and since $p^{(2k-1)}$ is a polynomial of degree at most 2, $p^{(2k-1)} \equiv 0$. It is sometimes useful to give a particular type of argument a name so that we can refer to it without repeating all the steps as we do, e.g., with a proof involving a "telescoping series." Thus we suggest that the previous argument is one in which we "Rolle up" the roots of derivatives of the polynomial p.

Now let us argue by induction. We have already demonstrated the case $k = 1$ since this is just Hermite interpolation. Thus suppose that we have demonstrated the lemma for $j = k - 1 \geq 1$ and that we now want to verify it for k. So consider a polynomial p of degree $2k + 1$ satisfying the conditions of the lemma. Our argument in which we "Rolle up" the roots of derivatives of p allows us to assert that the degree of p is at most $2k - 2$ since $p^{(2k-1)} \equiv 0$. But the data for the lemma for k include the data for $j = k - 1$, and since the degree of p is $\leq 2k - 2 < 2j + 1$, then $p \equiv 0$ by the induction hypothesis. QED

11.4.3 Numerical differentiation

Suppose we are given data and we want to compute the derivative of a function that the data represent. We can use interpolation as a general paradigm to compute any linear operator defined on functions. Suppose that D is a linear functional defined on $C^k(I)$, that is, a linear mapping $D : C^k(I) \to \mathbb{R}$ (cf. (11.31)). Suppose as well that we have a favorite interpolation scheme (Lagrange, Hermite, etc.) L_n that takes data $f \in \mathbb{R}^n$ and produces a polynomial $p = L_n f$ that represents the data f in some way. Then for data $f \in \mathbb{R}^n$, we define

$$D_n f = D(L_n f). \qquad\qquad (11.41)$$

Suppose now, by abuse of notation, we write $L_n f$ for a function $f \in C^k(I)$, where we mean that the data $f \in \mathbb{R}^n$ used to define L_n are taken from the function f in (11.34). Then we can compare the resulting approximation $D_n f$ with the exact Df:

$$Df - D_n f = Df - D(L_n f) = D(f - L_n f). \tag{11.42}$$

Thus the error $D - D_n$ is just D applied to the error in interpolation. So far, we have not considered the error in derivatives for interpolation, but this can be done in a fashion similar to what we did for the function values for interpolation error.

For $Df = f^{(k)}(x)$ for some $x \in I$, $D_n f$ provides an approximation to $f^{(k)}(x)$. Let us consider some examples. Suppose that $k = 1$ and $I = [0, 1]$. Suppose that L_n represents linear (Lagrange) interpolation at the points $x_0 = 0$ and $x_1 = 1$. Then $p = L_n f$ satisfies (cf. section 10.2.3)

$$p' = f(x_1) - f(x_0) = f[x_0, x_1], \tag{11.43}$$

and we note that p' is constant. Thus if $Df = f'(x)$, we have $D_n f = f(x_1) - f(x_0) = f[x_0, x_1]$ no matter which $x \in I$ we choose. On the other hand, if we take L_n to represent quadratic interpolation at $x_0 = 0$, $x_{1/2} = \frac{1}{2}$, and $x_1 = 1$, then $p = L_n f$ is quadratic and p' is a linear function. We can use the Newton divided difference representation (10.28) to determine the form of p':

$$p' = f[x_0, x_1]\omega_1' + f[x_0, x_1, x_{1/2}]\omega_2', \tag{11.44}$$

where $\omega_1(x) = (x - x_0) = x$ and $\omega_2(x) = (x - x_0)(x - x_1) = x(x - 1)$. Thus if we define $Df = f'(\frac{1}{2})$, then

$$f'(\tfrac{1}{2}) \approx D_n f = f[x_0, x_1] = f(x_1) - f(x_0) \tag{11.45}$$

since $\omega_2'(\frac{1}{2}) = 0$. Thus we get the same approximation for $f'(\frac{1}{2})$ using both linear and quadratic interpolants.

On the other hand, if we define $Df = f'(0)$, then we get a more complex expression for $D_n f$:

$$
\begin{aligned}
f'(0) &\approx D_n f \\
&= f[x_0, x_1]\omega_1'(0) + f[x_0, x_1, x_{1/2}]\omega_2'(0) = f[x_0, x_1] - f[x_0, x_1, x_{1/2}] \\
&= f[x_0, x_1] - (f[x_0, x_1] - f[x_0, x_{1/2}])/(x_1 - x_{1/2}) \qquad \text{[by (10.28)]} \\
&= f[x_0, x_1] - 2(f[x_0, x_1] - f[x_0, x_{1/2}]) = -f[x_0, x_1] + 2f[x_0, x_{1/2}] \\
&= -(f(x_1) - f(x_0)) + 2(f(x_{1/2}) - f(x_0)) \\
&= -3f(x_0) + 4f(x_{1/2}) - f(x_1).
\end{aligned}
\tag{11.46}
$$

Correspondingly, we see (e.g., by antisymmetry) that

$$f'(1) \approx f(x_0) - 4f(x_{1/2}) + 3f(x_1). \tag{11.47}$$

11.5 MORE READING

Applications of Chebyshev polynomials to solving differential equations is the topic of [18]. For information on the history and current research on Chebyshev polynomials, see [135]. For more detailed information about Lebesgue constants, see [147]. The Newton approach to interpolation can be applied to Hermite and other types of interpolation [42].

11.6 EXERCISES

Exercise 11.1 *Plot the error function w_n defined in (10.8) for points that are equally spaced (10.36) and for Chebyshev points.*

Exercise 11.2 *Prove the trigonometric identities*
$$\cos((n+1)t) = 2(\cos t)(\cos nt) - \cos((n-1)t). \qquad (11.48)$$
(Hint: use (11.6) to express $\cos kt = \frac{1}{2}(e^{ikt} + e^{-ikt})$ and then use this to expand the product $(\cos t)(\cos nt)$.)

Exercise 11.3 *Use the binomial theorem to expand*
$$(a+b)^k + (a-b)^k = \left(x + \sqrt{x^2-1}\right)^k + \left(x - \sqrt{x^2-1}\right)^k$$
and show that the square-roots disappear; then verify (11.5). (Hint: odd powers of the square-root terms cancel.)

Exercise 11.4 *Prove that the recursion relation (11.3) holds. (Hint: use (11.48). Note that $x = \cos t$ and interpret the other expressions in terms of the w's as functions of x.)*

Exercise 11.5 *The functions $T_k(x) = 2^{k-1}w_k(x)$ are called Chebyshev (often spelled Tchebyshev) polynomials when w_k is the error function corresponding to the Chebyshev points. Prove that these functions satisfy the following three-term recursion relation: $T_{k+1}(x) = 2xT_k(x) - T_{k-1}(x)$.*

Exercise 11.6 *Prove theorem 11.2. (Hint: use the second relation in (11.2) and observe that $\pm(x \pm \sqrt{x^2-1}) > 1$ and $1 > \pm(x \mp \sqrt{x^2-1}) > 0$ for $\pm x > 1$.)*

Exercise 11.7 *Prove that the functions $w_k(x)$ (and $T_k(x)$, cf. exercise 11.5) are orthogonal in the sense that*
$$\int_{-1}^{1} w_j(x)w_k(x)\sqrt{1-x^2}\,dx = 0 \qquad (11.49)$$
if $j \neq k$. (Hint: introduce the change of variables $x = \cos t$.)

Exercise 11.8 *Prove (11.24). (Hint: observe that*
$$\|L_n\|_\infty = \sup\left\{|(L_nf)(x)| \mid x \in I,\ \|f\|_\infty = 1\right\}, \qquad (11.50)$$
where I is the interval of interpolation.)

Exercise 11.9 *Verify figure 11.2 computationally.*

Exercise 11.10 *Let $x_1, \ldots, x_n$ be the Chebyshev points for $[-1,1]$. Consider the stretched set of points defined by*
$$\hat{x}_j = \left(\lambda + \frac{1-\lambda}{|x_1|}\right) x_j,$$
where $\lambda \in [0,1]$. When $\lambda = 1$, these are the regular Chebyshev points, and when $\lambda = 0$, we have $\hat{x}_1 = -1$ and $\hat{x}_n = 1$ [147]. Plot the Lebesgue function for various values of n and λ. Does $\lambda = \frac{1}{2}$ minimize the maximum norm of the Lebesgue function?

Exercise 11.11 *Suppose that $f \in C^0(I)$. Prove that there is an $x \in I$ such that $|f(x)| = \|f\|_{\infty,I}$.*

Exercise 11.12 *Consider mesh points where $x_1 = -1 + h^2$ and $x_i - x_{i-1} = h^2(1+h)^i$ for $i = 2, \ldots n-1$, where $n \approx \log(1 + 1/h)/\log(1+h)$. Reflect this set of points around the origin and add the origin as well. Compare polynomial interpolation on this set of points with the Chebyshev points.*

Exercise 11.13 *The Chebyshev points are not the optimal points for reducing the size of the Lebesgue function. The optimal points have the property that the Lebesgue function equi-oscillates in each subinterval [27, 43, 96], that is, between each pair of interpolation points x_i and x_{i+1}, there is a point ξ_i where $\lambda(\xi_i) = \|\lambda\|_\infty$ (at the ends, $\lambda(\pm 1) = \|\lambda\|_\infty$ as well). Determine the set of points x_i which minimize $\|\lambda\|_\infty$ for a given value of n.*

Exercise 11.14 *The estimate (11.30) implies that there is a constant E such that*

$$E \geq (\tfrac{1}{2}\pi \log n) - \|L_n\|_\infty \quad \forall n \tag{11.51}$$

so that we can define

$$E_0 = \sup \left\{ (\tfrac{1}{2}\pi \log n) - \|L_n\|_\infty \mid n \geq 1 \right\}. \tag{11.52}$$

Investigate the value of E_0. (By analogy with (18.6), we could call E_0 the Erdös number.)

Exercise 11.15 *The error polynomial ω_{n+1} for Chebyshev interpolation oscillates between ± 1, and in particular $|\omega_{n+1}(\pm 1)| = 1$. What is the derivative of ω_{n+1} at ± 1?*

Exercise 11.16 *Explain why (11.38) implies that $c = 0$ for the polynomial (11.39).*

Exercise 11.17 *Define $Df = f''(0)$ for $f \in C^2([-1,1])$. Let L_2 denote Lagrange interpolation by quadratics at $-1, 0, 1$. Determine the corresponding formula for $D_2 f = DL_2 f$. (Hint: see section 11.4.3.)*

Exercise 11.18 *Consider the error in Chebyshev approximation on a general interval $[a,b]$. Let $\hat{x}_i$ denote the Chebyshev points for $[-1,1]$ defined in (11.1) and define points x_i in $[a,b]$ by $x_i = a + \tfrac{1}{2}(b-a)(1+\hat{x}_i)$. Prove that*

$$\|f - L_n f\|_{\infty,[a,b]} \leq \frac{2}{(n+1)!} \left(\frac{b-a}{4} \right)^{n+1} \|f^{(n+1)}\|_{\infty,[a,b]}, \tag{11.53}$$

where L_n denotes Lagrange interpolation at the points x_i. (Hint: set $h = \tfrac{1}{2}(b-a)$ and define $\hat{f}(\hat{x}) = f(a+h(1+\hat{x}))$. Compare Lagrange interpolation on $[-1,1]$ with points $\hat{x}_i$, and on $[a,b]$ with points x_i. Observe that $\hat{f}^{(k)}(\hat{x}) = h^k f^{(k)}(a+h(1+\hat{x}))$.)

Exercise 11.19 *The Chebyshev polynomials T_n as defined in exercise 11.5 satisfy*

$$T_n(x) = \tfrac{1}{2}\left(\left(x + \sqrt{x^2 - 1}\right)^n + \left(x - \sqrt{x^2 - 1}\right)^n\right), \qquad (11.54)$$

in view of theorem 11.1. The Chebyshev polynomials of the second kind are denoted by $U_n(x)$ and can be defined by

$$U_n(x) = \frac{1}{2\sqrt{x^2 - 1}}\left(\left(x + \sqrt{x^2 - 1}\right)^{n+1} - \left(x - \sqrt{x^2 - 1}\right)^{n+1}\right). \quad (11.55)$$

Prove that $U_0(x) = 1$ and $U_1(x) = 2x$ and in general that

$$U_n(x) = \sum_{0 \le j \le n/2} \binom{n+1}{2j+1} (x^2 - 1)^j x^{n-2j} \qquad (11.56)$$

for all $n \ge 0$.

Exercise 11.20 *The Chebyshev polynomials of the second kind are defined in exercise 11.19. Prove the recursion relation*

$$U_{k+1}(x) = 2xU_k(x) - U_{k-1}(x), \ k \ge 1. \qquad (11.57)$$

Exercise 11.21 *The Chebyshev polynomials of the second kind are defined in exercise 11.54. Prove that $T'_n = nU_{n-1}$ for all $n \ge 1$, where T_n denotes the Chebyshev polynomial of the first kind defined in exercise 11.5.*

11.7 SOLUTIONS

Solution of Exercise 11.2. For $n = 0$, we have (note that $x_0 = 0$ in this case)

$$\cos((n + 1)\cos^{-1} x) = \cos(\cos^{-1} x) = x = x - x_0 = w_1(x). \qquad (11.58)$$

Let $t = \cos^{-1} x$. Then by a trigonometric identity,

$$\cos 2t = (2\cos^2 t) - 1 = 2x^2 - 1. \qquad (11.59)$$

Thus we know that $\cos 2t = \cos(2\cos^{-1} x)$ and $w_2(x)$ are both polynomials of degree 2 with the same roots. Matching the terms of order x^2, we find that $\cos 2t = 2w_2(x)$. This covers the case $n = 1$.

We can prove (11.48) by complex analysis; cf. (11.6):

$$\cos((n \pm 1)t) = \mathcal{R}e\, e^{(n\pm1)ti} = \mathcal{R}e\left(e^{nti}e^{\pm ti}\right)$$
$$= (\cos nt)\cos t \mp (\sin nt)\sin t, \qquad (11.60)$$

where $i = \sqrt{-1}$ and $\mathcal{R}e\, z$ denotes the real part of z. Then (11.48) follows by adding the plus and minus versions of (11.60). Similarly,

$$\sin((n \pm 1)t) = \mathcal{I}m\, e^{(n\pm1)ti} = \mathcal{I}m\left(e^{nti}e^{\pm ti}\right)$$
$$= (\sin nt)\cos t \pm (\cos nt)\sin t, \qquad (11.61)$$

where $\mathcal{I}m\,z$ denotes the imaginary part of z.

Using the induction hypothesis (11.48) translates to

$$\cos((n+1)t) = 2^n x w_n(x) - 2^{n-2} w_{n-1}(x). \tag{11.62}$$

This proves that $\cos((n+1)t) = \cos((n+1)\cos^{-1}x)$ is a polynomial in x of degree $n+1$.

Solution of Exercise 11.11. By definition, there is a sequence of points $x_i \in I$ such that $|f(x_i)| > \|f\|_{\infty,I} - 1/i$. Since I is closed, there is an accumulation point for the points x_i; that is, there is a subsequence x_{i_ℓ} such that $x_{i_\ell} \to x$ as $\ell \to \infty$. Since f is continuous, we have $f(x_{i_\ell}) \to f(x)$. Of course, we must have $|f(x)| \le \|f\|_{\infty,I}$. Thus

$$0 \le \|f\|_{\infty,I} - |f(x)| < |f(x_{i_\ell})| - |f(x)| + 1/i_\ell. \tag{11.63}$$

Letting $\ell \to \infty$ completes the proof since the right-hand side of (11.63) goes to zero.

Solution of Exercise 11.15. We can differentiate the expression (11.2) to get

$$\begin{aligned}
\frac{d}{dx}w_{n+1}(x) &= 2^{-n}\frac{d}{dx}\cos((n+1)\cos^{-1}x) \\
&= -2^{-n}(n+1)\left(\frac{d}{dx}\cos^{-1}x\right)\sin((n+1)\cos^{-1}x) \\
&= 2^{-n}(n+1)\frac{\sin((n+1)\cos^{-1}x)}{\sin(\cos^{-1}x)}.
\end{aligned} \tag{11.64}$$

Writing $x = \cos t$, we have

$$w'_{n+1}(\cos t) = 2^{-n}(n+1)\frac{\sin((n+1)t)}{\sin(t)}. \tag{11.65}$$

Therefore, l'Hôpital's rule implies that

$$\begin{aligned}
w'_{n+1}(1) &= 2^{-n}(n+1)\lim_{t\to 0}\frac{\sin((n+1)t)}{\sin(t)} \\
&= 2^{-n}(n+1)\lim_{t\to 0}\frac{(n+1)\cos((n+1)t)}{\cos(t)} \\
&= 2^{-n}(n+1)^2.
\end{aligned} \tag{11.66}$$

By symmetry, $w'_{n+1}(-1) = (-1)^n 2^{-n}(n+1)^2$.

Solution of Exercise 11.19. By definition, we have

$$U_n(x) = \frac{1}{2w(x)}\left((x+w(x))^{n+1} - (x-w(x))^{n+1}\right), \tag{11.67}$$

where we have substituted $w(x) = \sqrt{x^2-1}$ for simplicity. Using the binomial expansion, we have

$$
\begin{aligned}
U_n(x) &= \frac{1}{2w} \left(\sum_{j=0}^{n+1} \binom{n+1}{j} x^{n+1-j} w^j \right. \\
&\qquad \left. - \sum_{j=0}^{n+1} \binom{n+1}{j} x^{n+1-j} (-w)^j \right) \\
&= \frac{1}{2w} \sum_{j=0}^{n+1} \binom{n+1}{j} x^{n+1-j} w^j \left(1 - (-1)^j \right) \\
&= \frac{1}{w} \left(\sum_{0 \le \ell \le n/2} \binom{n+1}{2\ell+1} x^{n-2\ell} w^{2\ell+1} \right),
\end{aligned}
\tag{11.68}
$$

where we used the fact that $\left(1 - (-1)^j\right)$ is 0 for even j (and 2 for odd j) and made the substitution $j = 2\ell + 1$ for odd j. The limits on the summation are verified as follows: $0 \le j$ for j odd implies that $j \ge 1$, and so $\ell \ge 0$; for the upper limit, $j \le n+1$ iff $2\ell + 1 \le n+1$ iff $\ell \le n/2$. Thus

$$
\begin{aligned}
U_n(x) &= \sum_{0 \le \ell \le n/2} \binom{n+1}{2\ell+1} x^{n-2\ell} w^{2\ell} \\
&= \sum_{0 \le \ell \le n/2} \binom{n+1}{2\ell+1} x^{n-2\ell} (x^2 - 1)^\ell,
\end{aligned}
\tag{11.69}
$$

as claimed.

Chapter Twelve

Approximation Theory

> "The teaching of numerical analysis in a mathematics department poses a peculiar problem. At a time when the prime objectives in the instruction of most mathematical disciplines are rigor and logical coherence, many otherwise excellent textbooks in numerical analysis still convey the impression that computation is an art rather than a science, and that every numerical problem requires its own trick for its successful solution. It is thus understandable that many analysts are reluctant to take much interest in the teaching of numerical mathematics." (Peter Henrici in [80])

We now collect some important additional results about approximation theory. Not only are these results interesting in their own right, but the techniques of proof utilize novel concepts. We begin by considering the problem of finding the best approximation in the maximum norm. This is the problem that we addressed in section 1.3.2. We will see that the best approximation may be viewed as an adaptive interpolation process. Moreover, the mapping from a function to its best approximation is not linear, so we are forced outside the comfortable realm of linear operators.

We also present Bernstein's proof of Weierstrass' approximation theorem. The proof introduces a linear approximation operator, but unlike the Lagrange interpolant, this operator is not a projection. We contrast this type of approximation with least squares, which generates orthogonal polynomials. Finally, we compare all these with piecewise polynomial approximation which forms the basis of the finite element method [21, 110].

12.1 BEST APPROXIMATION BY POLYNOMIALS

We have seen that polynomial interpolation (especially using Chebyshev interpolation points) can be used to get good approximations of general functions, but we have yet to characterize what the very best approximation might be. It turns out that it is possible to do so, and it provides valuable insight into the approximation process. We will see that best approximation is necessarily adaptive in nature and that it is necessarily a nonlinear process (see exercise 12.1).

First, let us define what we mean by *best approximation*. For any $f \in C^0(I)$, define $d_n(f)$ by

$$d_n(f) = \inf \left\{ \|f - P\|_{\infty, I} \mid P \in \mathcal{P}_n \right\}, \tag{12.1}$$

where $\mathcal{P}_n$ denotes polynomials of degree n.

Definition 12.1 *Let $f \in C^0(I)$. Then $P \in \mathcal{P}_n$ is a best approximation to f provided that*

$$\|f - P\|_{\infty, I} = d_n(f). \tag{12.2}$$

As always, when the interval I is clear, we drop this from the subscripts on the norm. The existence of a best approximation is a simple matter of compactness. We always have $d_n(f) \le \|f\|_\infty$ by taking P in (12.1) to be the polynomial that is identically zero. If P is a best approximation, then it must satisfy

$$\|P\|_\infty \le \|P - f\|_\infty + \|f\|_\infty = d_n(f) + \|f\|_\infty \le 2\|f\|_\infty. \tag{12.3}$$

Therefore, it suffices to look for best approximation polynomials among the set of polynomials that satisfy $\|P\|_\infty \le 2\|f\|_\infty$. Since this set is closed and bounded, the continuous function $\phi(P) := \|P - f\|_\infty$ takes on its minimum on this set (see exercises 12.2 and 12.3).

What is somewhat surprising is that the best polynomial approximation is unique. For certain types of norms, this uniqueness is guaranteed, as we will see when we switch to an integral-based norm in section 12.3. But for the maximum norm, the uniqueness follows from a special alternation property that holds because we are working in one space dimension. We will prove the following result.

Theorem 12.2 *Let $f \in C^0(I)$. Then any best approximation $P \in \mathcal{P}_n$ to f satisfying (12.2) must also satisfy the alternation condition*

$$\pm(f - P)(\xi_j) = (-1)^j d_n(f) \quad \forall j = 0, \ldots, n+1, \tag{12.4}$$

where the points $\xi_0 < \xi_1 < \cdots < \xi_{n+1}$ lie in the interval I. Moreover, only one such polynomial can exist.

The expression $\pm(f - P)(\xi_j) = (-1)^j d_n(f)$ means that either

$$(f - P)(\xi_j) = (-1)^j d_n(f)$$

for all j, or

$$(P - f)(\xi_j) = (-1)^j d_n(f)$$

for all j. That is,

$$|f(\xi_j) - P(\xi_j)| = d_n(f) \tag{12.5}$$

for all j, and the signs of $(f - P)(\xi_j)$ alternate as j goes from 0 to 1 to 2, and so forth. The number of points ξ_j can be more than $n+1$. For example, the best constant approximation ($n = 0$) to $\sin \pi k x$ on $[-1, 1]$ is zero (see

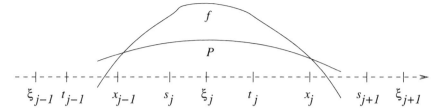

Figure 12.1 Notation for the alternation proof.

exercise 12.4), and there are $\mathcal{O}(k)$ such points for k large. The set of such points can even include open intervals.

The best polynomial approximation is quite different from the Lagrange interpolant for a fixed set of interpolation points, in that the norm of the best-approximation operator is uniformly bounded in the degree n of polynomials, whereas the norm of the Lagrange interpolant cannot be bounded, cf. (11.30). The inequality in (12.3) implies that the norm of the best-approximation operator is not greater than 2.

On the other hand, the best approximation does interpolate. One corollary of theorem 12.2 is that there exist points $x_0 < x_1 < \cdots < x_n$ in the interval I at which $f(x_j) = P(x_j)$ (see exercise 12.5). In particular, we have

$$\xi_j < x_j < \xi_{j+1} \tag{12.6}$$

for all $j = 0, \ldots, n$. That is, the best approximation $P \in \mathcal{P}_n$ to an arbitrary function $f \in C^0(I)$ interpolates f at $n + 1$ distinct points in the interval I. However, we have no information about the set of points. In this sense, the best approximation is an adaptive Lagrange interpolant. The specific interpolation points are adapted to the function being interpolated. It is easy to see as well that the mapping $f \to P$ cannot be a linear mapping (exercise 12.1).

Before we begin to prove the alternation property (12.4), let us consider an example, namely, the best approximation of x^{n+1} by polynomials of degree at most n. If we let P_n be the Lagrange interpolant of x^{n+1} at the Chebyshev points, then by (10.9) and (11.2) we know that

$$x_{n+1} - P_n(x) = \omega_{n+1}(x) = 2^{-n} \cos((n+1) \cos^{-1} x). \tag{12.7}$$

Thus we see explicitly that $x_{n+1} - P_n(x)$ has the claimed oscillation property (12.4), and by theorem 12.2, the Chebyshev-point Lagrange interpolant of x^{n+1} must be its best approximation.

Proof. To prove the property (12.4), we suppose that $P \in \mathcal{P}_n$ is any polynomial that satisfies (12.2). Define ξ_0 as

$$\xi_0 = \inf \left\{ x \in I \mid |f - P|(x) = d_n(f) \right\}. \tag{12.8}$$

The set on the right-hand side of (12.8) is not empty in view of exercise 11.11. Since both f and P are continuous, $|f - P|(\xi_0) = d_n(f)$. Define

$$\sigma_0 = (f - P)(\xi_0)/d_n(f) \ (= \pm 1). \tag{12.9}$$

Now suppose that we have defined $\xi_0, \ldots, \xi_k$ and $\sigma_0, \ldots, \sigma_k$ for $k \geq 0$. Then we define

$$\xi_{k+1} = \inf \left\{ x \in I \mid x > \xi_k \text{ and } (f - P)(x) = -\sigma_k d_n(f) \right\}, \qquad (12.10)$$

provided that the set on the right-hand side is not empty. In such a case, we define $\sigma_{k+1} = -\sigma_k$. Thus

$$(f - P)(\xi_{k+1}) = -\sigma_k d_n(f) = \sigma_{k+1} d_n(f). \qquad (12.11)$$

With this choice of σ's, we have

$$\sigma_j (f - P)(\xi_j) = d_n(f) \qquad (12.12)$$

for all j.

If the set on the right-hand side of (12.10) is empty, we stop the process. If we reach $k = n + 1$ as required by theorem 12.2, we also stop. Now we must show that if the process stops with $k < n + 1$, we must have a contradiction.

Note that (12.12) implies that

$$\sigma_j (f - P)(x) > d_n(f) - \epsilon$$

for x near ξ_j. In particular, there must be points s_j and t_j satisfying

$$\xi_{j-1} < s_j < \xi_j < t_j < \xi_{j+1} \qquad (12.13)$$

such that

$$\sigma_j (f - P)(x) \geq \tfrac{1}{2} d_n(f) \quad \text{for } x \in [s_j, t_j] \qquad (12.14)$$

(see figure 12.1). Moreover, we have $t_j \leq s_{j+1}$ because $f - P$ switches signs. In between, we can be assured that

$$M_j := \sup \left\{ |(f - P)(x)| \mid t_j \leq x \leq s_{j+1} \right\} < d_n(f). \qquad (12.15)$$

Let $M = \max_j M_j$. Define

$$x_j = \tfrac{1}{2}(t_j + s_{j+1}) \qquad (12.16)$$

and define a polynomial $Q \in \mathcal{P}_k$ by

$$Q(x) = \sigma_0 (x - x_1)(x - x_2) \cdots (x - x_k). \qquad (12.17)$$

Define $\beta = \|Q\|_\infty$ and let $\epsilon > 0$. Then there is a $q > 0$ such that for all j

$$\sigma_j Q(x) \geq q \qquad (12.18)$$

for $s_j < x < t_j$, so that

$$\begin{aligned}
\tfrac{1}{2} d_n(f) - \epsilon\beta \leq \sigma_j (f - P)(x) - \epsilon\beta &\leq \sigma_j (f - P - \epsilon Q)(x) \\
&\leq \sigma_j (f - P)(x) - \epsilon q \leq d_n(f) - \epsilon q
\end{aligned} \qquad (12.19)$$

for $s_j < x < t_j$. Therefore,

$$\|f - P - \epsilon Q\|_{\infty, [s_j, t_j]} \leq d_n(f) - \epsilon q, \qquad (12.20)$$

provided that $0 < \epsilon \leq \tfrac{3}{2} d_n(f)/(\beta + q)$. In the remaining regions,

$$|(f - P)(x)| \leq M < d_n(f), \qquad (12.21)$$

so that

$$|(f - P - \epsilon Q)(x)| \leq |(f - P)(x)| + \epsilon\beta \leq M + \epsilon\beta. \qquad (12.22)$$

Choosing $0 < \epsilon < (d_n(f) - M)/\beta$, we find that $P + \epsilon Q$ is a better approximation to f than P. If $k = n + 1$, there is no contradiction since the degree of $P + \epsilon Q$ is $n + 1$ in this case. But if $k < n + 1$, then $P + \epsilon Q \in \mathcal{P}_n$, and we have a contradiction to the optimality of P.

To prove uniqueness of the best approximation, we suppose that there are two polynomials $P, Q \in \mathcal{P}_n$ satisfying (12.2). Then so does the polynomial $R = \frac{1}{2}(P + Q)$ because, by the triangle inequality,

$$\begin{aligned}
\|f - R\|_{\infty,I} = \|f - \tfrac{1}{2}(P + Q)\|_{\infty,I} &= \|\tfrac{1}{2}(f - P) + \tfrac{1}{2}(f - Q)\|_{\infty,I} \\
&\leq \tfrac{1}{2}\|f - P\|_{\infty,I} + \tfrac{1}{2}\|f - Q\|_{\infty,I} = d_n(f).
\end{aligned} \qquad (12.23)$$

By the alternating condition (12.4), we conclude that there exist alternating $\sigma_j = \pm 1$ such that

$$\sigma_j(f - R)(\xi_j) = d_n(f) \quad \forall j = 0, \dots, n + 1, \qquad (12.24)$$

where the points $\xi_0 < \xi_1 < \cdots < \xi_{n+1}$ lie in the interval I. Therefore, for each $j = 0, \dots, n + 1$,

$$\tfrac{1}{2}\sigma_j(f - P)(\xi_j) + \tfrac{1}{2}\sigma_j(f - Q)(\xi_j) = d_n(f). \qquad (12.25)$$

We claim then (cf. exercise 12.6) that both

$$\sigma_j(f - P)(\xi_j) = d_n(f) \quad \text{and} \quad \sigma_j(f - Q)(\xi_j) = d_n(f). \qquad (12.26)$$

Otherwise, if, say, $\sigma_j(f - P)(\xi_j) < d_n(f)$, then we would have to have

$$\sigma_j(f - Q)(\xi_j) > d_n(f), \qquad (12.27)$$

contradicting the optimality of Q. But then

$$\sigma_j(f - P)(\xi_j) = \sigma_j(f - Q)(\xi_j), \qquad (12.28)$$

which implies that $P(\xi_j) = Q(\xi_j)$ for each $j = 0, \dots, n + 1$. Since $P, Q \in \mathcal{P}_n$, their equality at $n + 2$ points implies they must be equal. \hfill QED

12.2 WEIERSTRASS AND BERNSTEIN

The theorem of Weierstrass (see page 151) on the approximability of continuous functions by polynomials is often considered one of the main waypoints of basic analysis. Bernstein[1] developed an approximation scheme that can be used to prove Weierstrass' theorem constructively. Moreover, the Bernstein approximation introduces techniques of independent interest.

---

[1]Sergei Natanovich Bernstein (1880–1968), see page 65, was a student in both Paris and Göttingen and worked with both Picard (see page 258) and David Hilbert. Subsequently, he returned to Russia where he was required to complete additional graduate work in order to become qualified to be a professor "due to the conditions of life in tsarist Russia" [6].

12.2.1 Bernstein polynomials

The Bernstein polynomial $B_n f$ is defined by

$$B_n f(x) = \sum_{i=0}^{n} f(i/n)\beta_{i,n}(x), \tag{12.29}$$

where the Bernstein basis functions are defined by

$$\beta_{i,n}(x) = \binom{n}{i} x^i (1-x)^{n-i}. \tag{12.30}$$

Note that the basis functions $\beta_{i,n}$ are always nonnegative, so the Bernstein approximation B_n is monotone in the sense that $B_n f \geq 0$ whenever $f \geq 0$.

B_n is not an interpolation operator, but it does have certain special properties. For example, $B_n 1 = 1$; in other words,

$$\sum_{i=0}^{n} \beta_{i,n}(x) = 1 \quad \forall x \in [0, 1]. \tag{12.31}$$

The verification of (12.31) is just the binomial expansion:

$$(X + Y)^n = \sum_{i=0}^{n} \binom{n}{i} X^i Y^{n-i}, \tag{12.32}$$

applied with $X = x$ and $Y = 1 - x$. One consequence of (12.31) is a bound for operator B_n:

Lemma 12.3 *The Bernstein operator satisfies* $\|B_n\|_\infty = 1$.

Proof. Since the basis functions are nonnegative,

$$|B_n f(x)| \leq \sum_{i=0}^{n} |f(i/n)|\beta_{i,n}(x)$$
$$\leq \|f\|_\infty \sum_{i=0}^{n} \beta_{i,n}(x) = \|f\|_\infty \tag{12.33}$$

for any $x \in [0.1]$. Therefore $\|B_n\|_\infty \leq 1$. But since $B_n 1 = 1$, we must have equality.   QED

It is also the case that $B_n x = x$, i.e.,

$$\sum_{i=0}^{n} \frac{i}{n}\beta_{i,n}(x) = x \quad \forall x \in [0, 1] \tag{12.34}$$

(see exercise 12.7). However, B_n is not a projection in general. The Bernstein approximation to x^2 is $x^2 + x(1-x)/n$, that is,

$$\sum_{i=0}^{n} \frac{i^2}{n^2}\beta_{i,n}(x) = x^2 + \frac{x(1-x)}{n} \quad \forall x \in [0, 1] \tag{12.35}$$

(see exercise 12.7).

Although the Bernstein basis functions do not separate points in a way that allows B_n to be an interpolant, they are nevertheless quite local. The maximum of $\beta_{i,n}$ occurs at i/n (exercise 12.8), and the integral of $\beta_{i,n}$ is $(n+1)^{-1}$ for all i. Thus the Bernstein basis functions play the role of scaled approximate Dirac δ-functions. That is,

$$\int_0^1 f(x)\beta_{i,n}(x)\,dx \approx \frac{1}{n+1}f(i/n). \qquad (12.36)$$

12.2.2 Modulus of continuity

We need a way to measure the smoothness of a function that is subtle enough to be useful for any continuous function. The *modulus of continuity* $\omega(f;\delta)$ is such a measure.

Definition 12.4 *Let* $f \in C^0(I)$ *for some interval* I. *Then for all* $\delta > 0$,

$$\omega_I(f;\delta) = \sup\left\{|f(x) - f(y)| \mid x, y \in I \text{ and } |x - y| \le \delta\right\}. \qquad (12.37)$$

If the interval I is understood, we will drop the reference to it in the notation for the modulus of continuity and write $\omega_I(f;\delta)$ as $\omega(f;\delta)$. We can relate the modulus of continuity to other smoothness measures. For example, if $f \in C^1(I)$, then

$$\omega_I(f;\delta) \le \delta\|f'\|_{\infty,I} \qquad (12.38)$$

for any $\delta > 0$ (exercise 12.13). But the modulus of continuity is a more sensitive measure, as the following result shows.

Lemma 12.5 *Suppose that* I *is a closed, bounded interval. Then*

$$\lim_{\delta \to 0} \omega_I(f;\delta) = 0 \qquad (12.39)$$

for any $f \in C^0(I)$.

Proof. Any continuous function on a closed, bounded interval is uniformly continuous on that interval [141].   QED

The order of approximation for B_n is not optimal, but we do get convergence for any continuous function.

Theorem 12.6 *For any continuous function,*

$$\|f - B_nf\|_\infty \le C\omega(f;1/\sqrt{n}). \qquad (12.40)$$

Proof. By (12.31), we have for any $x \in [0,1]$,

$$\begin{aligned}
|f(x) - B_nf(x)| &= \left|\sum_{i=0}^n (f(x) - f(i/n))\beta_{i,n}(x)\right| \\
&\le \sum_{i=0}^n |f(x) - f(i/n)|\beta_{i,n}(x)
\end{aligned} \qquad (12.41)$$

since the Bernstein basis functions are nonnegative. We now break up the sum into two parts: one over points near x, and the other over points that are not close. Let $\delta > 0$ and define $J_x = \{j \mid |x - j/n| \le \delta\}$. By the definition of modulus of continuity and (12.31),

$$\sum_{i \in J_x} |f(x) - f(i/n)|\beta_{i,n}(x) \le \omega(f;\delta) \sum_{i \in J_x} \beta_{i,n}(x)$$

$$\le \omega(f;\delta) \sum_{i=0}^{n} \beta_{i,n}(x) = \omega(f;\delta). \tag{12.42}$$

Now suppose that i is not in J_x, so that $|x - i/n| > \delta$. For concreteness, let us suppose that $x > i/n$. Let $i/n = \xi_0 < \xi_1 < \cdots < \xi_k = x$ be a set of points such that $\xi_i - \xi_{i-1} \le \delta$ and k is as small as possible. This means that k is the smallest integer not less than $|x - i/n|/\delta$, and thus $k < 1 + |x - i/n|/\delta$. In this case,

$$|f(x) - f(i/n)| = \left| \sum_{j=1}^{k} f(\xi_i) - f(\xi_{i-1}) \right| \le \sum_{j=1}^{k} |f(\xi_i) - f(\xi_{i-1})| \tag{12.43}$$

$$\le k\omega(f;\delta) \le (1 + |x - i/n|/\delta)\omega(f;\delta).$$

If instead $x < i/n$, let $x = \xi_0 < \cdots < \xi_k = i/n$ and repeat the previous argument, so that (12.43) still holds. Therefore,

$$\sum_{i \notin J_x} |f(x) - f(i/n)|\beta_{i,n}(x) \le \omega(f;\delta) \sum_{i \notin J_x} (1 + |x - i/n|/\delta)\beta_{i,n}(x)$$

$$\le \omega(f;\delta)\left(1 + \sum_{i \notin J_x} (|x - i/n|/\delta)\beta_{i,n}(x)\right) \tag{12.44}$$

by (12.31). To estimate the last term, we note that for $i \notin J_x$, $1 < |x - i/n|/\delta$, so that $|x - i/n|/\delta < (|x - i/n|/\delta)^2$. Therefore,

$$\sum_{i \notin J_x} (|x - i/n|/\delta)\beta_{i,n}(x) \le \sum_{i \notin J_x} (|x - i/n|/\delta)^2 \beta_{i,n}(x)$$

$$\le \sum_{i=1}^{n} (|x - i/n|/\delta)^2 \beta_{i,n}(x). \tag{12.45}$$

$$= \delta^{-2} \sum_{i=1}^{n} (x - i/n)^2 \beta_{i,n}(x).$$

Using (12.31), (12.34), and (12.35), we can evaluate

$$\sum_{i=1}^{n} (x - i/n)^2 \beta_{i,n}(x) = x^2 - 2x \sum_{i=1}^{n} (i/n)\beta_{i,n}(x) + \sum_{i=1}^{n} (i/n)^2 \beta_{i,n}(x)$$

$$= x^2 - 2x^2 + x^2 + \frac{x(1-x)}{n} = \frac{x(1-x)}{n}. \tag{12.46}$$

Thus we have proved that

$$|f(x) - B_n f(x)| \le \omega(f;\delta)\left(2 + \frac{x(1-x)}{n\delta^2}\right). \tag{12.47}$$

Choosing $\delta = 1/\sqrt{n}$ completes the proof, with $C = 9/4$.   QED

12.3 LEAST SQUARES

Another way to define polynomial approximations is by least squares. This process is equivalent to expansion in orthogonal polynomials. The definition of such polynomials utilizes an inner-product structure on the linear space of square-integrable functions on an interval $I = [a, b]$. The work has already been done in section 5.4, so we just need to translate it into a new context. For simplicity, we continue to consider real-valued polynomials.

12.3.1 Polynomials as inner-product spaces

Define an inner product

$$(f, g) = \int_a^b f(x)g(x)\, w(x)dx, \qquad (12.48)$$

where $w > 0$ is some weight function. Then the associated norm is

$$\|f\|_2 = \sqrt{(f, f)} = \left(\int_a^b f(x)^2\, w(x)dx \right)^{1/2}. \qquad (12.49)$$

The condition

$$\int_a^b f(x)^2\, w(x)dx = 0$$

implies $f = 0$ under suitable integrability conditions on f, but we avoid full consideration of these issues. It is not hard to show that this holds (exercise 12.14) if $f \in V = C^0([a, b])$, but issues such as this provide motivation for studying the Lebesgue integral [142] in order to make these concepts rigorous for more general f. We have already seen the need to consider weights that vanish at the ends of the interval of integration in exercise 11.7, which shows that the Chebyshev polynomials ω_n are orthogonal with respect to the inner product (12.48) on the interval $[-1, 1]$ with the weight $w(x) = \sqrt{1 - x^2}$.

12.3.2 Orthogonal polynomials

We will construct polynomials that are orthonormal:

$$(P_i, P_j) = \int_a^b P_i(x)P_j(x)\, w(x)dx = \delta_{ij}, \qquad (12.50)$$

where P_i is a polynomial of degree i of the form

$$P_i(x) = a_i x^i + q_i(x), \qquad (12.51)$$

where $a_i \neq 0$ and the degree of $q_i(x)$ is $i - 1$. Notice that these conditions imply (exercise 12.15) that the P_i's are linearly independent. Thus the set $\{P_0, \ldots, P_n\}$ forms a basis for the space $\mathcal{P}_n$ of polynomials of degree n. For $i = 0$, it is trivial: $P_0 = 1/\sqrt{b - a}$.

  The construction of the orthogonal polynomials is immediate from the results in section 5.4, where we take the starting vectors to be $v^k = v^k(x) =$

x^{k-1}. Renumbering as necessary (starting at 0 instead of 1), the least-squares projection

$$L_n^S f = \sum_{i=0}^{n} (f, P_i) P_i \qquad (12.52)$$

is defined for any $f \in V$. Combining theorem 5.4 and lemma 5.4, we obtain

Theorem 12.7 *Given any $f \in V$,*

$$(f - L_n^S f, q) = 0 \qquad (12.53)$$

for all polynomials q of degree n, and

$$\|f - L_n^S f\|_2 = \min_{q \in \mathcal{P}_n} \|f - q\|_2. \qquad (12.54)$$

It is not hard to see that $L_n^S f$ in (12.52) is defined for any integrable function f and that theorem 12.7 holds for any square-integrable function f. However, we will stay with the more limited space $V = C^0(I)$ for the discussion here.

As before, there are some immediate corollaries. Suppose that $q \in \mathcal{P}_n$. Then $q = L_n^S q$ (L_n^S is a projection) because we must have $\|q - L_n^S q\| = 0$. Moreover,

$$\|f - L_n^S f\|_2 = 0 \qquad (12.55)$$

if and only if $f \in \mathcal{P}_n$.

Using the results above, the orthogonal polynomials can be defined by

$$P_{n+1} = \frac{1}{\|x^{n+1} - L_n^S x^{n+1}\|_2} \left(x^{n+1} - L_n^S x^{n+1} \right). \qquad (12.56)$$

The coefficient

$$a_{n+1} = 1/\|x^{n+1} - L_n^S x^{n+1}\|_2 \qquad (12.57)$$

is well-defined (and nonzero) because we must have

$$x^{n+1} - L_n^S x^{n+1} \neq 0$$

since $x^{n+1} \notin \mathcal{P}_n$. The scaling ensures that $(P_{n+1}, P_{n+1}) = 1$, and the orthogonality $(P_{n+1}, P_j) = 0$ is a consequence of (12.54).

12.3.3 Roots of orthogonal polynomials

First, we claim that the real roots of P_n are all simple. Suppose that

$$P_n(x) = (x - x_1)^2 r(x), \qquad (12.58)$$

where $r \in \mathcal{P}_{n-2}$. Then $P_n(x) r(x) = (x - x_1)^2 r(x)^2$, and the orthogonality of P_n implies that

$$0 = (P_n, r) = \int_a^b (x - x_1)^2 r(x)^2 w(x) \, dx. \qquad (12.59)$$

But this cannot happen unless $r \equiv 0$, which is impossible, as this would imply that $P_n \equiv 0$.

Second, all the real roots of P_n are within the interior of the interval $[a, b]$. To see this, enumerate all such roots in the interior of $[a, b]$ as $a < x_0 < x_1 < \cdots < x_m < b$ and define

$$q(x) = \pm(x - x_0)(x - x_1) \cdots (x - x_m). \tag{12.60}$$

These must all be simple, as they are real roots. Then $r = P_n q$ is of one sign in $[a, b]$. To see this, start the sign of q at the beginning (just to the left of x_0) with the same sign as P_n by adjusting the sign of q as necessary. At the next root, they both change sign, as both have simple roots there. So they stay of the same sign. This continues for all the roots, by induction. Thus $(P_n, q) > 0$. Since P_n is orthogonal to $\mathcal{P}_{n-1}$, we must have $m = n$.

Thus we have proved the following result.

Theorem 12.8 *The orthogonal polynomial P_n of degree n, defined equivalently by (12.56) and (12.57)) or by (12.50), has n simple roots in the interior of the interval $[a, b]$.*

Recall that the Chebyshev polynomials are orthogonal with respect to the weight $\sqrt{1 - x^2}$ on $[-1, 1]$. Thus it is natural to consider the roots of orthogonal polynomials as potential interpolation points, cf. exercise 12.16.

12.4 PIECEWISE POLYNOMIAL APPROXIMATION

The concept of piecewise approximation is simple. Suppose we have a subdivision of an interval

$$a = x_0 < x_1 < \cdots < x_n = b. \tag{12.61}$$

We can view each subinterval $[x_{j-1}, x_j]$ as independent and construct a particular approximation on it. In principle, these approximations could all be independent, but a common choice is to take them to be the same for each interval. For example, we might take linear Lagrange interpolation at the endpoints.

Once the local approximation is chosen, it may or may not be feasible (or of interest) to link them together. If we don't link them, we get a discontinuous piecewise approximation. For example, if we consider piecewise constant approximation, it necessarily must be discontinuous to be interesting. We can define $\Pi_n^0 f$ as a projection onto piecewise constants via

$$\Pi_n^0 f = \sum_{j=0}^{n-1} f(x_j)\phi_j, \tag{12.62}$$

where ϕ_j is the characteristic function of the interval $[x_{j-1}, x_j]$. Then it follows from the definition of the modulus of continuity (12.37) that

$$\|f - \Pi_n^0 f\|_{\infty, I} \leq \omega_I(f; \delta), \tag{12.63}$$

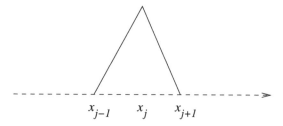

Figure 12.2 A picture of a typical basis function ϕ_j for continuous, piecewise linear interpolation; ϕ_j is zero outside the interval $[x_{j-1}, x_{j+1}]$.

where δ is defined by

$$\delta = \max\left\{x_j - x_{j-1} \mid j = 1, \ldots, n\right\}. \tag{12.64}$$

From (12.38), we also conclude that

$$\|f - \Pi_n^0 f\|_{\infty,I} \leq \delta \|f'\|_{\infty,I}, \tag{12.65}$$

provided that $f \in C^1(I)$.

It is also possible to define continuous piecewise linear approximation. Define basis functions ϕ_j for $i = 0, \ldots, n$ by the requirements that (see figure 12.1)

- $\phi_j \in C^0([a, b])$,

- ϕ_j is linear in each segment $[x_{k-1}, x_k]$ for $k = 1, \ldots, n$, and

- $\phi_j(x_k) = \delta_{jk}$ (Kronecker δ) for $k = 1, \ldots, n$.

By definition, a continuous, piecewise linear function is any function satisfying the first two conditions. We leave it as an exercise to see that any such function can be written as a linear combination of the ϕ_j's. Moreover, the ϕ_j's are linearly independent in view of the third condition.

The corresponding continuous, piecewise linear interpolant is defined by

$$\Pi_n^1 f = \sum_{j=0}^{n} f(x_j)\phi_j \tag{12.66}$$

for any $f \in C^0([a, b])$.

The interpolant Π_n^1 is a composite of Lagrange interpolants, so Π_n^1 is a projection since the Lagrange interpolant is a projection on each segment $[x_{j-1}, x_j]$ for $j = 1, \ldots, n$. Like the Bernstein approximation operator,

$$\|\Pi_n^1\|_{C^0 \to C^0} = 1 \tag{12.67}$$

because the basis functions are positive and

$$\sum_{j=0}^{n} \phi_j(x) = 1 \quad \forall x_0 \leq x \leq x_n \tag{12.68}$$

(see exercise 12.17).

Error estimates may be developed by considering the error on each segment separately. For example, it is elementary to show that for $k = 0, 1, 2$,

$$\|f - \Pi_n^1 f\|_\infty \leq c_k \delta^k \|f^{(k)}\|_\infty, \qquad (12.69)$$

where δ is defined by (12.64) (the case $k = 0$ is (12.67), with $c_0 = 1$; see exercise 12.19 for $k = 1$ and exercise 12.20 for $k = 2$).

12.5 ADAPTIVE APPROXIMATION

We have seen that best approximation by polynomials can be viewed as adaptive Lagrange interpolation. That is, the best approximant interpolates at points that depend on the function being approximated. This raises the question of whether adaptivity can be used to advantage with other types of approximations. The answer is decidedly yes, but the general subject is so large that we can give only a simple example based on piecewise constant approximation.

Observe that in all the examples considered in section 12.4, the measure of smoothness used for the function being approximated was always global. For example, (12.69) is the maximum norm of a derivative of f. But the modulus of continuity is also a global measure. Many functions of interest may have a localized behavior that is different from the general behavior. For example, consider $f(x) = \sqrt{x}$ on the interval $I = [0, 1]$. The derivative of f is not bounded, and its modulus of continuity is limited by its singularity at zero. For these reasons, we will consider instead a measure of smoothness that allows some localized singularities:

$$\|f\|_1 = \int_0^1 |f'(x)| \, dx. \qquad (12.70)$$

The subscript 1 denotes both that there is only one derivative and that only its first power (cf. (12.49)) is being integrated.

The expression in (12.70) is only a seminorm (see section 5.1.3). More seriously, it is not simple to express the right class of functions for which (12.70) is well-defined [21]. To bypass these issues, we make the simplifying assumption that f is differentiable on the open interval $]0, 1[$ with (12.70) finite. This allows functions of the form $f(x) = x^r$ for any $r > 0$. We can clearly generalize this concept to arbitrary finite intervals.

We propose to prove the following theorem [21].

Theorem 12.9 *Suppose that f is continuous on $[0, 1]$, that f is differentiable on the open interval $]0, 1[$, and that $\|f\|_1 < \infty$. Then there is a subdivision $0 = x_0 < x_1 < \cdots < x_n = 1$ such that*

$$\|f - \Pi_n^0 f\|_\infty \leq \frac{1}{n} \|f\|_1, \qquad (12.71)$$

where Π_n^0 denotes the piecewise constant interpolation defined in (12.62).

Proof. If $\|f\|_1 = 0$, then f is constant and $\Pi_n^0 f = f$ for any $n \geq 1$. So we assume that $\|f\|_1 > 0$. We introduce the auxiliary (continuous) function

$$\phi(t) = \frac{1}{\|f\|_1} \int_0^t |f'(x)| \, dx. \tag{12.72}$$

Then ϕ vanishes at $x = 0$ and is nondecreasing; moreover, $\phi(1) = 1$. Thus there are points x_j where $\phi(x_j) = j/n$, by the intermediate value theorem. If by chance $x_n < 1$, so that $\phi(t) \equiv 1$ for $t \in [x_n, 1]$, we simply redefine $x_n = 1$. By construction,

$$\frac{1}{\|f\|_1} \int_{x_{j-1}}^{x_j} |f'(x)| \, dx = \phi(x_j) - \phi(x_{j-1}) = \frac{1}{n}. \tag{12.73}$$

Thus it suffices to prove that for all j,

$$\|f - \Pi_n^0 f\|_{\infty, [x_{j-1}, x_j]} \leq \int_{x_{j-1}}^{x_j} |f'(x)| \, dx. \tag{12.74}$$

But for $x \in [x_{j-1}, x_j[$,

$$f(x) - \Pi_n^0 f(x) = f(x) - f(x_{j-1}) = \int_{x_{j-1}}^x f'(x) \, dx, \tag{12.75}$$

so (12.74) follows (note that $f(x) - \Pi_n^0 f(x) = 0$ for all $x = x_j$).   QED

Similar results hold for arbitrary finite intervals and for higher-order approximation, e.g., for Π_n^1 instead of Π_n^0 [21].

12.6 MORE READING

We have now seen five distinct types of approximations involving polynomials. The main features of these schemes are summarized in table 12.1. The significant observation is that there is no linear projection onto polynomials that has a norm uniformly bounded for all polynomial degrees n. This property is satisfied by piecewise linear approximation, but it can be shown [130] that indeed there can be no linear projection onto polynomials that has a norm uniformly bounded for all polynomial degrees n.

Approximation theory has been stimulated by a variety of influences. Polynomials are the most basic example of a function, so it is understandable that people wanted to know whether such simple functions could approximate general functions, as well as answers to other fundamental questions [34, 111, 137]. In addition to approximation problems from linear spaces, it is also possible to explore nonlinear spaces of functions [138].

12.7 EXERCISES

Exercise 12.1 Let $f_\pm \in C^0([-1, 1])$ be defined by $f_\pm(x) = \frac{1}{2} - |x \pm \frac{1}{2}|$ for $\pm x \leq 0$ and zero for $\pm x \geq 0$. Show that the best constant approximations to f_+, f_-, and $f_+ + f_-$ are all the same, and hence that the best approximations are not additive. (Hint: use exercise 12.4.)

Approximation type	Operator norm	Linear operator	Projection
Lagrange/Chebyshev	$\geq (\frac{1}{2}\pi \log n) - E$	Yes	Yes
Best approximation	≤ 2	No	Yes
Bernstein	1	Yes	No
Piecewise linear	1	Yes	Yes
Least squares	1	Yes	Yes

Table 12.1 Comparison of principal features of different approximation schemes. The top four are compared in the maximum norm; the top three involve polynomial approximation, whereas the fourth is piecewise polynomial. The fifth relates to approximation in the L^2-norm.

Exercise 12.2 *Prove that we can write*

$$d_n(f) = \inf \left\{ \|f - P\|_\infty \mid P \in \mathcal{P}_n, \ \|P\|_\infty \leq 2\|f\|_\infty \right\}, \qquad (12.76)$$

where d_n is defined in (12.1). (Hint: use (12.3).)

Exercise 12.3 *Fill in the remaining details of the existence proof for best-approximation polynomials. This will include answers to questions such as the following. Why can we view the set*

$$\left\{ P \in \mathcal{P}_n \mid \|P\|_\infty \leq 2\|f\|_\infty \right\} \qquad (12.77)$$

as a closed and bounded subset of $\mathbb{R}^{n+1}$? Why is $\phi(P) = \|f - P\|_\infty$ continuous when viewed as a function on $\mathbb{R}^{n+1}$? If you use a representation of $P \in \mathcal{P}_n$ in terms of some vector of coefficients $a \in \mathbb{R}^{n+1}$ (e.g., the coefficients of the representation of P as a sum of monomials), how do you relate the fact that there is an $\hat{a} = \min \phi(a)$ to having a polynomial P with the desired properties? That is, how do you make sure that the representation $P \leftrightarrow a$ is invertible?

Exercise 12.4 *Show that the best approximation of $f \in C^0(I)$ by a constant c is*

$$c = \tfrac{1}{2} \left(\inf \left\{ f(x) \mid x \in I \right\} + \sup \left\{ f(x) \mid x \in I \right\} \right). \qquad (12.78)$$

Exercise 12.5 *Show that the best approximation $P \in \mathcal{P}_n$ to $f \in C^0(I)$ satisfies $f(x_j) = P(x_j)$, where the points x_j satisfy*

$$\xi_0 < x_0 < \xi_1 < x_1 < \cdots < x_n < \xi_{n+1}. \qquad (12.79)$$

(Hint: apply the mean value theorem.)

Exercise 12.6 *Suppose that x_0 and x_1 are two real numbers such that $|x_i| \leq \frac{1}{2}$ for $i = 0, 1$ and such that $x_0 + x_1 = 1$. Prove that $x_0 = x_1 = \frac{1}{2}$.*

Exercise 12.7 *Prove (12.34) and (12.35). (Hint: differentiate (12.32) once for (12.34) and twice for (12.35) and rearrange terms.)*

Exercise 12.8 *Prove that the maximum of the functions $\beta_{i,n}(x)$ defined in (12.30) occurs at $x = i/n$ and determine its maximum value. (Hint: use the following formula due to Stirling[2]:*

$$e^n n!/n^n \approx \sqrt{2\pi n} \qquad (12.80)$$

for large n.)

Exercise 12.9 *Prove that the integral of $\beta_{i,n}$ defined in (12.30) is $1/(n+1)$ for all i.*

Exercise 12.10 *For a Lipschitz function, show that the Bernstein approximation error (12.40) is no bigger than $\sqrt{2}\lambda n^{-1/2}$, where λ is the Lipschitz constant on $[0, 1]$.*

Exercise 12.11 *Consider piecewise constant approximation on a uniform mesh of points i/n on $[0, 1]$. For a Lipschitz function, what is the best error estimate that you can give? Contrast this with exercise 12.10.*

Exercise 12.12 *(Discrete least squares.) Suppose that we gather data f_n associated with parameters x_n and that we want to depict these data as a function $f(x)$ with the property that $f(x_n) \approx f_n$. But now suppose that some of the x_n's are the same $(x_n = x_k$ for $n \neq k)$ but the f_n's are not the same! We can still construct a function that attempts to represent the data in a reasonable way. Define a polynomial P that minimizes*

$$\sum_n (P(x_n) - f_n)^2. \qquad (12.81)$$

Show that this minimization problem has a unique solution.

Exercise 12.13 *Suppose that $f \in C^1(I)$ for some interval I. Prove (12.38).*

Exercise 12.14 *Prove that if $f \in C^0(I)$ and $f \geq 0$ on I, then $\int_I f(x)\,dx = 0$ implies $\equiv 0$. (Hint: if $f(x) > 0$ for some, then $f(y) \geq \epsilon > 0$ for $y \in [x - \delta, x + \delta]$.)*

Exercise 12.15 *Show that the orthogonal polynomials (cf. (12.50)) are linearly independent.*

Exercise 12.16 *The Gauss points are the zeroes of orthogonal polynomials for the weight $w \equiv 1$ (see section 12.3.3). Investigate the size of the Lebesgue function (section 11.3) for the Gauss points for various values of n.*

Exercise 12.17 *Prove (12.68). (Hint: consider interpolating a constant and see what happens.)*

---

[2]James Stirling (1692–1770) was born in Scotland, near the town of Stirling, and entered Balliol College Oxford in 1711. He was proposed for membership of the Royal Society of London by Newton, to which he was elected in 1726.

Exercise 12.18 *Prove (12.67). (Hint: compare (12.33) and then use exercise 12.17.)*

Exercise 12.19 *Suppose that $f \in C^0(I)$ and that δ is defined in (12.64). Prove that*

$$\|f - \Pi_n^1 f\|_{\infty, I} \leq c\omega_I(f; \delta) \qquad (12.82)$$

for some constant c. Use this to prove (12.69) for $k = 1$. (Hint: see the piecewise constant case (12.63); use exercise 12.13.)

Exercise 12.20 *Prove (12.69) for $k = 2$. (Hint: in each interval $[x_{j-1}, x_j]$, the error $e = f - \Pi_n^1 f$ vanishes at the endpoints. Note that $e^{(2)} = f^{(2)}$. There must be some point $\xi \in [x_{j-1}, x_j]$ where $e'(\xi) = 0$ at which $|e|$ takes on its maximum value. Do a Taylor expansion around ξ.)*

Exercise 12.21 *For any set of vectors $v_1, \ldots, v_n$ in an inner-product space, the matrix with entries (v_i, v_j) is known as the Gram matrix (cf. the Gram-Schmidt process in section 5.4.3). Consider the inner-product space consisting of polynomials with inner product (12.48) on the interval $[0, 1]$. Prove that the Gram matrix in this case is the Hilbert matrix (4.14).*

12.8 SOLUTIONS

Solution of Exercise 12.7. The derivative of (12.32) with respect to X is

$$n(X + Y)^{n-1} = \sum_{i=1}^{n} \binom{n}{i} i X^{i-1} Y^{n-i}. \qquad (12.83)$$

Multiply by X and divide by n to get

$$X(X + Y)^{n-1} = \sum_{i=1}^{n} \binom{n}{i} \frac{i}{n} X^i Y^{n-i} = \sum_{i=0}^{n} \binom{n}{i} \frac{i}{n} X^i Y^{n-i}. \qquad (12.84)$$

Now set $X = x$ and $Y = 1 - x$ to obtain (12.34). Now differentiate (12.83) to get

$$n(n - 1)(X + Y)^{n-2} = \sum_{i=2}^{n} \binom{n}{i} i(i - 1) X^{i-2} Y^{n-i}. \qquad (12.85)$$

Multiply by X^2 and divide by n^2 to get

$$\begin{aligned}
\frac{n-1}{n} X^2 (X + Y)^{n-2} &= \sum_{i=2}^{n} \binom{n}{i} \frac{i(i-1)}{n^2} X^i Y^{n-i} \\
&= \sum_{i=0}^{n} \binom{n}{i} \frac{i(i-1)}{n^2} X^i Y^{n-i}.
\end{aligned} \qquad (12.86)$$

Setting $X = x$ and $Y = 1 - x$ and using (12.34), we have

$$\frac{n-1}{n}x^2 = \sum_{i=0}^{n} \binom{n}{i} \frac{i^2}{n^2} x^i (1-x)^{n-i} - \frac{x}{n}. \qquad (12.87)$$

Therefore,

$$\sum_{i=0}^{n} \binom{n}{i} \frac{i^2}{n^2} x^i (1-x)^{n-i} = \frac{n-1}{n}x^2 + \frac{x}{n} = x^2 + \frac{x - x^2}{n}, \qquad (12.88)$$

which verifies (12.35).

Solution of Exercise 12.8. We have $\beta_{i,n}(x) = cx^i(1-x)^{n-i}$, so

$$\begin{aligned}
\beta'_{i,n}(x) &= c\left(ix^{i-1}(1-x)^{n-i} - (n-1)x^i(1-x)^{n-i-1}\right) \\
&= cx^{i-1}(1-x)^{n-i-1}\left(i(1-x) - (n-i)x\right) \qquad (12.89) \\
&= cx^{i-1}(1-x)^{n-i-1}\left(i - nx\right).
\end{aligned}$$

The maximum value is thus

$$\begin{aligned}
\beta_{i,n}(i/n) &= \frac{n!}{i!(n-i)!}\frac{i^i(n-i)^{n-i}}{n^n} \\
&= \frac{n!}{n^n}\frac{i^i}{i!}\frac{(n-i)^{n-i}}{(n-i)!} = \frac{\phi(n)}{\phi(i)\phi(n-i)},
\end{aligned} \qquad (12.90)$$

where $\phi(n) := e^n n!/n^n$. By Stirling's formula (12.80), $\phi(n) \approx \sqrt{2\pi n}$ for large n. Thus if $0 < \epsilon \le i/n \le 1 - \epsilon$, then

$$\beta_{i,n}(i/n) \approx \frac{\sqrt{2\pi n}}{\sqrt{2\pi i}\sqrt{2\pi(n-i)}} = \frac{1}{\sqrt{2\pi n(i/n)(1-i/n)}}. \qquad (12.91)$$

Note that $\beta_{0,n}(0) = \beta_{n,n}(1) = 1$, and it appears that these are the largest values based on numerical computation (see figure 12.3).

Define b_n to be the piecewise linear function with values $\beta_{i,n}(i/n)$ at the mesh points i/n for $i = 0, \ldots, n$. This is the function plotted in figure 12.3 for three values of $n = 10, 100, 1000$. It is easy to see from figure 12.3 and the computations above that

$$\lim_{n\to\infty} \sqrt{n}\, b_n(x) = 1/\sqrt{2\pi x(1-x)}$$

for any $0 < x < 1$. This can be expressed in terms of a limit

$$\lim_{n\to\infty} \sqrt{n}\beta_{i_n,n}(i_n/n) = 1/\sqrt{2\pi x(1-x)},$$

where $x_n = i_n/n$ satisfies $\lim_{n\to\infty} x_n = x$ with $0 < x < 1$.

Solution of Exercise 12.9. The integral of the Bernstein polynomials can

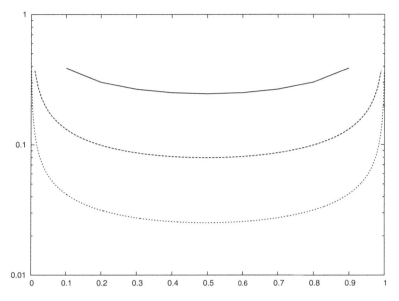

Figure 12.3 Values of $\beta_{i,n}$ as a function i for $n = 10, 100, 1000$. The horizontal axis is the scaled variable i/n.

be computed as follows. Suppose that $0 \leq i < n$. Then

$$\int_0^1 \beta_{i,n}(x)\, dx = \frac{n!}{i!(n-i)!} \int_0^1 x^i (1-x)^{n-i}\, dx$$

$$= \frac{n!}{(i+1)!(n-i)!} \int_0^1 \left(\frac{d}{dx} x^{i+1}\right) (1-x)^{n-i}\, dx$$

$$= -\frac{n!}{(i+1)!(n-i)!} \int_0^1 x^{i+1} \frac{d}{dx} (1-x)^{n-i}\, dx \qquad (12.92)$$

$$= \frac{n!}{(i+1)!(n-i-1)!} \int_0^1 x^{i+1} (1-x)^{n-i-1}\, dx$$

$$= \int_0^1 \beta_{i+1,n}(x)\, dx.$$

Since for $i = n$ we have

$$\int_0^1 \beta_{n,n}(x)\, dx = \int_0^1 x^n\, dx = \frac{1}{n+1}, \qquad (12.93)$$

all the integrals must have this value as well.

Solution of Exercise 12.19. In each interval $[x_{j-1}, x_j]$, $f - \Pi_n^1 f$ may be written as

$$f(x) - \Pi_n^1 f(x) = f(x) - (f(x_{j-1})\varphi_0(x) + f(x_j)\varphi_1(x)), \qquad (12.94)$$

where φ_i, $i = 0, 1$ denote the local basis functions. But since

$$\varphi_0(x) + \varphi_1(x) \equiv 1$$

for all $x \in [x_{j-1}, x_j]$, we have

$$
\begin{aligned}
f(x) - \Pi_n^1 f(x) &= f(x)(\varphi_0(x) + \varphi_1(x)) \\
&\quad - (f(x_{j-1})\varphi_0(x) + f(x_j)\varphi_1(x)) \\
&= \varphi_0(x)(f(x) - f(x_{j-1})) + \varphi_1(x)(f(x) - f(x_j)).
\end{aligned}
\tag{12.95}
$$

Since each φ_i has values only between 0 and 1,

$$
\begin{aligned}
|f(x) - \Pi_n^1 f(x)| &\leq \varphi_0(x)|f(x) - f(x_{j-1})| + \varphi_1(x)|f(x) - f(x_j)| \\
&\leq (\varphi_0(x) + \varphi_1(x))\,\omega_I(f; \delta) \\
&= \omega_I(f; \delta).
\end{aligned}
\tag{12.96}
$$

Since this holds for any $x \in [x_{j-1}, x_j]$ and for any j,

$$
\|f - \Pi_n^1 f\|_{\infty, I} \leq \omega_I(f; \delta).
\tag{12.97}
$$

Applying exercise 12.13 completes the proof, with $c = c_1 = 1$.

Solution of Exercise 12.21. Define polynomials $v_i(x) = x^{i-1}$. Then

$$
(v_i, v_j) = \int_0^1 x^{i+j-2}\, dx = \frac{1}{i+j-1},
\tag{12.98}
$$

in accord with (4.15).

Chapter Thirteen

Numerical Quadrature

> "We now recognize these schemes as examples of fixed-point (or functional) iteration. Similar schemes were previously proposed by James Gregory and communicated in letters to John Collins to solve the equations $b^n c + x^{n+1} = b^n x$ (8 November 1672) and $b^n c + x^{n+1} = b^{n-1}(b + c)x$ (2 April 1674)" [174].

The word *quadrature* is used in numerical analysis to denote approximate integration. We will see that some of the ideas predate the formal notions of the calculus. The most commonly used approaches involve polynomial interpolation as the basis.

Numerical quadrature may appear superficially as one of the simplest subjects covered so far. But we will also see that it introduces some of the deepest notions in analysis.

13.1 INTERPOLATORY QUADRATURE

The idea behind interpolatory quadrature is to define the approximate integral as the integral of an interpolant (or other approximant):

$$Qf = \int_a^b Lf(x)\,dx = \sum_{i=0}^n f(x_i) \int_a^b \phi_i(x)\,dx = \sum_{i=0}^n \alpha_i f(x_i), \qquad (13.1)$$

where the *quadrature coefficients* α_i are defined by

$$\alpha_i := \int_a^b \phi_i(x)\,dx. \qquad (13.2)$$

Here L denotes one of the operators we have constructed:

- Lagrange or Hermite interpolation (with the points chosen according to various objectives),

- piecewise polynomial interpolant (this is called a composite rule), or

- Bernstein (this is an unusual choice, but we will explore its properties briefly).

The basis functions ϕ_i are the basis functions for the Lagrange interpolation given in (11.33). Least-squares approximation does not lead to such a quadrature rule directly because it is defined in terms of an integral, but we will see that there is an intimate connection with Gaussian quadrature (section 13.1.3).

The quadrature error is easy to estimate for interpolatory quadrature:

$$Qf - \int_a^b f(x)\, dx = \int_a^b Lf(x) - f(x)\, dx. \tag{13.3}$$

Thus we have, for example,

$$\left| Qf - \int_a^b f(x)\, dx \right| \le (b-a)\|Lf - f\|_{\infty,[a,b]}, \tag{13.4}$$

so that we can apply estimates previously derived. For example, if L refers to Lagrange interpolation, then (10.9) implies that

$$\left| Qf - \int_a^b f(x)\, dx \right| \le \frac{(b-a)}{(n+1)!}\|f^{(n+1)}\|_{\infty,[a,b]}\|\omega_{n+1}\|_{\infty,[a,b]}, \tag{13.5}$$

where ω_k is defined in (10.8).

13.1.1 Newton-Cotes formulas

A Newton-Cotes[1] formula is based on choosing Lagrange interpolation with equally spaced points. There are two types of Newton-Cotes quadrature rules: open and closed. With the closed rules, the endpoints of the interval of integration are included as quadrature (interpolation) points. For example, the closed rule with two points is called the *trapezoidal rule*,

$$\int_a^b f(x)\, dx \approx \frac{b-a}{2}(f(a) + f(b)) =: Q_{\mathrm{TR}}f, \tag{13.6}$$

and the one with three points is called *Simpson's rule*[2]

$$\int_a^b f(x)\, dx \approx \frac{b-a}{6}(f(a) + 4f(\tfrac{1}{2}(b+a)) + f(b)) =: Q_{\mathrm{SR}}f. \tag{13.7}$$

The Newton-Cotes open rule with one point is known as the *midpoint rule*:

$$\int_a^b f(x)\, dx \approx (b-a)f(\tfrac{1}{2}(b+a)) =: Q_{\mathrm{MR}}f. \tag{13.8}$$

Error estimates can be derived directly from (13.5). For example, with the trapezoidal rule, $n = 1$ and $\omega_2(x) = (x-a)(x-b)$ has its maximum at $x = \tfrac{1}{2}(a+b)$, so that $\|\omega_2\|_{\infty,[a,b]} = \tfrac{1}{4}(b-a)^2$. Therefore,

$$\left| Q_{\mathrm{TR}}f - \int_a^b f(x)\, dx \right| \le \frac{(b-a)^3}{8}\|f^{(2)}\|_{\infty,[a,b]}. \tag{13.9}$$

---

[1]Roger Cotes (1682–1716) was nearly 40 years younger than Newton but became a close colleague and "is best known for his meticulous and creative editing of the second edition of Newton's *Principia*," done jointly with Newton [69].

[2]See pages 21 and 97 for information on Simpson.

Applying the same technique to Simpson's rule would imply an error of order $(b-a)^4$. However, for Simpson's rule, a better estimate can be obtained (exercise 13.1). Similarly, the midpoint rule has a higher-order of accuracy (the same as the trapezoidal rule) than would be implied by (13.5) (exercise 13.2). In section 13.1.2, we show how these rules, and others, can be treated in a uniform way. The key point is that a given quadrature rule can be derived from different approximation schemes, and we are free to pick the approximation scheme that produces the best error estimate. Not surprisingly, we will then see that the error in a quadrature rule is related to the problem of best approximation (section 12.1).

For all the Newton-Cotes rules, the quadrature coefficients (13.2) are proportional to the interval length $b-a$ and independent of the base point a (exercise 13.3). It is of interest to know whether the coefficients (13.2) are positive or not. For the open Newton-Cotes rules, they are not all positive, for example, for the rules with three, five, and six points. When the coefficients are negative, they are not suitable for certain applications. For example, it may be a requirement that the approximate integral of a nonnegative function be nonnegative.

13.1.2 Order of exactness

In general, one is interested in quadrature rules for weighted integrals of the sort we considered in section 12.3.2:

$$Qf = \int_a^b Lf(x)w(x)\,dx = \sum_{i=0}^n f(x_i) \int_a^b \phi_i(x)w(x)\,dx = \sum_{i=0}^n \alpha_i f(x_i).$$
(13.10)

The determining factor for error estimates for quadrature rules is not estimates for the interpolant as in (13.5) but rather is determined by their *order of exactness* together with a stability estimate.

Definition 13.1 *We say that a quadrature is exact for polynomials of degree k if*

$$Qp = \int_a^b p(x)w(x)\,dx$$
(13.11)

for all $p \in \mathcal{P}_k$.

Thus we see that (for $w \equiv 1$) Simpson's rule is exact for cubic polynomials (exercise 13.1), and the midpoint rule is exact for linear functions (exercise 13.2), as is trapezoidal rule.

Any quadrature rule may be viewed as a linear functional (cf. (11.31)), meaning a linear map from a vector space to the set of scalars, more specifically (exercise 13.4),

$$Q(f + cg) = Qf + cQg$$
(13.12)

for any continuous functions f and g and any scalar c. If a quadrature rule
(13.10) is exact for polynomials of degree k, then for any $p \in \mathcal{P}_k$,

$$
\left| Qf - \int_a^b f(x)w(x)\,dx \right| = \left| Q(f-p) - \int_a^b (f(x) - p(x))w(x)\,dx \right|
$$

$$
\leq |Q(f-p)| + \left| \int_a^b (f(x) - p(x))w(x)\,dx \right| \tag{13.13}
$$

$$
\leq \left(\sum_{i=1}^n |\alpha_i| \right) \|f - p\|_\infty + \int_a^b w(x)\,dx\, \|f - p\|_\infty = C\|f - p\|_\infty.
$$

Note that if all $\alpha_i > 0$, then

$$
\sum_{i=1}^n |\alpha_i| = \sum_{i=1}^n \alpha_i = \int_a^b w(x)\,dx, \tag{13.14}
$$

assuming that the quadrature rule is exact at least for constants. Thus we
have proved the following result, which reduces error estimates for quadrature to the previously studied problem of best approximation.

Theorem 13.2 *Suppose that w is a nonnegative, integrable weight function
and that the quadrature rule (13.10) is exact for polynomials of degree k.
Then there is a constant C such that for all $f \in C^0([a, b])$,*

$$
\left| Qf - \int_a^b f(x)w(x)\,dx \right| \leq C \min_{p \in \mathcal{P}_k} \|f - p\|_{\infty, [a,b]}. \tag{13.15}
$$

If all quadrature weights α_i are positive, then we may take

$$
C = 2 \int_a^b w(x)\,dx. \tag{13.16}
$$

13.1.3 Gaussian quadrature

In the Newton-Cotes formulas, the points are fixed at specified points. The
corresponding order of accuracy is approximately equal to the number of
points, with the proviso that the accuracy (cf. (13.9)) can increase by 1
because of symmetry, as occurs with Simpson's rule. One can pose the
quadrature problem as finding x_i's and α_i's so that

$$
\int_a^b p(x)w(x)\,dx = \sum_{i=0}^n \alpha_i p(x_i) \quad \forall p \in \mathcal{P}_k \tag{13.17}
$$

for k as large as possible. With the x_i's fixed, the system (13.17) is linear,
and the α_i's are just the integrals of the corresponding interpolation basis
functions. But if we allow the x_i's to be variables, we have the possibility
of getting exactness in (13.17) for a larger k, but at the expense of having a
nonlinear system to solve for the x_i's.

Gaussian quadrature may be defined by taking the points x_i such that we
get a formula exact for as high a degree as possible. Stated as a system of

equations, it is highly nonlinear. With n values of x_i's and n values of α_i's, we might expect to integrate a polynomial of degree $2n - 1$ exactly since the dimension of $\mathcal{P}_{2n-1}$ is $2n$.

By symmetry, the midpoint rule gives the optimal solution for $n = 1$ when $w \equiv 1$. We propose in exercise 13.5 to solve this problem for $n = 2$. However, proceeding in this way for higher degrees would be tedious. Fortunately, if we take the x_i's to be the roots of the orthogonal polynomial P_n, all is well. First, we know the roots are in the interval in question and that they are distinct (see section 12.3.3). We will refer to these roots as the *Gauss points* and let L_n^G be Lagrange interpolation at these points. We denote by Q_n^G the corresponding interpolatory quadrature rule.

Suppose that $f \in \mathcal{P}_{2n-1}$. Then $f - L_n^G f$ vanishes at the roots of P_n, so we can write $f - L_n^G f = P_n q$, where $q \in \mathcal{P}_{n-1}$. Therefore,

$$Q_n^G f - \int_a^b f(x)\,dx = \int_a^b L_n^G f(x) - f(x)\,dx = \int_a^b P_n(x)q(x)\,dx = 0 \quad (13.18)$$

because P_n is orthogonal to $\mathcal{P}_{n-1}$.

Fortuitously, the coefficients α_i are positive. Let $f(x) = P_n(x)^2/(x - x_i)^2$. By (13.18), since the degree of f is $2n - 2$,

$$\alpha_i f(x_i) = Q_n^G f = \int_a^b f(x)\,dx > 0 \quad (13.19)$$

since f is positive except at the x_j's, where it vanishes for all $j \neq i$. We also have $f(x_i) \neq 0$ since P_n has only a simple zero there. But we also know that $f(x) > 0$ for x near x_i (f is the square of a function that is not zero near x_i), so we must have $f(x_i) > 0$ as well. Since α_i is the quotient of positive terms, it must be positive.

Thus we have proved the following result.

Theorem 13.3 *Let x_i be the roots of the orthogonal polynomial P_n (the Gauss points) and let the quadrature coefficients be defined by (13.2) for the corresponding Lagrange interpolation basis functions. Then the resulting Gaussian quadrature (13.10) is exact for polynomials of maximum degree $2n - 1$.*

13.1.4 Hermite quadrature

Any approximation scheme can be used to create a quadrature rule via the recipe

$$Q_{\text{gen}} f = \int_a^b G_n f(x) w(x)\,dx \quad (13.20)$$

for a general approximation operator G_n of the form (11.34). Hermite interpolation (section 11.4.2) is one example that introduces a new ingredient. In this case we have

$$Q_\text{H} f = \frac{b-a}{2}(f(a) + f(b)) + \frac{(b-a)^2}{12}(f'(a) - f'(b)), \quad (13.21)$$

where the coefficients can be verified by various means. One approach of course is to evaluate the integrals of the basis functions in (13.2). For $a = 0$ and $b = 1$, the basis functions are $\phi_0(x) = 1 - 3x^2 + 2x^3$ and $\phi_1(x) = x(1 - x)^2$ for the value and derivative nodes at $x = 0$. The corresponding basis functions at $x = 1$ are $\phi_0(1 - x)$ and $-\phi_1(1 - x)$. Note the Q_H is again a linear functional, but now it is defined only on C^1-functions.

Consider the interpolation scheme implied in lemma 11.6. This suggests that there is a quadrature rule of the form

$$Q_k^{\text{EM}} f = \frac{b - a}{2}(f(a) + f(b)) + \sum_{i=1}^{k} c_i (b - a)^{2i}(f^{(2i-1)}(a) - f^{(2i-1)}(b)) \quad (13.22)$$

that is exact for polynomials of degree $2k+1$ and defined on C^{2k-1}-functions. Here $c_1 = \frac{1}{12}$, in keeping with the case $k = 1$ in which (13.22) is just the Hermite quadrature. This is basis of the Euler-Maclaurin[3] formula (13.25); we will see that the coefficients c_i can be identified in general (section 13.3).

13.1.5 Composite rules

There are two ways to think of deriving *composite rules*. First, we start with a subdivision of the interval $a = \xi_0 < \xi_1 < \cdots < \xi_n = b$. We then apply one of the previously discussed methods to each interval $[\xi_{i-1}, \xi_i]$ for $i = 1, \ldots, n$. For example, if we apply the trapezoidal rule to each interval, we obtain the rule

$$\tfrac{1}{2} h_1 f(a) + \sum_{i=1}^{n-1} \tfrac{1}{2}(h_i + h_{i+1}) f(\xi_i) + \tfrac{1}{2} h_n f(b), \quad (13.23)$$

where $h_i = \xi_i - \xi_{i-1}$ for $i = 1, \ldots, n$. The same quadrature rule arises from (13.1) if we define L to be continuous piecewise linear interpolation (section 12.4) using the points ξ_i.

It is interesting to consider the composite trapezoidal rule on a regular subdivision: $\xi_j = a + jh$, where $h = (b - a)/n$. The quadrature rule then is

$$h\left(\tfrac{1}{2} f(a) + \sum_{i=1}^{n-1} f(\xi_i) + \tfrac{1}{2} f(b)\right). \quad (13.24)$$

Except for the endpoints, this is a very simple rule, and yet it is very powerful, as we will see shortly.

Suppose we consider the composite version of (13.22). The odd-order derivative terms all cancel in the intermediate intervals, and we obtain the

---

[3] Colin Maclaurin (1698–1746) entered the University of Glasgow in 1709 at the age of 11 and was awarded an M.A. at age 14. By 1717 he was a professor. He was significant for his clarification of the ideas of Newton, who supported the appointment of Maclaurin to the University of Edinburgh in 1725 [163].

quadrature rule

$$h\left(\tfrac{1}{2}f(a) + \sum_{i=1}^{n-1} f(\xi_i) + \tfrac{1}{2}f(b)\right)$$
$$+ \sum_{i=1}^{k} c_i h^{2i}(f^{(2i-1)}(a) - f^{(2i-1)}(b)) \tag{13.25}$$

that is exact for polynomials of degree $2k + 1$. It is again defined only on C^{2k-1} functions. We can think of this as the trapezoidal rule with endpoint corrections. This formula is attributed to Euler and Maclaurin. We will provide an alternate derivation that identifies the coefficients c_i (cf. section 13.3).

We can make the composite trapezoidal rule even simpler for periodic functions. For simplicity, let us assume that $a = 0$ and $b = 1$ and that f is 1-periodic. Then $f(0) = f(1)$, and all the derivative corrections cancel, so (13.25) simplifies further to

$$\frac{1}{n} \sum_{i=1}^{n} f(i/n). \tag{13.26}$$

Theorem 13.4 *Suppose that f is a 1-periodic function. Then the trapezoidal rule (13.26) is exact to any order; that is, if $f \in C^{2k+1}([0,1])$, then*

$$\left| \int_0^1 f(x)\,dx - \frac{1}{n} \sum_{i=1}^{n} f(i/n) \right| \le C_k n^{-2k-1} \| f^{(2k+1)} \|_\infty \tag{13.27}$$

for any value of $n \ge 1$, where C_k is a constant that depends only on k.

We postpone the proof of this theorem, as it is a simple corollary of the Peano kernel theorem; cf. section 13.2. As an example of the use of the trapezoidal rule for a periodic function, we consider the integral [41]

$$\int_0^1 \frac{dt}{1 + \tfrac{1}{2}\sin(2\pi t)}. \tag{13.28}$$

Computational results are shown for various values of n in table 13.1.

13.2 PEANO KERNEL THEOREM

There is a general abstract result due to Peano[4] that gives a representation of the error for a wide class of numerical approximations. The error in quadrature is a typical example. Consider the setup in theorem 13.2 and define

$$Ef = Qf - \int_a^b f(x)w(x)\,dx. \tag{13.29}$$

---

[4]Giuseppe Peano (1858–1932) is best known for his contributions to the foundations of mathematics. But he also did research on numerical analysis [127].

n	Integral	Error
3	1.15384615384615	8.5×10^{-4}
5	1.15469613259669	4.4×10^{-6}
7	1.15470051566839	2.3×10^{-8}
9	1.15470053**8**26218	1.2×10^{-10}
11	1.15470053837**8**65	6.0×10^{-13}

Table 13.1 Errors in computing the integral (13.28) via the trapezoidal rule with n points. The exact answer is 1.15470053837925, which is obtained with $n = 13$ and does not change for larger n. The bold face digits are the first incorrect digits for each n.

Note that $EP = 0$ for all polynomials of degree k, where k is the order of exactness of Q, and that E is linear,

$$E(f + cg) = Ef + cEg, \qquad (13.30)$$

as long as the same is true of Q, since this holds for the integral. In particular, $Ef = E(f - P)$ for any polynomial P of degree k.

Recall Taylor's theorem with integral remainder (7.81):

$$f(x) - P_k(x) = \frac{1}{k!} \int_a^x (x - t)^k f^{(k+1)}(t) \, dt, \qquad (13.31)$$

where P_k is the Taylor polynomial

$$P_k(x) = \sum_{j=0}^k \frac{f^{(j)}(a)}{j!} (x - a)^j. \qquad (13.32)$$

Let us use the notation $(X)_+$ to mean X if $X \geq 0$ and 0 if $X \leq 0$. Then we can rewrite (13.31) as

$$f(x) - P_k(x) = \frac{1}{k!} \int_a^b (x - t)_+^k f^{(k+1)}(t) \, dt. \qquad (13.33)$$

Since E is linear, we have

$$Ef = E(f - P) = \frac{1}{k!} E \left[\int_a^b (x - t)_+^k f^{(k+1)}(t) \, dt \right] \qquad (13.34)$$
$$= \frac{1}{k!} \int_a^b E \left[(x - t)_+^k \right] f^{(k+1)}(t) \, dt.$$

The last equality may seem like a leap of faith, and in any case the notation needs to be made more precise. Define

$$\phi(x) = \int_a^b (x - t)_+^k f^{(k+1)}(t) \, dt \qquad (13.35)$$

for $x \in [a, b]$. Then (13.33) says that $f - P_k = (k!)^{-1} \phi$, so $Ef = (k!)^{-1} E\phi$. Similarly, define a one-parameter family of functions $\psi_t^k(x) = (x - t)_+^k$ for $x \in [a, b]$ and let

$$K(t) = E\psi_t^k. \qquad (13.36)$$

Then we claim that

$$E\phi = \int_a^b K(t) f^{(k+1)}(t)\, dt. \tag{13.37}$$

The proof of (13.37) relies on the linearity of E and the linearity of the integration process. For example, this can be verified by approximating the integral by Riemann sums (exercise 13.6). Thus we have proved the following.

Theorem 13.5 *Suppose that the quadrature Q is linear, exact of order k, and defined on $C^{k+1}([a,b])$. Then the error E defined by (13.29) satisfies*

$$Ef = \frac{1}{k!} \int_a^b K(t) f^{(k+1)}(t)\, dt, \tag{13.38}$$

where K is defined by (13.36).

The function K is called the *Peano kernel* for this error relation. We can provide an error estimate using the Peano kernel:

$$|Ef| \le \frac{1}{k!} \int_a^b |K(t)|\, dt\, \|f^{(k+1)}\|_{\infty,[a,b]}, \tag{13.39}$$

which can be compared with (13.5) (see exercise 13.7).

There is a not so small missing detail here, namely, whether the function defined in (13.36) is integrable in an appropriate sense (exercise 13.6). Nevertheless, let us try to develop some general rules about the Peano kernels. It may not be clear why K is well-defined at all since it involves the application of Q to the function ψ_t^k, which is not so smooth. For $t \le x$, $\psi_t^k \equiv 0$, and so the kth derivative of ψ_t^k is discontinuous at $x = t$. However, it is easy to see that $\psi_t^k \in C^{k-1}(\mathbb{R})$ and

$$\begin{aligned}
K'(t) &= \lim_{h \to 0} h^{-1}\left(K(t+h) - K(t)\right) = \lim_{h \to 0} h^{-1}\left(E\psi_{t+h}^k - E\psi_t^k\right) \\
&= \lim_{h \to 0} h^{-1} E\left(\psi_{t+h}^k - \psi_t^k\right) = E \lim_{h \to 0} h^{-1}\left(\psi_{t+h}^k - \psi_t^k\right) \\
&= -kE\left[(x-t)_+^{k-1}\right] = -kE\psi_t^{k-1},
\end{aligned} \tag{13.40}$$

provided that E is well-defined for functions in C^{k-2}. By definition, $\psi_t^0(x)$ is the Heavyside function that is 0 for $x < t$ and 1 for $x > t$.

When $t = a$, $\psi_a^k(x) = x^k$ on $[a,b]$, so we have $K(a) = 0$ because Q is exact of order k. Similarly, when $t = b$, $\psi_b^k \equiv 0$ on $[a,b]$, so again $K(b) = 0$. Therefore, (13.38) implies that

$$K^{(i)}(a) = K^{(i)}(b) = 0 \tag{13.41}$$

for $i = 0, 1, \dots, k-1-m$, provided that Qf is well-defined for $f \in C^m([a,b])$. In the case of the Hermite quadrature rule (13.21), we have $m = 1$.

Now let us see if we can figure out what K might look like in examples. Let us start with Q = midpoint rule on $[0,1]$, which is exact for polynomials of degree $k = 1$. In this case, the statement is

$$Ef = f(\tfrac{1}{2}) - \int_0^1 f(t)\, dt = \int_0^1 K_{\mathrm{MR}}(t) f^{(2)}(t)\, dt. \tag{13.42}$$

The quadrature rule $Qf = f(\frac{1}{2})$ is well-defined for $f \in C^0$, so we conclude from (13.40) that $K_{MR} \in C^0$ and that K'_{MR} is defined for $x \neq \frac{1}{2}$ and bounded. Thus we can integrate by parts to find

$$Ef = f(\tfrac{1}{2}) - \int_0^1 f(t)\,dt = -\int_0^1 K_{MR}^{(1)}(t)f^{(1)}(t)\,dt. \qquad (13.43)$$

We can integrate by parts again, but we have to be careful since K_{MR} is not C^1. However, the only point where K_{MR} fails to be smooth is $x = \frac{1}{2}$, and so we can break the integral into two parts and integrate by parts again. To make a long story short, we find that

$$K_{MR}(t) = -\begin{cases} \frac{1}{2}t^2 & t \leq \frac{1}{2} \\ \frac{1}{2}(t-1)^2 & t \geq \frac{1}{2}. \end{cases} \qquad (13.44)$$

We leave as exercise 13.8 verification that this K_{MR} satisfies (13.42) for all $f \in C^2$. Similarly, it is not hard to see (exercise 13.7) that the kernel for the trapezoidal rule is

$$K_{TR}(t) = \tfrac{1}{2}t(1-t) \qquad (13.45)$$

and the kernel for Hermite quadrature (13.21) is

$$K_H(x) = -\tfrac{1}{24}x^2(1-x)^2. \qquad (13.46)$$

We will consider the form of the general kernels K_k^{EM} for the Euler-Maclaurin quadrature subsequently.

If we make a simple change of variables in the integration, the Peano kernel changes in a predictable way. Suppose that $\widehat{K}$ denotes the Peano kernel for the interval $[0, 1]$. Then the kernel for the interval $[a, a + h]$ is

$$K(a + ht) = h^k \widehat{K}(t), \qquad (13.47)$$

where k is the order of exactness.

For the Euler-Maclaurin formula (13.25), we have

$$h\left(\tfrac{1}{2}f(a) + \sum_{i=1}^{n-1} f(\xi_i) + \tfrac{1}{2}f(b)\right) + \sum_{i=1}^{k} c_i h^{2i}(f^{(2i-1)}(a) - f^{(2i-1)}(b))$$

$$= \int_a^b f(x)\,dx + h^{2k+1} \sum_{i=0}^{n-1} \int_0^1 K_k^{EM}(x)f^{(2k+1)}(a + h(i + x))\,dx. \qquad (13.48)$$

This completes the proof of theorem 13.4. The kernels K_k^{EM} are related to the Bernoulli polynomials [41, 100].

13.3 GREGORIE-EULER-MACLAURIN FORMULAS

Gregorie[5] developed a formula for numerical integration that predated, or at least was contemporary with, the work of Newton on calculus. This

---

[5] James Gregorie (1638–1675), a.k.a. James Gregory, a Scottish mathematician and astronomer, was successively professor at the University of St. Andrews and the University of Edinburgh. He had "a reputation among his peers second only to that of Newton" [162].

formula is also related to the formula (13.25) attributed later to Euler and Maclaurin. The Gregorie formula has been utilized in codes for solving partial differential equations [16]. The following derivation of these formulas provides an application of operator calculus.

13.3.1 More operator calculus

In order to compute the coefficients arising in the Euler-Maclaurin formula (13.25), we make a small detour to develop further the technology regarding operators on function spaces that we began in section 9.3.1. We have seen many such operators so far, but we now treat them as abstractions in which we will view them much like a point in the complex plane. We make a formal analogy between functions of a complex variable and corresponding functions of operators. We begin with an example.

Let $h > 0$ be fixed. We define the difference operator Δ by

$$\Delta f(x) = f(x + h) - f(x). \tag{13.49}$$

This operator makes sense for any $f \in C^0(\mathbb{R})$, but we will often restrict the operators in this discussion to the set of polynomials (a dense subset of C^0 at least on finite intervals; cf. section 12.2). We have also used the notation D for the derivative operator (7.17), i.e.,

$$Df(x) = f'(x) \tag{13.50}$$

in the one-dimensional case. This operator is no longer defined on all of C^0, so we restrict it always to polynomials. More precisely, we define the vector space (exercise 13.11) $\mathcal{P}_\infty$ by

$$\mathcal{P}_\infty = \cup_{k=0}^\infty \mathcal{P}_k. \tag{13.51}$$

Then both Δ and D map $\mathcal{P}_\infty \to \mathcal{P}_\infty$. Note that although $\mathcal{P}_\infty$ is infinite-dimensional, each $P \in \mathcal{P}_\infty$ has a finite degree.

It is not surprising that we could find a formal relationship between D and Δ. The Taylor expansion (see exercise 7.4)

$$f(x + h) = \sum_{k=0}^\infty \frac{h^k f^{(k)}(x)}{k!} = \sum_{k=0}^\infty \frac{(hD)^k f(x)}{k!} \tag{13.52}$$

(valid for any polynomial f) leads to the relationship

$$\Delta f(x) = \sum_{k=1}^\infty \frac{(hD)^k f(x)}{k!} = \sum_{k=1}^\infty \frac{(hD)^k}{k!} f(x) \quad \forall f \in \mathcal{P}_\infty. \tag{13.53}$$

For any polynomial f, the sum in (13.53) is finite, so there are no convergence issues. The function represented by the series in (13.53) is familiar since we can write

$$\zeta(z) = \sum_{k=1}^\infty \frac{z^k}{k!} = e^z - 1. \tag{13.54}$$

Replacing z by hD formally, we obtain

$$\boxed{\Delta f(x) = \zeta(hD)f(x).} \tag{13.55}$$

We now explain how to make this rigorous.

We now generalize the operator calculus derived in section 9.3.1 for matrices, especially (9.48), to operators on polynomials. Of course, this is not exactly a generalization since any operator on a finite-dimensional vector space can be written as a matrix. But we want to use the calculus for operators on $\mathcal{P}_\infty$ which is infinite-dimensional, so it makes sense to approach the theory more abstractly.

We know that for any linear operator T, it makes sense to talk about powers of T, e.g., $T^2 f = T(Tf)$, and $T^k f = T(T^{k-1} f)$ is defined by induction. We again define $T^0 = I$, where I denotes the identity operator $If = f$ for all f. Thus any polynomial $p(z) = \sum_{k=1}^n c_i z^i$ can be applied to T to get $p(T)$ by summing all the monomials T^k with appropriate scalar coefficients:

$$p(T)f = \sum_{k=1}^n c_i T^i f. \tag{13.56}$$

It also makes sense to talk about the infinite sum $\zeta(D)f$ for any polynomial f since it involves only finitely many terms in the sum in (13.54). Thus we have proved the following result.

Lemma 13.6 *For any polynomial $f \in \mathcal{P}_\infty$, we have*

$$\Delta f(x) = \zeta(hD)f(x) = e^{hD}f(x) - f(x) \quad \forall x \in \mathbb{R}, \tag{13.57}$$

where $\zeta(z) = e^z - 1$.

lemma 13.6 provides an explicit relationship between Δ and D:

$$\Delta = e^{hD} - I, \tag{13.58}$$

where I is the identity operator (and h is the parameter in the definition of Δ). This uses the following fact that we leave as exercise 13.12:

$$\sum_{k=0}^\infty (c_k + b_k)T^k f = \sum_{k=0}^\infty c_k T^k f + \sum_{k=0}^\infty b_k T^k f \quad \forall f \in \mathcal{P}_\infty. \tag{13.59}$$

To make this a bit more formal, we need to say what operators T are allowed and show at least that $T = hD$ is one of them. Note that both $T = D$ and $T = \Delta$ have the property that the degree of Tf is 1 less than the degree of f for any polynomial f. Thus both D and Δ are in the following set of operators:

$$\mathcal{T} = \left\{ T : \mathcal{P}_\infty \to \mathcal{P}_\infty \mid \forall f \in \mathcal{P}_\infty \, \exists \hat{k} < \infty \text{ such that } T^{\hat{k}} f \equiv 0 \right\}. \tag{13.60}$$

Note that if $T^{\hat{k}} f \equiv 0$, then $T^k f \equiv 0$ for all $k > \hat{k}$ as well. Thus the set $\mathcal{T}$ comprises the operators for which infinite expressions like (13.59) always reduce to finite expressions for any given $f \in \mathcal{P}_\infty$. Restricting to this set greatly simplifies convergence arguments.

13.3.2 Product formula

The summation rule (13.59) for functions of operators is elementary, but the corresponding rule for products is more subtle. Suppose that $\mu(z) = \sum_{k=0}^{\infty} c_k z^k$ and $\nu(z) = \sum_{k=0}^{\infty} b_k z^k$ are power series that converge for $|z| < \epsilon$ for some $\epsilon > 0$. Define $\upsilon(z) = \mu(z)\nu(z)$, which has the power series $\upsilon(z) = \sum_{k=0}^{\infty} a_k z^k$, where

$$a_k = \sum_{i=0}^{k} c_i b_{k-i} \tag{13.61}$$

(see exercise 13.13).

Lemma 13.7 *Let $T \in \mathcal{T}$. If μ and ν are power series as above and $\upsilon(z) = \mu(z)\nu(z)$, then $\upsilon(T)f = \mu(T)(\nu(T)f)$ for any polynomial f.*

Proof. Let $g = \nu(T)f = \sum_{k=0}^{\hat{k}} b_k T^k f$, where $\hat{k}$ is chosen depending on $f \in \mathcal{P}_{\infty}$ according to the defining property of (13.60). In particular, we conclude that $g \in \mathcal{P}_{\infty}$, and we can then write

$$\mu(T)g = \sum_{k=0}^{\hat{k}} c_k T^k g, \tag{13.62}$$

where we have increased the value of $\hat{k}$ using (13.60) as necessary. Multiplying the two expressions gives the desired result.   QED

13.3.3 Inverse operators

Now we consider finding inverses in the operator calculus. There is a new ingredient in that in addition to power series, we need to add the symbol $1/z$ to the set of functions that we can apply to an operator $T \in \mathcal{T}$. Since 1 is the symbol for the identity operator and z is the symbol for the operator itself, then $1/z$ should be the symbol for the inverse since $z(1/z) = 1$, which is consistent with our calculus for products of power series. We already saw this when we applied the expression

$$(1 - z)^{-1} = \sum_{k=0}^{\infty} z^k \tag{13.63}$$

to write the inverse of the matrix $I - M$ as

$$(I - M)^{-1} = \sum_{k=0}^{\infty} M^k \tag{13.64}$$

for a convergent matrix M (section 8.1). We now apply this idea to compute an expansion for $\eta(z) = 1/\zeta(z)$, where ζ was defined in (13.54).

First, note that (see exercise 13.14)

$$\eta(z) = 1/\zeta(z) = \tfrac{1}{2}(-1 + \coth \tfrac{1}{2}z) \tag{13.65}$$

and recall that the hyperbolic cotangent has the expansion

$$\frac{1}{2}\coth\frac{1}{2}z = \sum_{k=0}^{\infty} \frac{B_{2k}}{(2k)!}z^{2k-1}$$

$$= (1/z) + \frac{1}{12}z - \frac{1}{720}z^3 + \cdots,$$

(13.66)

where B_n is the nth Bernoulli[6] number (cf. exercises 3.11 and 13.16). We will show that $\eta(hD)$ is the inverse of $\zeta(hD)$, in the sense that for any polynomial f, we have $f = \zeta(hD)\eta(hD)f$.

Lemma 13.8 *Suppose that* $\zeta(z) = \sum_{k=1}^{\infty} c_k z^k$ *and that*

$$\zeta(z)^{-1} = 1/z + \sum_{k=0}^{\infty} b_k z^k,$$

in the sense that both series converge for $|z| < \epsilon$ *and*

$$\zeta(z)\left(1/z + \sum_{k=0}^{\infty} b_k z^k\right) = 1$$

(13.67)

for all z *in* $0 < |z| < \epsilon$. *Suppose that* $T \in \mathcal{T}$ *has a right inverse* $R\colon TRf = f$ *for all* $f \in \mathcal{P}_\infty$. *Then the operator* U *defined by* $U = R + \sum_{k=0}^{\infty} b_k T^k$ *satisfies* $\zeta(T)Uf = f$ *for any polynomial* f.

Note that we require that the operator $R: \mathcal{P}_\infty \to \mathcal{P}_\infty$, but we do not require that $R \in \mathcal{T}$.

Proof. The expression $\zeta(z)\left(1/z + \sum_{k=0}^{\infty} b_k z^k\right) = 1$ means that

$$\zeta(z)\nu(z) = 1 - \mu(z),$$

(13.68)

where $\mu(z) = \sum_{k=1}^{\infty} c_k z^{k-1}$ and $\nu(z) = \sum_{k=0}^{\infty} b_k z^k$. In particular, $Uf = Rf + \nu(T)f$.

We have $\zeta(z) = \mu(z)z$, so that

$$\zeta(T)(Uf) = \mu(T)(TUf) = \mu(T)(f + T\nu(T)f)$$

(13.69)

for any $f \in \mathcal{P}_\infty$. But rewriting (13.68) gives $1 - \mu(z) = \zeta(z)\nu(z) = \mu(z)z\nu(z)$, so that $1 = \mu(z)(1 + z\nu(z))$, as required. QED

Combining lemmas 13.6 and 13.8 shows that the following holds.

Corollary 13.9 *For any polynomial* $f \in \mathcal{P}_\infty$, *we have*

$$f(x) = \Delta\eta(hD)f(x),$$

(13.70)

where $\eta(z) = 1/\zeta(z) = (e^z - 1)^{-1} = \frac{1}{2}(-1 + \coth\frac{1}{2}z)$.

---

[6]See page 49.

13.3.4 The Euler-Maclaurin formula

To understand corollary 13.9, we need to interpret the meaning of the term $1/z$ when D is substituted for z. The inverse of differentiation is integration:

$$D^{-1}f(x) = \int^x f(s)\,ds, \tag{13.71}$$

but the inverse is not uniquely defined. That is, define

$$\mathcal{I}_c f(x) = \int_c^x f(s)\,ds \tag{13.72}$$

for any constant c, and then we have $D\mathcal{I}_c f = f$ for any polynomial f. What this means is that $\mathcal{I}_c$ is a right inverse for D for any c, and thus we can write $I = \zeta(D)\eta(D)$, but the interpretation of $\eta(D)\zeta(D)$ is problematic. The formal interpretation of $\eta(hD)$ is then

$$\eta(hD) = h^{-1}\mathcal{I}_c - \tfrac{1}{2}I + \tfrac{1}{12}hD - \tfrac{1}{720}(hD)^3 + \cdots . \tag{13.73}$$

Fortunately,

$$\Delta\mathcal{I}_c f(x) = \int_x^{x+h} f(s)\,ds, \tag{13.74}$$

which is independent of the value of c, so we have

$$f(x) = h^{-1}\int_x^{x+h} f(s)\,ds - \tfrac{1}{2}\Delta f(x) + \tfrac{1}{12}h\Delta Df(x) - \tfrac{1}{720}h^3\Delta D^3 f(x) + \cdots . \tag{13.75}$$

Reorganizing (13.75), we find

$$\int_x^{x+h} f(s)\,ds = \tfrac{1}{2}h(f(x+h) + f(x)) - \tfrac{1}{12}h^2\Delta Df(x) + \tfrac{1}{720}h^4\Delta D^3 f(x) + \cdots . \tag{13.76}$$

Summing this as we do for a composite quadrature rule, we find

$$\int_a^b f(s)\,ds = h\sum_{i=1}^{n-1} f(a+ih) + \tfrac{1}{2}h(f(a) + f(b))$$
$$- \tfrac{1}{12}h^2\left(f'(b) - f'(a)\right) + \tfrac{1}{720}h^4\left(f^{(3)}(b) - f^{(3)}(a)\right) + \cdots . \tag{13.77}$$

where $b = a + nh$. Note the appearance of the trapezoidal rule (cf. (13.27)) in the middle of (13.77), which is the *Euler-Maclaurin formula* (13.25). The coefficients c_i in (13.25) are those in the power series expansion of $z\eta(z) = z/(e^z - 1)$, i.e.,

$$\frac{z}{e^z - 1} = \frac{z}{2}(-1 + \coth \tfrac{1}{2}z) = 1 - \tfrac{1}{2}z - \sum_{i=1}^{\infty} c_i z^{2i}. \tag{13.78}$$

The numbers $B_{2i} = (2i)!c_i$ are known as the Bernoulli numbers (cf. (13.66)): $c_1 = 1/12$, $c_2 = -1/720$, $c_3 = 1/30240$, $c_4 = -1/1209600$, and so forth.

13.3.5 Euler's constant γ

Let us derive the result that

$$\boxed{A_n = \sum_{i=1}^{n} \frac{1}{i} \approx \gamma + \log n,} \tag{13.79}$$

where $\gamma = 0.57721 \cdots$ is Euler's constant. We see that A_n is very close to the trapezoidal rule for the integral

$$\log n = \int_1^n \frac{dx}{x} \approx \frac{1}{2} + \frac{1}{2n} + \sum_{i=2}^{n-1} \frac{1}{i}. \tag{13.80}$$

Set $f(x) = 1/x$. Then (by induction) $f^{(k)}(x) = (-1)^k k! x^{-k-1}$. In particular, $f^{(2i-1)}(x) = -(2i-1)! x^{-2i}$. Therefore, the Euler-Maclaurin formula gives

$$\log(n/m) = \int_m^n \frac{dx}{x} \approx \sum_{i=m}^{n} \frac{1}{i} - \frac{1}{2m} - \frac{1}{2n}$$
$$- \sum_{i=1}^{k} \frac{B_{2i}}{2i} \left(m^{-2i} - n^{-2i} \right) + \mathcal{O}\left(m^{-2k-2} \right), \tag{13.81}$$

where the numbers B_{2i} are the Bernoulli numbers (exercises 3.11 and 13.16). Thus we can write

$$\sum_{i=1}^{n} \frac{1}{i} = \sum_{i=1}^{m-1} \frac{1}{i} + \sum_{i=m}^{n} \frac{1}{i} = \sum_{i=1}^{m-1} \frac{1}{i} + \log(n/m)$$
$$+ \frac{1}{2m} + \frac{1}{2n} + \sum_{i=1}^{k} \frac{B_{2i}}{2i} \left(m^{-2i} - n^{-2i} \right) + \mathcal{O}\left(m^{-2k-2} \right). \tag{13.82}$$

Thus we find (e.g., by letting $n \to \infty$) that

$$\gamma = \sum_{i=1}^{m-1} \frac{1}{i} - \log m + \frac{1}{2m} + \sum_{i=1}^{k} \frac{B_{2i}}{2i} m^{-2i} + \mathcal{O}\left(m^{-2k-2} \right). \tag{13.83}$$

In table 13.2, we give the results of applying this algorithm for various values of k and m computed using floating-point arithmetic. Note the effect of round-off error for $m = 1000$ and $k = 4$ (see section 18.1.1).

13.3.6 Gregorie's quadrature

The quadrature rule of Gregorie can now be realized as a simple application of the Euler-Maclaurin formula [169]. The derivatives at the ends of the intervals are replaced by suitable difference quotients that approximate the derivatives [66]; cf. section 11.4.3. Because the first few Bernoulli numbers are quite small, and the trapezoidal rule is very efficient to compute (since the coefficients are all the same), the Gregorie rules can be quite useful in applications [16].

m	$k = 0$	$k = 2$	$k = 4$
10	0.576383160974208	0.577215660974208	0.577215664900795
20	0.577007383589691	0.577215664839691	0.577215664901532
100	0.577207331651528	0.577215664901528	0.577215664901532
1000	0.577215581568205	0.577215664901530	0.577215664901530

Table 13.2 Computation of the Euler constant γ using (13.83) for various values of k and m. The first incorrect digit is shown in boldface. The cases $k = 4$ and $m = 20, 100$ agree with the exact value to the digits shown.

For example, if we take the approximation (11.46) (which we have to scale by a factor of 2 to account for the interval size), for the derivative in (13.22) for $k = 1$, we get

$$\int_a^b p(s)\, ds \approx Q_{\text{GEM}} = h \sum_{i=1}^{n-1} p(a + ih) + \tfrac{1}{2}h(p(a) + p(b))$$

$$+ \tfrac{1}{12}h\left(-\tfrac{3}{2}p(a) + 2p(a + h) - \tfrac{1}{2}p(a + 2h)\right) \quad (13.84)$$

$$- \left(\tfrac{3}{2}p(b) - 2p(b - h) + \tfrac{1}{2}p(b - 2h)\right)\Big).$$

For $n = 1$, we obtain Simpson's rule. Thus the quadrature rule is exact for cubics in this case. For larger n, it is clear that the quadrature rule is exact for quadratics because the difference approximation (11.46) is exact for quadratics. The symmetry of the quadrature rule around the midpoint $m = \tfrac{1}{2}(a+b)$ implies that $Q_{\text{GEM}}(x-m)^3 = 0$, the exact result for the integral. Thus Q_{GEM} is exact for cubics for all n.

In figure 13.1, the use of (13.84) to approximate the integral of x^4 on $[0, 1]$ is depicted. The slope of the error curve confirms that the approximation error is proportional to n^{-4}, as would be expected from the Peano kernel theorem, cf. the error expression for the Euler-Maclaurin formula (13.48). For the periodic function in (13.28), the errors for Q_{GEM} are similar and therefore much larger than would be obtained using the trapezoidal rule Q_{TR} given in (13.36) without any endpoint corrections, as is reflected in table 13.1. Thus for periodic functions, the simple trapezoidal rule Q_{TR} given in (13.36) is more accurate than the formally more accurate rule Q_{GEM} defined in (13.84).

13.4 OTHER QUADRATURE RULES

Any type of approximation naturally leads to a quadrature rule. Here we briefly discuss two that are related to approximation techniques studied earlier.

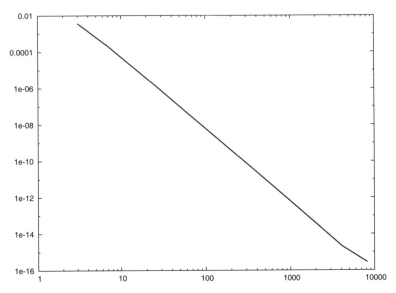

Figure 13.1 Errors (vertical axis) in computing the integral of x^4 on $[0, 1]$ via the
Gregorie rule (13.84) with n points (horizontal axis).

13.4.1 Chebyshev quadrature

A natural quadrature rule can be associated with interpolation at the Chebyshev points (11.1). This quadrature is called the "first rule" of Fejér[7] whereas
Fejér's "second rule" uses instead the interior extrema of the Chebyshev
polynomials,

$$x_j = \cos(j\pi/n), \quad 0 < j < n. \tag{13.85}$$

The closely related rule, which includes the points in (13.85) for $j = 0$ and
$j = n$ (that is, $x = \pm 1$), is known as the Clenshaw-Curtis rule. The latter
rule is popular for several reasons, including the fact that it is often as
accurate as Gaussian quadrature with the same number of points as well
as the availability of an algorithm that, in effect, computes the quadrature
weights very efficiently [159].

13.4.2 Bernstein quadrature

We have seen that any linear approximation scheme can generate an interesting quadrature rule. The same holds for the Bernstein approximation.
However, this does not yield a new quadrature rule but rather an interpretation of a variant of the trapezoidal rule. In particular, the Bernstein

---

[7]Leopold (Lipót) Fejér (1880–1959) was a student of Schwarz (see page 73) and had
a remarkable list of advisees including Paul Erdös, George Pólya, Marcel Riesz, Gabor
Szegö, and John von Neumann.

quadrature is of the form

$$Q_n^B f := \sum_{j=0}^{n} \left(\int_0^1 B_{j,n}(x)\, dx \right) f(j/n) = \frac{1}{n+1} \sum_{j=0}^{n} f(j/n). \qquad (13.86)$$

What is striking is that theorem 12.6 implies that

$$\lim_{n\to\infty} Q_n^B f = \int_0^1 f(x)\, dx, \qquad (13.87)$$

for any $f \in C^0([0,1])$. It is easy to compare the quadrature rule (13.86) with the trapezoidal rule: the coefficients at the end differ by $\mathcal{O}\left(n^{-1}\right)$ and the coefficients in the middle differ by $\mathcal{O}\left(n^{-2}\right)$, but the quadrature points are the same. In particular, this allows us to show that the (composite) trapezoidal rule also converges to the integral, as in (13.87) for any $f \in C^0([0,1])$ (see exercise 13.19).

13.5 MORE READING

The book [41] provides a comprehensive introduction to numerical quadrature. See [100] as well for information on the Bernoulli polynomials which play the role of Peano kernels in the Euler-Maclaurin formula, as well as other applications.

13.6 EXERCISES

Exercise 13.1 *Prove that Simpson's rule is exact for cubics (hint: use symmetry). Explain how to modify (13.3) to reflect the extra accuracy of Simpson's rule. (Hint: consider a Hermite-type interpolation involving the derivative at the midpoint and show that the corresponding basis function has integral zero, so that there is no corresponding α.)*

Exercise 13.2 *Prove that the midpoint rule is exact for linear polynomials (hint: use symmetry). Explain how to modify (13.3) to reflect the extra accuracy of the midpoint rule. (Hint: consider a Hermite-type interpolation involving the derivative at the midpoint and show that the corresponding basis function has integral zero, so that there is no corresponding α.)*

Exercise 13.3 *Show that the coefficients α_i in (13.2) for the Newton-Cotes quadrature are linearly proportional to the interval length $b - a$:*

$$\alpha_i = (b-a)\alpha_i^1, \qquad (13.88)$$

where the the coefficients α_i^1 correspond to the integration rules with $a = 0$ and $b = 1$. (Hint: show that the corresponding Lagrange basis functions satisfy a similar type of scaling when you map $[a,b] \to [0,1]$.)

Exercise 13.4 *Prove that any quadrature rule Qf of the form (13.10), i.e., $Qf = \sum_{i=0}^{n} \alpha_i f(x_i)$, is a linear functional on continuous functions.*

Exercise 13.5 *Suppose that $a = -1$ and $b = 1$. Consider the problem (13.17) for $n = 2$. Determine the optimal points $x_i = \pm\xi$ by solving the 4×4 system of equations. (Hint: use symmetry, the representation (13.2), and the observation in exercise 13.1 to eliminate as many variables as possible.)*

Exercise 13.6 *Give conditions on K such that (13.37) is valid. (Hint: write the integral as a limit of finite sums, e.g., (13.87), and show how E applied to the sum is the sum of E applied to the individual terms.)*

Exercise 13.7 *Prove that the error for the trapezoidal rule on $[-1, 1]$ satisfies*

$$\int_{-1}^{1} f(t)\, dt - \tfrac{1}{2}\left(f(-1) + f(1)\right) = \int_{0}^{1} \tfrac{1}{2}(t^2 - 1) f^{(2)}(t)\, dt. \qquad (13.89)$$

Use this to determine if the error estimate (13.9) is sharp.

Exercise 13.8 *Verify that the K defined in (13.44) satisfies (13.42) for all $f \in C^2$.*

Exercise 13.9 *Verify that the rule (13.23) arises as an example of interpolatory quadrature (13.1) if we define L to be continuous piecewise linear interpolation (section 12.4) using the points ξ_i.*

Exercise 13.10 *Derive the composite midpoint rule by summing the midpoint rule for each interval and show that it corresponds to interpolatory quadrature (13.1) if we define L to be piecewise constant interpolation (section 12.4) at the midpoints of each interval $[\xi_{i-1}, \xi_i]$ for $i = 1, \ldots, n$.*

Exercise 13.11 *Show that $\mathcal{P}_\infty$ defined in (13.51) is a vector space. Show that it also has a ring structure defined by pointwise multiplication.*

Exercise 13.12 *Prove that (13.59) is valid for any $T \in \mathcal{T}$ (see (13.60)). Apply (13.59) to justify the expression (13.58) by showing that the additive decomposition rule is valid:*

$$\zeta(hD) = e^{hD} - hD. \qquad (13.90)$$

Verify that the operator e^{hD} is well-defined on $\mathcal{P}_\infty$. (Hint: use Taylor's theorem; cf. exercise 7.4.)

Exercise 13.13 *Prove that (13.61) defines a power series υ that is convergent for $|z| < \epsilon$ provided this holds for μ and ν.*

Exercise 13.14 *Prove (13.65). (Hint: write the definition of the hyperbolic cotangent and simplify.)*

Exercise 13.15 *Prove that the power series in (13.78) converges for $|z| < 2\pi$. (Hint: ignoring the removable singularity $z = 0$, the smallest zeros of the denominator $e^z - 1$ are $z = \pm 2\pi i$.)*

Exercise 13.16 *The* generating function *for the Bernoulli numbers is the function*

$$\chi(z) = z/(e^z - 1), \tag{13.91}$$

in the sense that

$$\chi(z) = \sum_{k=0}^{\infty} \frac{B_k}{k!} z^k. \tag{13.92}$$

Show that the power series (13.92) has no odd-order terms of degree 3 and higher. (Hint: consider the function

$$\phi(z) = (z/(e^z - 1)) - 1 + \tfrac{1}{2}z \tag{13.93}$$

and show that $\phi(-z) = \phi(z)$ for all z.)

Exercise 13.17 *Prove the following analog of the Euler-Maclaurin formula in which the trapezoidal rule is replaced by the midpoint rule: for any polynomial p,*

$$\begin{aligned}
\int_a^b p(s)\,ds = {}& h \sum_{i=0}^{n-1} p(a + (i + \tfrac{1}{2})h) \\
& + \sum_{i=1}^{\infty} \hat{c}_i h^{2i} \left(p^{(2i-1)}(b) - p^{(2i-1)}(a) \right),
\end{aligned} \tag{13.94}$$

where $h = (b-a)/n$ and the coefficients $\hat{c}_i$ are also related to the Bernoulli numbers [163].

Exercise 13.18 *Determine the Peano kernel K_2 for the Euler-Maclaurin quadrature formula for $k = 2$ on the interval $[-1, 1]$. (Hint: write*

$$K_2(x) = (6!)^{-1}(1+x)^3(1-x)^3 + \alpha(1+x)^2(1-x)^2$$

and determine the value of α that ensures that $K_2^{(3)}(\pm 1) = 0$. Integrate by parts to verify the required formula.)

Exercise 13.19 *Prove that the trapezoidal rule*

$$\frac{1}{n}\left(\tfrac{1}{2}f(0) + \tfrac{1}{2}f(1) + \sum_{j=1}^{n-1} f(j/n) \right) \tag{13.95}$$

converges to $\int_0^1 f(x)\,dx$ as $n \to \infty$ for any $f \in C^0([0,1])$. Determine the rate of convergence in terms of the modulus of continuity of f. (Hint: compare exercise 12.19).

13.7 SOLUTIONS

Solution of Exercise 13.1. In view of exercise 13.3, it suffices to take $a = 0$ and $b = 1$. By symmetry,

$$\phi_0(x) = \phi_2(1-x) = 2(x - \tfrac{1}{2})(x - 1) = 2x^2 - 3x + 1, \qquad (13.96)$$

and thus

$$\alpha_2 = \alpha_0 = \int_0^1 2(x - \tfrac{1}{2})(x - 1)\, dx = \tfrac{2}{3} - \tfrac{3}{2} + 1 = \tfrac{1}{6}. \qquad (13.97)$$

Similarly, $\phi_1(x) = 4x(1-x) = -4x^2 + 4x$, and

$$\alpha_1 = \int_0^1 -4x^2 + 4x\, dx = -\tfrac{4}{3} + 2 = \tfrac{2}{3}. \qquad (13.98)$$

To see that this rule is exact for cubics, observe that the cubic

$$\phi_3(x) = x(x - \tfrac{1}{2})(x - 1) \qquad (13.99)$$

is antisymmetric around $x = \tfrac{1}{2}$, and thus $\int_0^1 \phi_3(x)\, dx = 0$.

Since any cubic may be written as $\beta\phi_3 + q$, where q is quadratic, Simpson's rule computes its integral exactly. If we define a Hermite-type interpolation involving Lagrange interpolation at $0, \tfrac{1}{2}, 1$ and derivative interpolation at $\tfrac{1}{2}$, then the corresponding basis functions will just involve simple additions of ϕ_3 and not change the quadrature coefficients α. However, the interpolation error (13.3) will involve an interpolant that is exact for cubics.

Solution of Exercise 13.16. Consider the difference

$$
\begin{aligned}
\phi(-z) - \phi(z) &= \frac{-z}{e^{-z} - 1} - 1 - \tfrac{1}{2}z - \frac{z}{e^z - 1} + 1 - \tfrac{1}{2}z \\
&= \frac{-z}{e^{-z} - 1} - \frac{z}{e^z - 1} - z \qquad (13.100) \\
&= -z\left(\frac{1}{e^{-z} - 1} + \frac{1}{e^z - 1} + 1\right).
\end{aligned}
$$

But a common denominator yields (cf. exercise 2.12)

$$
\begin{aligned}
\frac{1}{e^{-z} - 1} + \frac{1}{e^z - 1} &= \frac{e^z - 1 + e^{-z} - 1}{(e^{-z} - 1)(e^z - 1)} \\
&= \frac{e^z + e^{-z} - 2}{1 - e^{-z} - e^z + 1} = -1. \qquad (13.101)
\end{aligned}
$$

Therefore, $\phi(-z) - \phi(z) = 0$ for all z.

Chapter Fourteen

Eigenvalue Problems

"For his work on acceleration of the PageRank algorithm Gene [Golub] received Google stock; he donated most of these funds to found the Paul and Cindy Saylor Chair at the University of Illinois"–from *Gene H. Golub Biography* by Chen Greif. Gene Howard Golub (1932–2007) passed away shortly before his 19th (leap year) birthday, 29 February 2008, which was celebrated as the Gene Golub Around the World Day.

Eigenvalues and eigenvectors arise in many situations. We have seen the fundamental role of the spectral radius (6.8) in determining the convergence rate of iterative methods (chapter 8). Similarly, the convergence properties of the conjugate gradient method (cf. theorem 9.9) are encoded in the eigenvalues. These give only two hints of the crucial role of eigenvalues and why it is natural to seek efficient algorithms to compute the eigenvalues of matrices.

We begin by giving some further examples of eigenproblems for motivation. Then we present results that allow estimation of the location of eigenvalues by simple computations. We then explain fundamental limitations of the eigenvalue problem which show that, in general, only approximate algorithms are feasible, unlike the problem of solution of linear systems for which "exact" algorithms like Gaussian elimination and conjugate gradients are available. Finally, we discuss the Hessenberg factorization, which simplifies the eigenproblem substantially.

14.1 EIGENVALUE EXAMPLES

We have seen several examples of families of matrices, e.g., (4.20) and (8.2), for which the eigenvalues may be of interest. We explain in one case why this is so. We also give other examples to provide some additional motivation.

14.1.1 Mechanical resonance

Many physical models are described by differential equations. For a two-point boundary value problem, a finite difference approximation produces a linear system of the form (4.20). The model (4.20) is a reasonable approximation to the deflection of a string under modest loading. Thus the frequency of sounds can be determined as the eigenvalues of mechanical systems [123].

But far more complex physical systems, approximated by either finite difference or finite element methods, are modeled by matrices with structures very similar to (4.20), with the matrix size n limited only by the size of computer memory. Finding the deflection x of a bridge subject to a force f would require solving $A x = f$ for such a system. But finding the frequencies of vibration [14] require the eigenvalues of A.

14.1.2 Quality rankings

Suppose you want to quantify connectivity of related objects based on the quality of the relationships among the objects. For example, we might want to measure personal connections so that we can target advertising to people who influence influential people. To see how this might be done, suppose we had access to all cell phone data. We can rank people based on whether their numbers are kept in another person's cell phones for "one-touch" dialing.

It is reasonable to say that you are more connected if the people who keep your number are also highly connected. One way to implement this is to define your connectivity rating to be a fixed multiple of the sum of all the ratings of the people who list you. Since this definition is circular, we have to write an equation.

Let $\mu > 0$ denote a parameter to be picked later. Let $\mathbf{A} = (a_{ij})$ denote the matrix with the property that $a_{ij} = 1$ if and only if the jth person keeps the ith person's phone number, and zero otherwise. Then the connectivity ratings x_i of the ith person can be determined from other connectivity ratings (x_j) by the relationship

$$x_i = \mu \sum_{\{j \mid a_{ij} \neq 0\}} x_j. \tag{14.1}$$

This just says that the ith rating is proportional to the sum of the ratings of the people that connect to the ith person, with a constant of proportionality given by μ. By the definition of the matrix $\mathbf{A} = (a_{ij})$, we find for all i that

$$\lambda x_i = \sum_{\{j \mid a_{ij} \neq 0\}} x_j = \sum_j a_{ij} x_j = (\mathbf{A}X)_i, \tag{14.2}$$

where $\lambda = 1/\mu$. This says that x and λ form an eigenpair: $\mathbf{A}x = \lambda x$.

If $\mathbf{A}$ is irreducible (section 8.3.4), then there is a nonnegative eigenpair (theorem 8.18), so that the ratings are all of the same sign. In many cases, only relative, not absolute, ratings are of interest. In this case ratings, like eigenvectors, can be subjected to arbitrary scalings by a constant factor.

The above example is similar to the problem of determining *link relevance* for search engines for the World Wide Web. Instead of phone numbers, links between web pages generate A, and a similar model can be derived. Current web search engines[1] compute the corresponding eigenvalue problem for the entire web periodically, with several billions of web pages ranked currently.

---

[1] The PageRank algorithm of Google is similar but more involved, cf. [25, 90, 117].

For such a matrix size, it is significant that the matrices are quite sparse. A typical web page might link to only a few dozen other web pages. However, the location of the nonzero entries is quite arbitrary in this case, not leading to an obvious banded structure.

14.1.3 Not so sparse eigenvalue problems

So far, all the examples given have sparse matrices. To avoid the impression that this is the main application area, we note that iterative eigenvalue methods are often employed with success on nonsparse matrices as well. For example, the Lanczos algorithm has been used to solve the eigenvalue problem associated with electronic structure prediction for large atoms [49].

14.2 GERSHGORIN'S THEOREM

Although computing eigenvalues of a matrix is quite difficult, getting useful bounds on their location is sometimes quite easy. As we have seen (corollary 3.3), the eigenvalues of a triangular matrix are displayed on the diagonal. A theorem of Gershgorin[2] extends this observation using a type of perturbation argument known as a *homotopy method*.

To start the discussion, let us imagine describing where eigenvalues are *not* to be found. The set of λ that are not eigenvalues of a matrix A are those for which $A - \lambda I$ is invertible. We know that a matrix that is diagonally dominant is invertible, by corollary 8.9. Thus if

$$|a_{ii} - \lambda| > \sum_{i \neq j = 1}^{n} |a_{ij}| \tag{14.3}$$

for all $j = 1, \ldots, n$, then λ is not an eigenvalue of A. In other words, each eigenvalue λ of A must satisfy

$$|a_{ii} - \lambda| \leq \sum_{i \neq j = 1}^{n} |a_{ij}| \tag{14.4}$$

for some $j \in \{1, \ldots, n\}$. This says that each eigenvalue must be in a disk of radius

$$r_i = \sum_{i \neq j = 1}^{n} |a_{ij}| \tag{14.5}$$

in the complex plane that is centered at the diagonal entry a_{ii}. We can call such a disk $D(a_{ii}, r_i)$. Therefore, all the eigenvalues are in $\cup_{i=1}^{n} D(a_{ii}, r_i)$. This is the main gist of Gershgorin's theorem, but if the disks do not all overlap, we can say something more.

---

[2]Semyon Aranovich Gershgorin (1901–1933) is credited with being the first to state the theorem in print, but it is said that the result, and its connection to diagonal dominance, was known earlier: "It seems that the circles theorem was known to Schur" [143] (see page 81 for more on Schur).

Theorem 14.1 *Suppose that $\cup_{i=1}^n D(a_{ii}, r_i)$ consists of k distinct connected components C_j, where $C_j \cap C_{j'} = \emptyset$ if $j \neq j'$ and*

$$\cup_{i=1}^n D(a_{ii}, r_i) = C_1 \cup \cdots \cup C_k, \qquad (14.6)$$

where each C_j is the union of l_j disks. Then there are exactly l_j eigenvalues (counting multiplicity) in each connected component C_j.

Proof. We just need to establish the number of eigenvalues in each component. Define a one-parameter family of matrices

$$A^s = (1 - s)\,\text{diag}(A) + sA, \qquad (14.7)$$

where $s \in [0, 1]$ and $\text{diag}(A)$ denotes the diagonal matrix with diagonal entries matching those of A. When $s = 0$, $A^s = \text{diag}(A)$ and the eigenvalues are a_{ii}; when $s = 1$, $A^s = A$. Note that $\text{diag}(A^s) = \text{diag}(A)$ for all s. By the first part of the theorem, the eigenvalues of A^s lie in the disks

$$\cup_{i=1}^n D(a_{ii}, sr_i) = C_1^s \cup \cdots \cup C_{k^s}^s. \qquad (14.8)$$

When $s = 0$, the number of components k^0 is the number of distinct coefficients a_{ii}, and each $C_i^0 = \{a_{jj}\}$ for some j since the centers of the disks do not change with s. The number of distinct components can decrease as disks merge, but it cannot increase.

The matrix coefficients of A^s are continuous in s, and this means that the eigenvalues λ_i^s depend continuously on s, because of the following lemma.

Lemma 14.2 *The eigenvalues of a matrix depend continuously on the coefficients.*

We will prove this result in section 14.2.2.

There is a set of discrete points $0 = s_0 < s_1 < \cdots < s_r \leq 1$ with the property that the number k_s of components in (14.8) does not vary in the intervals $s_{i-1} \leq s < s_i$ for $i = 1, \ldots, r$. That is, the values s_i denote the merge points of the monotonically growing components. Fix i for the moment. Suppose that for some $s < s_i$, an eigenvalue that started in one of the components $C_j^{s_{i-1}}$ reaches the exterior of the component. Reducing the value of s if necessary, we can ensure that the eigenvalue is in the complement of all the components since they are separated by a finite distance, in contradiction to the first part of Gershgorin's theorem. So if there are m_j eigenvalues in $C_j^{s_{i-1}}$, then there must be m_j eigenvalues in C_j^s for all $s < s_i$. If two components merge at s_i, all we can say is that the eigenvalues in each separate component for $s < s_i$ are in the union at $s = s_i$, again by continuity.   QED

Since we use the notion of diagonal dominance to define the Gershgorin disks, we could also use the notion of generalized diagonal dominance (section 8.3.1) and obtain more general results [167]. Instead, we leave such ideas to exercise 14.1, and we consider more complex geometric estimates.

14.2.1 Ovals of Cassini

It is possible to locate eigenvalues even more precisely than by using the Gershgorin disks. Define the Cassini ovals

$$\mathcal{C}_{ij} = \left\{\lambda \in \mathbb{C} \mid |\lambda - a_{ii}||\lambda - a_{jj}| \leq r_i r_j\right\} \tag{14.9}$$

for $i, j = 1, \ldots, n$, where the r_i's are defined by (14.5). The following is a theorem of A. Brauer.[3]

Theorem 14.3 *The eigenvalues of A are contained in $\cup_{i,j=1}^{n} \mathcal{C}_{ij}$.*

Proof. Consider an eigenpair $A x = \lambda x$ with $x \neq 0$. For any index j, we find

$$(\lambda - a_{jj})x_j = \sum_{l \neq j} a_{jl} x_l. \tag{14.10}$$

Let i be an index such that

$$|x_i| = \|x\|_\infty, \tag{14.11}$$

so that applying (14.10) for any j we have

$$|\lambda - a_{jj}||x_j| = \left|\sum_{l \neq j} a_{jl} x_l\right| \leq \sum_{l \neq j} |a_{jl}||x_i| = r_j |x_i|, \tag{14.12}$$

by using (14.11). Then (14.12) yields for all j,

$$|\lambda - a_{jj}| \leq r_j |x_i|/|x_j|. \tag{14.13}$$

Now let $j \neq i$ be an index such that

$$|x_j| \geq |x_k| \quad \text{for all } k \neq i. \tag{14.14}$$

There are two cases to consider. It might be that $x_j = 0$, which means that $x_k = 0$ for all $k \neq i$. Computing, we find for $k \neq i$ that

$$0 = \lambda x_k = (A x)_k = \sum_{l=1}^{n} a_{kl} x_l = a_{ki} x_i. \tag{14.15}$$

Thus $a_{ki} = 0$ for all $k \neq i$, so a repeat of the calculation (14.15) with $k = i$ says that $\lambda = a_{ii}$. This trivially implies that $\lambda \in \mathcal{C}_{ik}$ for all k.

Now suppose that $x_j \neq 0$. Estimating (14.10) using (14.14), we find

$$|\lambda - a_{ii}||x_i| \leq \sum_{l \neq i} |a_{il}||x_j| = r_i |x_j|, \tag{14.16}$$

yielding the inequality

$$|\lambda - a_{ii}| \leq r_i |x_j|/|x_i|. \tag{14.17}$$

Multiplying (14.13) and (14.17), we conclude that $\lambda \in \mathcal{C}_{ij}$. QED

---

[3]Alfred Theodor Brauer (1894–1985) was a student of Issai Schur and Erhard Schmidt, as was his younger brother Richard Dagobert Brauer (1901–1977).

The location of eigenvalues via Cassini ovals is at least as precise as using the Gershgorin disks, that is,

$$\boxed{\cup_{i,j=1}^{n} \mathcal{C}_{ij} \subset \cup_{i=1}^{n} D(a_{ii}, r_i).} \qquad (14.18)$$

To prove this, suppose that $\lambda \notin D(a_{ii}, r_i) \cup D(a_{jj}, r_j)$. Then

$$|\lambda - a_{ii}| > r_i \quad \text{and} \quad |\lambda - a_{jj}| > r_j, \qquad (14.19)$$

so that $|\lambda - a_{ii}| \, |\lambda - a_{jj}| > r_i r_j$ and $\lambda \notin \mathcal{C}_{ij}$. Thus if $\lambda \notin \cup_{i=1}^{n} D(a_{ii}, r_i)$, then $\lambda \notin \cup_{i,j=1}^{n} \mathcal{C}_{ij}$, which is just the set-theoretic complement of (14.18).

To compare the two predictions of Cassini and Gershgorin, consider the family of matrices

$$\begin{pmatrix} 1 & -t \\ t & -1 \end{pmatrix} \qquad (14.20)$$

whose eigenvalues are $\lambda_{\pm} = \pm\sqrt{1 - t^2} \approx \pm(1 - \frac{1}{2}t^2)$. The Gershgorin estimate would predict only that $|\lambda_{\pm} \mp 1| \leq t$, whereas the Cassini estimate is more accurate:

$$|1 \mp \lambda_{\pm}| \leq t^2 / |1 \pm \lambda_{\pm}|, \qquad (14.21)$$

which can be used to show that the Cassini ovals have a diameter that is $\mathcal{O}(t^2)$ (exercise 14.2).

The Cassini ovals are known to be optimal in the sense that for a given set of numbers a_{ii} and r_i, there is a matrix A with these diagonals and off-diagonal absolute row sums having an eigenvalue λ for any $\lambda \in \cup_{i,j=1}^{n} \mathcal{C}_{ij}$ [167, 168].

Locating eigenvalues is related to proving invertibility, in the sense that the latter is equivalent to $\lambda = 0$ not being an eigenvalue. Thus the following corollary holds.

Corollary 14.4 *An $n \times n$ matrix A is invertible if*

$$|a_{ii} a_{jj}| > r_i r_j \qquad (14.22)$$

for all $i, j = 1, \ldots, n$, where the r_i's are defined by (14.5).

14.2.2 Eigenvalue continuity

The continuity of the eigenvalues of a matrix, as a function of the entries, can be proved by brute force. First, we can write the characteristic polynomial in terms of the coefficients (exercise 14.3), and this polynomial is clearly continuous with respect to the coefficients. Then we can show that the zeros of a polynomial are continuous functions of the coefficients of the polynomial. This is the content of appendices A, B, and K in [122]. However, this dependence is not very smooth. Consider the roots of $p(x) = x^n + a$. One of them is the real root $x = \sqrt[n]{a}$ (for $a > 0$), which is only Hölder-continuous of order $1/n$ at $a = 0$; see exercise 14.5 and appendix K in [122].

Instead of reproducing the proof in [122], we take a completely different approach advocated recently [116]. It relies on the Schur decomposition and the fact that unitary matrices form a compact set. This is easy to see since $\|U\|_2 = 1$ for any unitary matrix (exercise 6.5). Thus, in whatever norm you like (theorem 5.3), the unitary matrices form a bounded set in $\mathbb{C}^{n^2}$. The limit of unitary matrices is also unitary (exercise 14.6), so that the unitary matrices form a compact set in $\mathbb{C}^{n^2}$.

When we identify the eigenvalues of a matrix with the diagonal entries of the triangular factor in the Schur decomposition, there is a great deal of possible ambiguity due to simple permutations of indices since a permutation is itself a unitary matrix (exercise 14.7). We need a way to talk about eigenvalues that reflects their basic properties. The diagonal of a matrix is naturally a vector, but since the order of the eigenvalues does not matter, the concept of vector is not right. On the other hand, a discrete set correctly eliminates the ambiguity of permutations of indices, but it eliminates too much, namely, the multiplicity of the eigenvalues. So we introduce a new space $\mathbb{L}^n$ to encapsulate the right properties. First, we define a relation $\mathcal{R}$ on $\mathbb{C}^n$ as follows:

$$x\mathcal{R}y \quad \text{iff} \quad x = \Pi y \text{ for some permutation } \Pi. \tag{14.23}$$

Define $\mathbb{L}^n$ to be the set of equivalence classes (exercise 14.8) of $\mathcal{R}$ in $\mathbb{C}^n$.

We also need a way to measure closeness for sets of eigenvalues to deal with this ambiguity.

Definition 14.5 *Define a nonnegative function* $\mathcal{D}(x,y)$ *on* $\mathbb{C}^n \times \mathbb{C}^n$ *by*

$$\mathcal{D}(x,y) = \min\left\{\|x - \Pi y\|_\infty \mid \Pi \text{ is a permutation matrix}\right\}. \tag{14.24}$$

Note that, for any n, there is a finite set of permutations, so we can assert that the minimum is attained in (14.24); that is, there is a permutation matrix Π such that $\mathcal{D}(x,y) = \|x - \Pi y\|_\infty$. The function d is symmetric $(\mathcal{D}(x,y) = \mathcal{D}(y,x))$ and satisfies a triangle inequality

$$\mathcal{D}(x,y) \leq \mathcal{D}(x,w) + \mathcal{D}(w,y) \tag{14.25}$$

for all $w, x, y \in \mathbb{C}^n$ (exercise 14.9). But for $[x] \in \mathbb{L}^n$, $\mathcal{D}(x,y) = 0$ for all $y \in [x]$. In fact, $x\mathcal{R}y$ iff $\mathcal{D}(x,y) = 0$.

There is a natural extension of the function $\mathcal{D}$ to the space $\mathbb{L}^n$:

$$\mathcal{D}([x],[y]) = \mathcal{D}(x,y) \tag{14.26}$$

(but you have to check that this is well-defined, exercise 14.10). Thus $(\mathbb{L}^n, \mathcal{D})$ forms a metric space [141]. The following result makes lemma 14.2 precise.

Lemma 14.6 *Suppose that* $A^k \to A$ *as* $k \to \infty$. *Apply the Schur decomposition to each matrix:*

$$(U^k)^\star A^k U^k = T^k \quad \text{and} \quad U^\star A U = T, \tag{14.27}$$

where T *and each* T^k *are upper-triangular. Then*

$$\mathcal{D}(\text{diag}(T^k), \text{diag}(T)) \to 0 \text{ as } k \to \infty,$$

where $\text{diag}(M)$ *denotes the vector corresponding to the diagonal of* M.

Proof. Since the unitary matrices are compact, we can pick a subset of the indices k such that (after renaming this subsequence k again) $U^k \to \hat{U}$ as $k \to \infty$, where $\hat{U}$ is unitary. With this subsequence, we have

$$T^k = (U^k)^\star A^k U^k \to \hat{U}^\star A \hat{U}. \tag{14.28}$$

The limit of upper-triangular matrices is necessarily upper-triangular; thus we can write

$$T^k \to \hat{T} = \hat{U}^\star A \hat{U}, \tag{14.29}$$

where $\hat{T}$ is upper-triangular. Since the eigenvalues of A^k are the diagonal entries of T^k, (14.29) implies that the eigenvalues of A^k converge to the diagonal entries of $\hat{T}$. But since $\hat{T} = \hat{U}^\star A \hat{U}$, these are eigenvalues of A (with some ordering). That is, $\mathcal{D}(\text{diag}(\hat{T}), \text{diag}(T)) = 0$.

This would be the end of the proof of continuity, except for the fact that we have established this only for a subsequence of the original sequence. However, we can also establish this more generally. If we take any subsequence of the original sequence A^k, we can pick a subsequence of this subsequence for which convergence of eigenvalues occurs. In each case, the ordering of the eigenvalues on the diagonal of the limiting triangular matrix may be different, but the set of values must be the same: the eigenvalues of A. As a result, we have proved convergence of the original sequence (exercise 14.11).

QED

14.3 SOLVING SEPARATELY

We have viewed the eigenproblem as a nonlinear system (section 7.2.4) in the variables (x, λ). But it is a special one, in that the expression $A x - \lambda x$ is bilinear. Thus, if we know the eigenvalue λ, we just have to solve the linear system $(A - \lambda I)x = 0$ for some nonzero x. We showed in section 3.4.4 that this can be carried out with LU factorization using appropriate pivoting, at least when the kernel of $A - \lambda I$ is one-dimensional. (The case when the dimension is higher can be handled similarly; cf. section 3.4.3.)

On the other hand, if by chance we have an eigenvector x, we can also determine the eigenvalue easily. There are several ways to do this, but a convenient one is based on the Rayleigh[4] quotient

$$\lambda^R(x) = \frac{x^\star A x}{x^\star x}, \tag{14.30}$$

defined for any $0 \neq x \in \mathbb{C}^n$. Thus if $A x = \lambda x$, then

$$\lambda^R(x) = \frac{x^\star A x}{x^\star x} = \frac{x^\star \lambda x}{x^\star x} = \lambda. \tag{14.31}$$

We will look at the Rayleigh quotient in more detail in section 15.1.1.

---

[4]John William Strutt, 3rd Baron Rayleigh (1842–1919) received the Nobel prize in physics in 1904 for his study of gases and the discovery of argon. He was a student of Stokes (page 118).

14.4 HOW NOT TO EIGEN

The eigenproblem sounds simple, at least for small matrix sizes: just find the roots of the characteristic equation. For the 2×2 case, we have

$$p_2(\lambda) = \det(A - \lambda I)$$

$$= \det \begin{pmatrix} a_{11} - \lambda & a_{12} \\ a_{21} & a_{22} - \lambda \end{pmatrix} \tag{14.32}$$

$$= (a_{11} - \lambda)(a_{22} - \lambda) - a_{12}a_{21}$$

$$= \lambda^2 - \lambda(a_{11} + a_{22}) + a_{11}a_{22} - a_{12}a_{21}.$$

The roots of a quadratic are also easy to determine:

$$\lambda = \tfrac{1}{2} \left(a_{11} + a_{22} \pm \sqrt{(a_{11} + a_{22})^2 - 4(a_{11}a_{22} - a_{12}a_{21})} \right)$$

$$= \tfrac{1}{2} \left(a_{11} + a_{22} \pm \sqrt{a_{11}^2 + a_{22}^2 - 2a_{11}a_{22} + 4a_{12}a_{21}} \right) \tag{14.33}$$

$$= \tfrac{1}{2} \left(a_{11} + a_{22} \pm \sqrt{(a_{11} - a_{22})^2 + 4a_{12}a_{21}} \right).$$

It is a bit more cumbersome to write down the characteristic polynomial of a 3×3 matrix and then the equation for the roots of a cubic. Similarly, we can still imagine writing the characteristic polynomial of a 4×4 matrix and then the equations for the roots of a quartic, but only in a smaller font.

But there the process stops, not just because the formulas get too messy or unstable. For some polynomials of degree 5 and higher, there is no general formula for their roots in terms of a finite sequence of steps involving radicals and ordinary algebraic operations. This result, due originally to Abel,[5] can be seen [149] by considering the equation

$$\lambda^5 - a\lambda + b = 0. \tag{14.34}$$

Some polynomial roots can be written in terms of such formulas (e.g., if they can be factored into lower-order polynomials whose roots would be determined by simple formulas), but Galois[6] characterized the polynomials whose roots cannot be written in terms of such formulas.

On the other hand, one has to temper this discussion by what it means to compute a "radical" since this is still a complicated computational issue. We saw in (1.2) an effective way to compute square roots, but it requires some attention. Similarly, a solution to

$$\lambda^5 - \lambda + b = 0 \tag{14.35}$$

is called a *Bring radical*[7] (or *ultraradical*), and one could easily develop algorithms to solve for $\lambda = \lambda(b)$ (exercise 2.17). If the ultraradical is allowed

---

[5]See the beginning of chapter 8.

[6]The short life of Évariste Galois (1811–1832) "stands as a symbol of precocious mathematical genius, misunderstood, disturbed, an object of persecution by the authorities of the time and particularly by the principal French mathematicians, who did not appreciate the depth and value of his work" [155]. The work leading to what we now call Galois theory was repeatedly rejected for publication in his lifetime.

[7]Erland Samuel Bring (1736–1798) was a Swedish mathematician who discovered a way to reduce a general quintic to the form (14.34).

as a basic step, then there is a formula for quintic roots in general [2]. We leave as exercise 14.12 verification in the case that the quintic takes the form (14.34).

However, in addition to the difficulty of finding a formula for the roots of a polynomial, there is also the complexity of determining the coefficients of $p_n(\lambda) = \det(A - \lambda I)$, which can be factorial in n if we choose the wrong approach (exercise 14.3). So we cannot simply form the characteristic polynomial in a naïve way and then find its roots, at least for large n.

One solution to this problem is to first reduce the matrix via algebraic operations similar to Gaussian elimination to a form where the determinant can be evaluated efficiently. Such a form is the Hessenberg[8] form (section 14.5).

14.5 REDUCTION TO HESSENBERG FORM

We know from the work of Abel and Galois (section 14.4) that we cannot expect to find an algorithm that can, by a finite sequence of operations, reduce a matrix to triangular form via similarity transformations. Likewise, we know that every matrix cannot be diagonalized. If we take the Jordan canonical form as guide, we might guess that it would be possible to achieve a relaxed goal, to reduce a matrix to nearly triangular form, with, say, just one extra subdiagonal. Such a form is called the *Hessenberg form*. More precisely, consider the following definition analogous to definition 3.1.

Definition 14.7 *A matrix $B = (b_{ij})$ is called upper-Hessenberg (respectively, lower-Hessenberg) if $b_{ij} = 0$ for all $j > i + 1$ (respectively, $i > j + 1$).*

We have seen before how to perform matrix decompositions to produce zeros in required places. Both the Schur decomposition and the triangular factorization produced by Gaussian elimination are of this type. In this case, we will see that an algorithm can be applied that is very similar to the Gram-Schmidt process (5.45). The original algorithm was proposed by Lanczos[9] for Hermitian matrices and then generalized by Arnoldi[10] to arbitrary matrices. Since the general algorithm applies as well to the special case

---

[8]Karl Hessenberg (1904–1959) demonstrated the decomposition in his thesis [124] and worked for most of his career as an engineer for A.E.G., the German "general electric" company. He applied for a U.S. patent on "electric valve circuits" on 1 July 1939 (just two months before the invasion of Poland), which was issued (#2,356,589) on 22 August 1944 during the liberation of Paris.

[9]Cornelius Lanczos (1893–1974) studied with Fejér (page 220) in Budapest but then wrote a thesis in mathematical physics and later was Einstein's assistant in Berlin. Lanczos moved to Purdue University in 1932, and in the United States his interests turned computational. Prompted by McCarthy's investigations of political sympathies, Lanczos moved to Ireland in 1954 [118].

[10]Walter Edwin Arnoldi (1917–1995) was employed throughout his career at United Aircraft Corporation (later United Technologies) where he obtained several patents including U.S. Patent #3,144,317 for a freezing process to remove carbon dioxide from the air.

and gives the same result, we prefer to refer to the two cases as the Lanczos-Arnoldi algorithm. Note, however, that a Hermitian Hessenberg matrix is tridiagonal, so there can be a substantial computational simplification in the Hermitian case.

14.5.1 Lanczos-Arnoldi algorithm

The full Hessenberg decomposition takes the form

$$Q^{\star}AQ = H, \tag{14.36}$$

where Q is unitary and H is upper-Hessenberg. We derive the Hessenberg form iteratively as follows:

$$AQ^k = Q^{k+1}H^k, \tag{14.37}$$

where Q^i is an $n \times i$ matrix and H^k is a $k+1 \times k$ matrix. Let us write

$$Q^k = \begin{pmatrix} q^1 & \cdots & q^k \end{pmatrix}. \tag{14.38}$$

For Q to be unitary, we need the q's to be orthonormal:

$$\|q^i\|_2 = 1 \text{ for all } i \text{ and } (q^i)^{\star}q^j = 0 \text{ for } i \neq j. \tag{14.39}$$

The kth column of (14.37) can be written

$$Aq^k = \sum_{j=1}^{k+1} h_{jk}q^j. \tag{14.40}$$

The Lanczos-Arnoldi algorithm allows an arbitrary (normalized) initial vector q^1 satisfying $\|q^1\|_2 = 1$. We will see later how this choice affects the quality of the decomposition. For $k = 1$, we need to construct q^2 and scalars h_{11} and h_{21} such that

$$Aq^1 = h_{11}q^1 + h_{21}q^2. \tag{14.41}$$

If by chance q^1 is an eigenvector (the choice in the Schur decomposition; cf. section 6.2.3), then we choose h_{11} to be the eigenvalue (e.g., take $h_{11} = \lambda^R(q^1)$ using the Rayleigh quotient (14.30)) and set $h_{12} = 0$ and choose q^2 arbitrarily such that $\|q^2\|_2 = 1$. In this case, the algorithm essentially starts over.

If Aq^1 and q^1 are not collinear, we define $h_{21} = \|Aq^1 - h_{11}q^1\|_2$ (necessarily, $h_{21} > 0$, cf. exercise 14.15) and

$$q^2 = h_{21}^{-1}(Aq^1 - h_{11}q^1), \tag{14.42}$$

which satisfies $\|q^2\|_2 = 1$. Note that, so far, we have not specified h_{11}. To create a unitary matrix, we need $(q^1)^{\star}q^2 = 0$, and if we multiply (14.42) by $(q^1)^{\star}$, we see that this means $h_{11} = (q^1)^{\star}Aq^1$.

In general, (14.40) provides an algorithm to generate the q's and h's inductively. Suppose we have defined orthonormal vectors $q^1, \ldots, q^k$. Define

$$h_{jk} = (q^j)^{\star}Aq^k, \quad j = 1, \ldots, k. \tag{14.43}$$

Rewrite (14.40) as

$$r = Aq^k - \sum_{j=1}^{k} h_{jk}q^j = h_{k,k+1}q^{k+1}. \tag{14.44}$$

In view of (14.43), $(q^i)^\star r = 0$ for $i = 1, \ldots, k$. With r defined in this way, we define

$$h_{k,k+1} = \|r\|_2 \quad \text{and} \quad q^{k+1} = h_{k,k+1}^{-1}r. \tag{14.45}$$

If by chance $r = 0$, then we have $AQ^k = \hat{H}^k Q^k$, where $\hat{H}^k$ is the $k \times k$ matrix where we omit the last row of H^k. In this case, we can restart the process by choosing q^{k+1} to be an arbitrary vector of norm 1 that is orthogonal to $q^1, \ldots, q^k$. Of course, when we reach $k = n$, we must have $r = 0$ because the q's form an orthonormal basis.

14.5.2 Optimality of Lanczos-Arnoldi

The quality of the Hessenberg matrix H in (14.36) depends on how small the off-diagonal terms $h_{k,k+1}$ are. We show that the Lanczos-Arnoldi algorithm minimizes these values among certain choices.

Suppose (as is the generic case) that q^1 is not an eigenvector of A and that the residual r in (14.44) does not vanish. Then by (14.42) and (14.44) (and induction), q^{k+1} is a linear combination of the first $k + 1$ Krylov vectors (see exercise 14.16)

$$q^1, Aq^1, \ldots A^k q^1. \tag{14.46}$$

That is, we can write $q^{k+1} = P_k(A)q^1$ for some polynomial P_k of degree k. More precisely, provided that $h_{k,k+1} \neq 0$, we can write

$$h_{k,k+1}q^{k+1} = P_k(A)q^1 = A^k q^1 + \hat{P}_{k-1}(A)q^1, \tag{14.47}$$

where $\hat{P}_{k-1}$ is a polynomial of degree $k - 1$ or less. Thus P_k is a monic polynomial (i.e., the coefficient of the term of order k is 1). Define $\mathcal{P}_k^M$ to be the set of monic polynomials of degree k.

Lemma 14.8 *Suppose that q^1 is not an eigenvector of A and that the residual r in (14.44) does not vanish. Let P_k be the monic polynomial (14.47) generated by the Lanczos-Arnoldi process. Then*

$$(P_k(A)q^1, Q(A)q^1)_I = 0 \tag{14.48}$$

for all polynomials Q of degree $k - 1$. Thus

$$h_{k,k+1} = \|P_k(A)q^1\|_2 = \min \left\{ \|Q(A)q^1\|_2 \mid Q \in \mathcal{P}_k^M \right\} \tag{14.49}$$

for $k = 1, \ldots, n$.

Proof. From (14.47), we see that $0 = (q^j)^\star P_k(A)q^1$ for all $j = 1, \ldots, k$. But since $q^1, \ldots, q^{k+1}$ are orthonormal, it must also be the case that the $k + 1$

Krylov vectors (14.46) can be expressed in terms of $q^1, \ldots, q^{k+1}$. That is, the $k+1$ Krylov vectors (14.46) have to span a $(k+1)$-dimensional space, that is, the one spanned by $q^1, \ldots, q^{k+1}$ (exercise 14.17). Therefore, we conclude that $(A^j q^1)^\star P_k(A) q^1 = 0$ for all $j = 1, \ldots, k$. This proves (14.48). To prove (14.49), expand the expression $\| (P_k(A) + tQ(A)) q^1 \|_2^2$ for any scalar t and polynomial Q of degree $k-1$; cf. section 12.3 and in particular (12.54). QED

Suppose that A has a complete set of eigenvectors $X^1, \ldots, X^n$. Write $q^1 = \sum_{j=1}^n a_j X^j$, where $X^1, \ldots, X^n$. Then by (9.63), the orthogonality (14.48) becomes

$$0 = (P_n(A)q^1, Q(A)q^1)_I = \sum_{j=1}^n P_n(\lambda_j) Q(\lambda_j) a_j^2 \tag{14.50}$$

for any polynomial Q of degree less than n. Choosing Q_i such that $Q_i(\lambda_j) = \delta_{ij}$ for $i, j = 1, \ldots, n$, we see that $P_n(\lambda_j) = 0$ for all j provided that none of the coefficients a_j vanish. Thus P_n is a constant multiple of the characteristic polynomial of A. Since P_n is monic, they must be equal.

Similarly, (9.63) implies that

$$h_{k,k+1}^2 = \sum_{j=1}^n P(\lambda_j)^2 a_j^2 = \| P_k(A) q^1 \|_2^2$$

$$= \min \left\{ \| Q(A) q^1 \|_2^2 \mid Q \in \mathcal{P}_k^M \right\} \tag{14.51}$$

$$= \min \left\{ \sum_{j=1}^n Q(\lambda_j)^2 a_j^2 \mid Q \in \mathcal{P}_k^M \right\}.$$

In the case where A is Hermitian, the eigenvalues lie in an interval

$$I = [\lambda_{\min}, \lambda_{\min} + 2\Lambda]$$

for some $\Lambda > 0$. Thus (14.51) implies

$$h_{k,k+1}^2 \leq \| Q \|_{\infty, I}^2 \sum_{j=1}^n a_j^2 = \| Q \|_{\infty, I}^2 \| q^1 \|_2^2 \tag{14.52}$$

for any $Q \in \mathcal{P}_k^M$. For example, we can take $Q(x) = x^n - L_{n-1}(x^n)$, and L_{n-1} denotes Chebyshev interpolation on I. Thus (exercise 11.18) we have

$$h_{k,k+1} \leq 2 \left(\frac{\Lambda}{2} \right)^n. \tag{14.53}$$

We leave as exercise 14.18 formulation and proof of a version of lemma 14.8 that covers the case where the residual r in (14.44) vanishes and the algorithm is restarted.

14.6 MORE READING

The classic text [122] was referenced in section 14.2.2 but is also of general interest, as is the monograph [172]. There are several more recent texts as well, cf. [118].

14.7 EXERCISES

Exercise 14.1 *Use the notion of generalized diagonal dominance (see section 8.3.1) to define generalized Gershgorin disks of the form [167] $D(a_{ii}, r_i^x)$, where r_i^x is the weighted sum*

$$r_i^x = \sum_{i \neq j=1}^{n} |a_{ij}| x_j / x_i \tag{14.54}$$

for any vector x with nonzero entries.

Exercise 14.2 *Prove that the Cassini ovals have a diameter that is $\mathcal{O}(t^2)$ for the matrices (14.20). (Hint: first use Gershgorin's theorem to get the bound $|\lambda_\pm \mp 1| \leq t$ and use this to bound the denominator in (14.21) from below via the triangle inequality.)*

Exercise 14.3 *Write a code to determine the values of the characteristic polynomial of a general $n \times n$ matrix A for a given value of λ using induction on n. That is, define $p(A, n, \lambda) = A_{11} - \lambda$ for $n = 1$ and for $n \geq 2$,*

$$p(A, n, \lambda) = \sum_{i=1}^{n} (A_{ii} - \lambda) p(e(A, i), n - 1, \lambda), \tag{14.55}$$

where the matrix function $e(A, i)$ eliminates the ith row and column of A. Test this code for various matrices A for which the characteristic polynomial is known and study its performance as a function of n.

Exercise 14.4 *Write a code to determine the coefficients of the characteristic polynomial of a general $n \times n$ matrix as a polynomial in the variable λ using induction on n. That is, define $c(a, n, \lambda) = [a, -1]$ for $n = 1$, corresponding to the representation of $p_1(\lambda) = a - \lambda$. Determine the data structures needed to form the required iteration. Test this code for various matrices a for which the characteristic polynomial is known and study its performance as a function of n.*

Exercise 14.5 *The notion of Hölder continuity of order $\alpha > 0$ generalizes Lipschitz continuity (2.9):*

$$|g(x) - g(y)| \leq \lambda |x - y|^\alpha. \tag{14.56}$$

In particular, Hölder continuity of order $\alpha = 1$ is the same as Lipschitz continuity. Prove that the zeros of a polynomial of degree n are Hölder-continuous of order $\alpha = 1/n$.

Exercise 14.6 *Prove that the limit of unitary matrices is also unitary. (Hint: just prove that $U^\star U = \lim (U^k)^\star U^k$ if $U = \lim U^k$.)*

Exercise 14.7 *A permutation matrix Π is a matrix such that $\Pi_{\sigma(i), \sigma(j)} = \delta_{i,j}$ (Kronecker δ) for a permutation σ of $\{1, \ldots, n\}$. Prove that a permutation matrix Π is unitary and that $\|\Pi\|_\infty = 1$.*

Exercise 14.8 *Show that the relation $\mathcal{R}$ in (14.23) is reflexive, symmetric, and transitive. (Hint: use the facts that a product of permutations is a permutation and that the inverse of a permutation is a permutation; cf. exercise 14.7).*

Exercise 14.9 *Prove that the function $\mathcal{D}$ defined in (14.24) is symmetric ($\mathcal{D}(x,y) = \mathcal{D}(y,x)$) and satisfies the triangle inequality (14.25) for all $w, x, y \in \mathbb{C}^n$.*

Exercise 14.10 *Consider the equivalence relation $\mathcal{R}$ in (14.23), with the corresponding equivalence classes denoted by $[x]$. Suppose that $x^1, x^2 \in [x]$. Prove that $\mathcal{D}(x^1, y) = \mathcal{D}(x^2, y)$ for any $y \in \mathbb{C}^n$. (Hint: write $x_2 = \Pi x_1$ and explain and exploit the expression $\mathcal{D}(x_2, y) \leq \|x_2 - \Pi P y\|_\infty$ for any permutation P to show that $\mathcal{D}(x_2, y) \leq \mathcal{D}(x_1, y)$.)*

Exercise 14.11 *Suppose that there is a real number x and a sequence of real numbers x_n with the property that for any subsequence x_{n_j}, there is a further subsequence $x_{n_{j_k}}$ that converges to x. Prove that the full sequence must converge to x. (Hint: lack of convergence of the full sequence would imply that there is a subsequence that avoids an open ball around x. But if this subsequence has a subsequence converging to x, we have a contradiction.)*

Exercise 14.12 *Suppose that $\lambda = \beta(b)$ denotes a solution to (14.35). Show that the roots of equation (14.34) can be written in the form*

$$\lambda = a^{1/4} \beta\left(a^{-5/4} b\right). \tag{14.57}$$

For simplicity, assume that $a > 0$.

Exercise 14.13 *Write a code to determine the values of the characteristic polynomial of an $n \times n$ Hessenberg matrix A for a given value of λ using induction on n. That is, define $p(A, n, \lambda) = A_{11} - \lambda$ for $n = 1$ and for $n \geq 2$,*

$$p(A, n, \lambda) = \sum_{i=1}^{n} (A_{ii} - \lambda)p(e(A, i), n - 1, \lambda), \tag{14.58}$$

where the matrix function $e(A, i)$ eliminates the ith row and column of A. Test this code for various matrices A for which the characteristic polynomial is known and study its performance as a function of n.

Exercise 14.14 *Write a code to determine the coefficients of the characteristic polynomial of an $n \times n$ Hessenberg matrix as a polynomial in the variable λ using induction on n. That is, define $c(a, n, \lambda) = [a, -1]$ for $n = 1$, corresponding to the representation of $p_1(\lambda) = a - \lambda$. Determine the data structures needed to form the required iteration. Test this code for various matrices a for which the characteristic polynomial is known and study its performance as a function of n.*

Exercise 14.15 *Suppose that x is a nonzero vector that is not an eigenvector of A. Show that $\|Ax - tx\|_2 \neq 0$ for any scalar t.*

Exercise 14.16 *Prove by induction on k that q^{k+1} can be written in terms of the vectors in (14.46).*

Exercise 14.17 *Suppose that n vectors $x^1, \ldots, x^n$ are orthonormal and are spanned by the n vectors $y^1, \ldots, y^n$. Prove that the vectors $y^1, \ldots, y^n$ have to be linearly independent.*

Exercise 14.18 *Formulate and prove a version of lemma 14.8 that covers the case where the residual r in (14.44) vanishes and the algorithm is restarted. (Hint: write H in block form and show that lemma 14.8 holds for each block.)*

14.8 SOLUTIONS

Solution of Exercise 14.9. To prove symmetry, write $\mathcal{D}(y, x) = \|y - \Pi x\|_\infty$ and define a new permutation $P = \Pi^{-1}$, which is well-defined because permutations are unitary (exercise 14.7). Then by the definition (14.24),

$$\mathcal{D}(x, y) \leq \|x - Py\|_\infty = \|P(\Pi x - y)\|_\infty \leq \|\Pi x - y\|_\infty = \mathcal{D}(y, x) \quad (14.59)$$

since the norm of a permutation matrix is 1 (exercise 14.7). Reversing the roles of x and y yields the reverse inequality, so they must be equal.

To prove (14.25), write $\mathcal{D}(x, w) = \|x - \Pi w\|_\infty$ and $\mathcal{D}(w, y) = \|w - Qy\|_\infty$. Define a new permutation $P = \Pi Q$. By definition,

$$
\begin{aligned}
\mathcal{D}(x, y) \leq \|x - Py\|_\infty &\leq \|x - \Pi w\|_\infty + \|\Pi w - Py\|_\infty \\
&= \mathcal{D}(x, w) + \|\Pi(w - Qy)\|_\infty \\
&\leq \mathcal{D}(x, w) + \|w - Qy\|_\infty = \mathcal{D}(x, w) + \mathcal{D}(w, y)
\end{aligned}
\quad (14.60)
$$

since the norm of a permutation matrix is 1 (exercise 14.7).

Solution of Exercise 14.12. Let $B = a^{-5/4}b$. If λ is defined by (14.57), we have $\lambda = a^{1/4}\beta(B)$, and thus

$$
\begin{aligned}
\lambda^5 - a\lambda + b &= a^{5/4}\beta(B)^5 - aa^{1/4}\beta(B) + b \\
&= a^{5/4}\left(\beta(B)^5 - \beta(B)\right) + b \\
&= a^{5/4}(-B) + b = 0
\end{aligned}
\quad (14.61)
$$

because (14.35) implies that $\beta(B)^5 - \beta(B) = -B$.

Chapter Fifteen

Eigenvalue Algorithms

> "Perhaps situations exist where highly sensitive eigenvalues of nonnormal operators are of genuine physical significance, but they are outnumbered by situations where eigenvalues are mistakenly investigated when a deeper analysis is properly called for." [160]

The number of different algorithms for computing eigenvalues and eigenvectors is extensive [81]. Here we will focus on one technique that is at once simple and fundamental, the power method. We will see that the power method is primarily a method for approximating eigenvectors, with various ways of generating associated eigenvalue approximations. One of these, using the Rayleigh quotient, has special properties. We also consider an important variant of the power method, called inverse iteration. Combining inverse iteration with the Rayleigh quotient for computing the eigenvalue approximation gives an algorithm called Rayleigh quotient iteration. This has a surprising (third-order) rate of convergence.

We also consider the singular value decomposition in this chapter, and we compare it with other factorizations we have seen before.

15.1 POWER METHOD

The power method approximates a single eigenpair by an iterative technique similar to fixed-point iteration. The basic idea is already encoded in theorem 6.11, which states that $A^n \to 0$ iff the eigenvalues of A are less than 1 in modulus. Thus we already have a way to test the size of the eigenvalues. Suppose that $Ax = \lambda x$ and that we scale A by a fixed factor: $B = \alpha A$. Then

$$Bx = \alpha A x = \alpha \lambda x, \tag{15.1}$$

so that $\alpha \lambda$ is an eigenvalue of the scaled matrix. So if $B^n \to 0$, then $|\alpha \lambda| < 1$. By scaling in the right way, we could obtain an algorithm for determining the size of the largest eigenvalue of A; cf. exercise 15.1. Note that an additive shift $B = A + \alpha I$ also has a simple effect on the eigenvalues (exercise 15.2).

What is not obvious from theorem 6.11 is that the basic iterations in chapter 8 also provide information on an eigenvector as well as an eigenvalue. We modify (8.8) by dropping f, and we consider an iterative process of the form

$$x^{(k+1)} = A\,x^{(k)}, \tag{15.2}$$

where we will discuss later the effect of the choice of the starting vector $x^{(0)}$. We know that $x^{(k)}$ may blow up if $\rho(A) > 1$, and it will tend to zero if $\rho(A) < 1$.

More generally, we can predict the expected behavior of the iteration (15.2) by starting with $x^{(0)} = x$, where $x \neq 0$ is an eigenvector with eigenvalue $\lambda = re^{i\theta}$. Then (by induction)

$$x^{(k)} = \lambda^k x = r^k e^{ik\theta} x. \tag{15.3}$$

Thus the iterates grow (or decay) like r^k, and they change direction because of the multiplication by $e^{ik\theta}$. Thus there are two types of scaling we need to do. One of them is to moderate the growth (or decay) of the vector sizes, but in addition, we need to be attentive to the directional change. For example, if $\theta = \pi$ and $r = 1$, then the vectors satisfy $x^{(k)} = (-1)^k x$. Only if $\theta = 0$ can we ignore the direction.

Fortunately, we will see that we need to worry only about the "size" scaling as an integral part of the algorithm. The directional scaling can be added later. Thus we modify (15.2) by introducing a scaling:

$$y^{(k)} = A\,x^{(k)},$$
$$x^{(k+1)} = \frac{1}{\|y^{(k)}\|_2} y^{(k)}. \tag{15.4}$$

The choice of the norm in the definition of x^{k+1} is somewhat arbitrary, but we have made a convenient choice that both simplifies the discussion and is computationally beneficial. The main point is that the normalization ensures that $\|x^{(k+1)}\|_2 = 1$. Note that since we have not included any directional information about the vectors, we do not expect the vectors $x^{(k)}$ themselves to converge in general. However, we can use the vectors $x^{(k)}$ and $y^{(k)}$ to generate convergent eigenvalues. There are different ways to do this, but a convenient one is based on the Rayleigh quotient (14.30):

$$\lambda_k = \lambda^R(x^{(k)}). \tag{15.5}$$

15.1.1 Rayleigh quotient

To begin with, let us investigate some properties of the Rayleigh quotient (14.30), namely,

$$\lambda^R(x) = \frac{x^\star A\,x}{x^\star x}, \tag{15.6}$$

defined for any $0 \neq x \in \mathbb{C}^n$.

First, (14.31) says that the Rayleigh quotient is a fixed-point operator, in the sense that if $A\,x = \alpha x$ for $x \neq 0$, then $\lambda^R(x) = \alpha$. Second, the Rayleigh quotient is independent of scaling,

$$\lambda^R(\alpha x) = \frac{(\alpha x)^\star A(\alpha x)}{(\alpha x)^\star (\alpha x)} = \frac{|\alpha|^2 x^\star A\,x}{|\alpha|^2 x^\star x} = \lambda^R(x), \tag{15.7}$$

for any complex number $\alpha \neq 0$. Third, the Rayleigh quotient is bounded. If we define $\alpha = \max\{|a_{ij}| \mid i, j = 1, \ldots, n\}$, then we see that $|\lambda^R(x)| \leq \alpha$. Therefore, all the iterates λ_k defined in (15.5) are also bounded: $|\lambda_k| \leq \alpha$. Thus at least a subsequence of the λ's will converge. Finally, let us establish the continuity of the Rayleigh quotient.

Lemma 15.1 *Suppose that x and ϵ are two vectors in $\mathbb{C}^n$ such that $\|x\|_2 > \|\epsilon\|_2$. Then*

$$|\lambda^R(x + \epsilon) - \lambda^R(x)| \leq \frac{6\|A\|_2\|x\|_2\|\epsilon\|_2}{(\|x\|_2 - \|\epsilon\|_2)^2}. \tag{15.8}$$

In particular, λ^R is Lipschitz-continuous on $\{x \in \mathbb{C}^n \mid \|x\|_2 > 0\}$.

Proof. We write

$$\begin{aligned}
\lambda^R(x + \epsilon) - \lambda^R(x) &= \frac{1}{\|x + \epsilon\|_2^2}\left((x+\epsilon)^\star A(x+\epsilon) - \frac{\|x+\epsilon\|_2^2}{\|x\|_2^2}x^\star A\,x\right)\\
&= \frac{1}{\|x + \epsilon\|_2^2}\left(\epsilon^\star A\,x + x^\star A\epsilon + \epsilon^\star A\epsilon + \left(1 - \frac{\|x+\epsilon\|_2^2}{\|x\|_2^2}\right)x^\star A\,x\right)\\
&= \frac{1}{\|x + \epsilon\|_2^2}\left(\epsilon^\star A\,x + x^\star A\epsilon + \epsilon^\star A\epsilon - \left(\frac{x^\star\epsilon + \epsilon^\star x + \epsilon^\star\epsilon}{\|x\|_2^2}\right)x^\star A\,x\right).
\end{aligned} \tag{15.9}$$

Observe that for all $y, w \in \mathbb{C}^n$,

$$|y^\star Aw| \leq \|y\|_2\|Aw\|_2 \leq \|y\|_2\|w\|_2\|A\|_2, \tag{15.10}$$

by (5.12) and (6.2). Therefore,

$$\begin{aligned}
|\lambda^R(x + \epsilon) - \lambda^R(x)| &\leq \frac{\|A\|_2\left(4\|x\|_2\|\epsilon\|_2 + 2\|\epsilon\|_2^2\right)}{\|x + \epsilon\|_2^2}\\
&\leq \frac{6\|A\|_2\|x\|_2\|\epsilon\|_2}{\|x + \epsilon\|_2^2}
\end{aligned} \tag{15.11}$$

since $\|\epsilon\|_2 < \|x\|_2$. But by exercise 5.16, we have

$$\frac{1}{\|x + \epsilon\|_2} \leq \frac{1}{\|x\|_2 - \|\epsilon\|_2}. \tag{15.12}$$

Combining (15.12) with (15.11) completes the proof.   QED

15.1.2 Back to the power method

In the power method, we defined x^k and y^k by (15.4) and λ_k by (15.5). Thus

$$\lambda_k = \lambda^R(x^k) = \frac{(x^k)^\star A\,x^k}{(x^k)^\star x^k} = \frac{(x^k)^\star y^k}{(x^k)^\star x^k} = (x^k)^\star y^k, \tag{15.13}$$

where $\lambda^R(x)$ denotes the Rayleigh quotient (15.6). Here we have dropped the parentheses around the iteration indices (superscripts) on the x's and y's to simplify the notation. Note that we have allowed for the possibility that the vectors x^k may be complex, and we must check that $x^k \neq 0$, or rather that $y^{k-1} \neq 0$. Of course, this algorithm continues only if $y^{k-1} \neq 0$. If by chance we find $y^{k-1} = 0$ at some point, we have found a null vector (x^{k-1}) for A and thus an eigenpair with $\lambda = 0$. On the other hand, it may happen that $\lambda_k = 0$ at some stage, i.e., that $(x^k)^\star A x^k = 0$. If A is symmetric and positive definite, this cannot happen at all (exercise 15.3), and if x^k is close to an eigenvector x whose eigenvalue λ is not zero, it also will not happen since $\lambda^R(x^k) \approx \lambda^R(x) = \lambda$, as we will see.

15.1.3 Eigenvector convergence

Regarding convergence of the x^k's, we return to (15.3). We can see that at best we would expect convergence of the sequence $(\overline{\lambda}/|\lambda|)^k x^k$. Let us state a basic convergence result analogous to exercise 2.1.

Lemma 15.2 *Suppose that the power method iteration (15.4) proceeds, with $A x^k \neq 0$ and $\lambda_k \neq 0$, and converges, that is,*

$$\lambda_k \to \lambda \neq 0, \qquad (\overline{\lambda}/|\lambda|)^k x^k \to x, \tag{15.14}$$

as $k \to \infty$. Then $A x = \lambda x$.

Proof. First, we see by using (15.7) that

$$\lambda = \lim_{k \to \infty} \lambda_k = \lim_{k \to \infty} \lambda^R(x^k) = \lim_{k \to \infty} \lambda^R((\overline{\lambda}/|\lambda|)^k x^k) = \lambda^R(x) \tag{15.15}$$

because the Rayleigh quotient (15.6) is continuous (lemma 15.1) on the unit sphere

$$\left\{ y \in \mathbb{C}^n \mid \|y\|_2 = 1 \right\}. \tag{15.16}$$

Since $\|x^k\|_2 = 1$ for each k, $\|x\|_2 = 1$ (cf. exercise 15.5). It suffices to show that $A x = \alpha x$ for some complex α because (14.31) then implies that $\lambda = \lambda^R(x) = \alpha$.

Note that the scaled convergence of x^k implies scaled convergence of y^k:

$$\lim_{k \to \infty} (\overline{\lambda}/|\lambda|)^k y^k = \lim_{k \to \infty} (\overline{\lambda}/|\lambda|)^k A x^k = A \lim_{k \to \infty} (\overline{\lambda}/|\lambda|)^k x^k = A x. \tag{15.17}$$

In particular, we find that (cf. exercise 15.5)

$$\lim_{k \to \infty} \|y^k\|_2 = \lim_{k \to \infty} \|(\overline{\lambda}/|\lambda|)^k y^k\|_2 = \|A x\|_2. \tag{15.18}$$

But now recalling the definition of x^{k+1} in terms of y^k, we find

$$A x = \lim_{k \to \infty} (\overline{\lambda}/|\lambda|)^k y^k = \lim_{k \to \infty} (\overline{\lambda}/|\lambda|)^k \|y^k\|_2 x^{k+1}$$

$$= \lim_{k \to \infty} \|y^k\|_2 \left(|\lambda|/\overline{\lambda}\right) \left(\overline{\lambda}/|\lambda|\right)^{k+1} x^{k+1} \tag{15.19}$$

$$= \|A x\|_2 \left(|\lambda|/\overline{\lambda}\right) x = \|A x\|_2 \left(\lambda/|\lambda|\right) x = \alpha x,$$

where we used exercise 15.5 regarding the limit of products. QED

This result says that it is always a reasonable idea to apply the power method because if it converges, it converges to an eigenpair. This also leads us to a special case of the Perron-Frobenius theorem (section 8.3.4).

Theorem 15.3 *Suppose that M is a nonnegative matrix and that the power method converges to an eigenpair (x, λ) satisfying $Mx = \lambda x$, starting with a nonnegative initial vector x^0. Then $\lambda \geq 0$ and $x \geq 0$.*

The result is obvious because all the iterates x^k are nonnegative. Since x is nonnegative, so is Mx, and thus λ must be also. We address convergence of the power method in section 15.1.4, and those results combined with theorem 15.3 give a proof of theorem 8.18 under certain conditions. We leave the proof of the general form of theorem 8.18 to further reading [13, 168].

15.1.4 Power method convergence

We now turn to the question of when we can anticipate that the power method will converge. Suppose that A is diagonalizable, that is, $A = B^{-1}MB$, where M is diagonal with diagonal entries $\mu_1, \ldots, \mu_n$. Suppose, moreover, that there is an eigenvalue that is largest in modulus and renumber the indices so that it is μ_n:

$$|\mu_n| > |\mu_i| \quad \forall i \neq n. \tag{15.20}$$

Define

$$x = B^{-1}E_n, \tag{15.21}$$

where E_n is the unit vector with zeros in each entry except the nth. Note that $ME_n = \mu_n E_n$ and

$$\begin{aligned} Ax &= (B^{-1}MB)(B^{-1}E_n) = B^{-1}ME_n \\ &= B^{-1}\mu_n E_n = \mu_n B^{-1}E_n = \mu_n x. \end{aligned} \tag{15.22}$$

To see how the power method converges, we return to the original concept (15.2) and define

$$X^k = AX^{k-1} \quad \forall k \geq 1, \tag{15.23}$$

where we take $X^0 = x^0$. Then by induction we have

$$X^k = A^k x^0 \quad \forall k \geq 0. \tag{15.24}$$

We claim that we can write

$$x^k = \|X^k\|_2^{-1} X^k \tag{15.25}$$

for all k, where x^k is generated by the algorithm (15.4). It is true for $k = 0$, so we proceed by induction:

$$\begin{aligned} x^{k+1} &= \|Ax^k\|_2^{-1} Ax^k \quad \text{[by (15.4)]} \\ &= \|A(\|X^k\|_2^{-1}X^k)\|_2^{-1} A(\|X^k\|_2^{-1}X^k) \quad \text{[by (15.25) for k]} \\ &= \|X^k\|_2 \|A(X^k)\|_2^{-1} \|X^k\|_2^{-1} A(X^k) \\ &= \|AX^k\|_2^{-1} AX^k = \|X^{k+1}\|_2^{-1} X^{k+1}, \end{aligned} \tag{15.26}$$

which verifies (15.25) for $k + 1$. Therefore, by (15.7), we conclude that

$$\lambda_k = \lambda^R(x^k) = \lambda^R(X^k). \tag{15.27}$$

Now we consider the asymptotics of the sequence X^k. First, we note the fact that

$$A^k = B^{-1} M^k B, \tag{15.28}$$

which can be proved by induction, see (6.30) and exercise 15.6. Because of our assumption (15.20) about the eigenvalues of A (which are the diagonal entries of M) we see that

$$\lim_{k \to \infty} (\mu_n)^{-k} M^k = \lim_{k \to \infty}
\begin{pmatrix}
(\mu_1/\mu_n)^k & 0 & \cdots & 0 & 0 \\
& & \vdots & & \\
0 & 0 & \cdots & (\mu_{n-1}/\mu_n)^k & 0 \\
0 & 0 & \cdots & 0 & 1
\end{pmatrix} \tag{15.29}$$

$$= \begin{pmatrix}
0 & 0 & \cdots & 0 & 0 \\
& & \vdots & & \\
0 & 0 & \cdots & 0 & 0 \\
0 & 0 & \cdots & 0 & 1
\end{pmatrix} = E_n E_n{}^{\star}.$$

But (15.24), (15.28), and (15.21) imply that

$$(\mu_n)^{-k} X^k = (\mu_n)^{-k} A^k x^0 = B^{-1} (\mu_n)^{-k} M^k B x^0$$
$$\to B^{-1} E_n E_n{}^{\star} B x^0 = (E_n{}^{\star} B x^0) x = ((Bx)^{\star} B x^0) x = \alpha x \tag{15.30}$$

as $k \to \infty$, where $\alpha = (Bx)^{\star} B x^0 = x^{\star} B^{\star} B x^0 = (x, x^0)_{B^{\star}B}$.

In addition to convergence, we can also establish a rate. Define

$$\hat{\rho} = \max \{ |\mu_i|/|\mu_n| \mid i \neq n \} < 1. \tag{15.31}$$

Then

$$\|(\mu_n)^{-k} M^k - E_n E_n{}^{\star}\|_\infty \leq \hat{\rho}^k, \tag{15.32}$$

where $\|D\|_\infty$ denotes the operator norm associated with the maximum norm, which happens to be the same as the maximum absolute entry for a diagonal matrix D (see exercise 8.2). Therefore,

$$\|(\mu_n)^{-k} X^k - \alpha x\|_\infty \leq \|B^{-1}\|_\infty \|B\|_\infty \|x^0\|_\infty \hat{\rho}^k. \tag{15.33}$$

The product $\kappa_\infty(B) := \|B^{-1}\|_\infty \|B\|_\infty$ is often called the *condition number* of the matrix B (with respect to the maximum norm); cf. (9.74).

Suppose that $\alpha = x^{\star} B^{\star} B x^0 = (x, x^0)_{B^{\star}B}$. By (15.6), (15.7), and (15.27), we know that

$$\lambda_k = \lambda^R(\mu_n^{-k} X^k) \to \lambda^R(\alpha x) = \lambda^R(x) = \mu_n. \tag{15.34}$$

Applying (15.33) and (15.8), we find that

$$|\lambda_k - \mu_n| \leq C \hat{\rho}^k, \tag{15.35}$$

provided that k is sufficiently large. More precisely, there are constants k_0 and C such that (15.35) holds for all $k \geq k_0$. First, pick k_0 large enough that

$$\|B^{-1}\|_\infty \|B\|_\infty \|x^0\|_\infty \hat{\rho}^{k_0} \leq \tfrac{1}{2} \|\alpha x\|_\infty. \tag{15.36}$$

This allows us to erase the denominator on the right-hand side of (15.8), replacing the factor 6 in the numerator by 24. Determining the remaining ingredients in C is left as an exercise.

Write $\mu_n = Re^{i\theta}$ with $R > 0$. Then

$$\lim_{k\to\infty} R^{-k}\|X^k\|_2 = \lim_{k\to\infty} \|\mu_n^{-k}X^k\|_2 = \|\alpha x\|_2. \tag{15.37}$$

Therefore, by (15.25),

$$\begin{aligned} e^{-ik\theta}x^k &= e^{-ik\theta}\|X^k\|_2^{-1}X^k \\ &= \mu_n^{-k}(R^k/\|X^k\|_2)X^k \to \|\alpha x\|_2^{-1}\alpha x. \end{aligned} \tag{15.38}$$

Therefore, we have proved the following result.

Theorem 15.4 *Suppose that $A = B^{-1}MB$ is a diagonalizable $n \times n$ matrix whose eigenvalues μ_i satisfy (15.20). Suppose that the starting vector x_0 satisfies*

$$\alpha = x^\star B^\star Bx^0 = (x, x^0)_{B^\star B} \neq 0, \tag{15.39}$$

where x is the eigenvector (15.21) of A corresponding to μ_n. Then (15.35) holds for the eigenvalue convergence, where $\hat{\rho}$ is defined in (15.31), and

$$\left\|(|\mu_n|/\mu_n)^k x^k - \|\alpha x\|_2^{-1}\alpha x\right\| \leq C\hat{\rho}^k \tag{15.40}$$

for k sufficiently large, where x is the eigenvector of A corresponding to μ_n.

15.1.5 Power method limitations

Unfortunately, the power method does not work universally. If the eigenvalues largest in complex modulus occur as a conjugate pair, which happens frequently for a real matrix, then the power method will oscillate, as seen by considering the matrix

$$A = \begin{pmatrix} 0 & 1 \\ -1 & 0 \end{pmatrix}, \tag{15.41}$$

which has the property that $A^{4n+j} = A^j$ for $j = 1, 2, 3, 4$ and n any positive integer. In particular, $A^2 = -I$, $A^3 = -A$, $A^4 = I$, $A^5 = A$, etc. If we define $X^{(n)} = A^k X^{(0)}$, we have $X^{(4n+2)} = -X^{(0)}$ for all n, whereas $X^{(4n)} = X^{(0)}$. Note that if $X^{(0)}$ is normalized, then so are all the subsequent vectors. Thus convergence does not occur. We leave as exercise 15.7 characterization of what happens when we apply the Rayleigh quotient to the vectors $X^{(n)}$.

15.1.6 Defective matrices

When A is defective, the behavior is more complex, but the power method still converges when there is a single largest eigenvalue. This is easily seen by considering the fact that

$$\begin{pmatrix} \lambda & 1 \\ 0 & \lambda \end{pmatrix}^k = \lambda^{k-1}\begin{pmatrix} \lambda & k \\ 0 & \lambda \end{pmatrix} \tag{15.42}$$

for any positive integer k. Thus if we apply the power method starting with the vector

$$x^0 = \begin{pmatrix} 0 \\ 1 \end{pmatrix}, \tag{15.43}$$

we get the sequence of vectors

$$X^{(k)} = \lambda^{k-1} \begin{pmatrix} k \\ \lambda \end{pmatrix} = k\lambda^{k-1} \begin{pmatrix} 1 \\ \frac{\lambda}{k} \end{pmatrix}. \tag{15.44}$$

Thus we get convergence of the normalized x^k to the eigenvector $\begin{pmatrix} 1 \\ 0 \end{pmatrix}$ at a rate of λ/k. On the other hand

$$\begin{pmatrix} \lambda & 1 \\ 0 & \lambda \end{pmatrix}^k \begin{pmatrix} 1 \\ 0 \end{pmatrix} = \lambda^k \begin{pmatrix} 1 \\ 0 \end{pmatrix} \tag{15.45}$$

for all k. Thus it is easy to see that the power method converges to the correct eigenvector for any starting vector (exercise 15.8), but the convergence is no longer exponential in general.

The matrix in (15.42) is an example of what we might call a *Jordan matrix* A which is equal to λ on the diagonal, equal to 1 on the superdiagonal $(a_{i,i+1} = 1)$, and 0 elsewhere:

$$\begin{pmatrix} \lambda & 1 & 0 & \cdots & 0 & 0 & 0 \\ 0 & \lambda & 1 & \cdots & 0 & 0 & 0 \\ \cdot & & \cdot & \cdots & & \cdot & \cdot \\ 0 & 0 & 0 & \cdots & 0 & \lambda & 1 \\ 0 & 0 & 0 & \cdots & 0 & 0 & \lambda \end{pmatrix}. \tag{15.46}$$

The general form of the powers of a Jordan matrix A of order n has the form

$$(A^k)_{i,i+j} = \lambda^{k-j} \binom{k}{j} \tag{15.47}$$

for all $i = 1, \ldots, n$ and $j = 0, \ldots, \min\{n - i, k\}$ (exercise 15.11). Thus $A^k x$ can be computed as follows. Let m be the smallest index such that $x_j = 0$ for all $j > m$. If $x_n \neq 0$, then set $m = n$.

If $m = 1$, then x is an eigenvector of A, so $A^k x = \lambda^k x$ for all k. So suppose that $m \geq 2$. Then for $k > 2n$,

$$(A^k x)_1 = \sum_{j=0}^{m-1} \lambda^{k-j} \binom{k}{j} x_{j+1}$$

$$= \lambda^{k-m+1} \binom{k}{m-1} \left(x_m + \sum_{j=0}^{m-2} \epsilon_j^{(k)} x_{j+1} \right), \tag{15.48}$$

where the error terms $\epsilon_j^{(k)}$ are given by

$$\epsilon_j^{(k)} = \lambda^{m-j-1} \binom{k}{j} \Big/ \binom{k}{m-1} = \lambda^{m-j-1} \frac{(m-1)!(k-m+1)!}{j!(k-j)!}$$

$$= \lambda^{m-j-1} \frac{(m-1)\cdots(j+1)}{(k-j)\cdots(k-m+2)} = \prod_{i=1}^{m-j-1} \lambda \frac{j+i}{k+1-j-i}. \tag{15.49}$$

Estimating the terms in the fractions, we find

$$|\epsilon_j^{(k)}| \leq \left(\frac{\lambda(m-1)}{k-m+2}\right)^{m-1-j} \to 0 \qquad (15.50)$$

as $k \to \infty$ since $m - 1 - j \geq 1$. Therefore,

$$\lambda^{m-j-1}\binom{k}{m-1}^{-1}(A^k x)_1 \to x_m \quad \text{as } k \to \infty. \qquad (15.51)$$

Similarly, for $i \geq 2$,

$$(A^k x)_i = \sum_{j=0}^{m-i} \lambda^{k-j}\binom{k}{j}x_{j+i}$$

$$= \lambda^{-k+m-1}\binom{k}{m-1}\left(\sum_{j=0}^{m-i}\epsilon_j^{(k)}x_{j+1}\right). \qquad (15.52)$$

Therefore, for $i \geq 2$,

$$\lambda^{-k+m-1}\binom{k}{m-1}^{-1}(A^k x)_i \to 0 \quad \text{as } k \to \infty. \qquad (15.53)$$

Lemma 15.5 *Suppose that A is the $n \times n$ Jordan matrix shown in (15.46) and let $x \neq 0$ be an arbitrary initial vector. Let m be the smallest index such that $x_j = 0$ for all $j > m$. If $x_n \neq 0$, then set $m = n$. If $m = 1$, then $A^k x = \lambda^k x$ for all k. If $m \geq 2$, the power method starting with the initial vector x converges to the eigenvector $E = (1, 0, \ldots, 0)^T$, and the error satisfies*

$$\left|\lambda^{-k+m-1}\binom{k}{m-1}^{-1}A^k x - x_m E\right| \leq \|x\|_\infty \sum_{j=0}^{m-2}\left(\frac{|\lambda|(m-1)}{k-m+2}\right)^{m-1-j} \qquad (15.54)$$

for $k > 2n$. If also $k > (1 + |\lambda|)n$, then

$$\left|\lambda^{-k+m-1}\binom{k}{m-1}^{-1}A^k x - x_m E\right| \leq \|x\|_\infty C k^{-1}, \qquad (15.55)$$

where C is a constant that depends only on n and λ.

A general matrix resulting from the Jordan decomposition has several Jordan blocks, but the analysis is similar. As long as there is a single μ_r with largest complex modulus, that is, the eigenvalues satisfy the following analog of (15.20),

$$|\mu_r| > |\mu_i| \quad \forall i \neq r, \qquad (15.56)$$

then the power method converges provided there is a suitable starting vector (exercise 15.14). The convergence behavior is determined by the Jordan block having the eigenvalue largest in magnitude.

15.2 INVERSE ITERATION

The power method is effective in determining the largest eigenvalue of a matrix, but with a slight variation it can be used to find any eigenvalue. If we want to find an eigenvalue close to some value $\mu \in \mathbb{C}$, we can apply the power method to $B = (A - \mu I)^{-1}$. If $Bx = \lambda x$, then $(A - \mu I)x = \lambda^{-1}x$, so that $\mu + \lambda^{-1}$ is an eigenvalue of A (see exercise 15.2). If λ is the largest eigenvalue of $(A - \mu I)^{-1}$, then $\mu + \lambda^{-1}$ is the eigenvalue of A closest to μ. In particular, if $\mu = 0$ then λ will be the eigenvalue with the smallest modulus. Of course, applying the power method to B requires solving systems $(A - \mu I)x = y$ successively. For this reason, the algorithm is called *inverse iteration*, and is defined as follows:

$$(A - \mu I)y^k = x^k,$$

$$x^{k+1} = \frac{1}{\|y^k\|_2}y^k, \quad \text{and} \tag{15.57}$$

$$\lambda_{k+1} = \lambda^R(x^{k+1}).$$

We can imagine applying inverse iteration at any stage in the process of approximating eigenvalues. Even if we seek the largest eigenvalue of A, if we have found a good guess $\mu = \lambda_k$, then we can apply inverse iteration, that is, the power method for $B = (A - \lambda_k I)^{-1}$, to refine our estimate. We will pursue this idea in detail in section 15.2.2. But what is disconcerting about this process is that, as λ_k becomes a better approximation to an eigenvalue, $A - \lambda_k I$ becomes more nearly singular.

We have seen that the eigenvalue problem for an $n \times n$ matrix can be viewed as a system of nonlinear equations in $n + 1$ variables (7.53). In that case, we saw that the system was equivalent to (7.59) and (7.60), with a different scaling relating x^k and y^k. These equations are very similar to those in inverse iteration (15.57), and we saw that the linear system was nonsingular provided the eigenvalue was simple. But we can look at inverse iteration more directly, as follows.

15.2.1 The nearly singular system

Suppose for simplicity that the eigenvalue of interest is $\lambda = 0$, so that the eigenvector x is a null vector of A. Inverse iteration involves solving an equation of the form

$$(A + \epsilon I)x^\epsilon = f, \tag{15.58}$$

where we take $f \approx x$ and hope that x^ϵ is an even better approximation to x. It may seem a strange choice to take $f \approx x$ to look for a null vector, but in fact $x^\epsilon = \epsilon^{-1}x$ is an exact solution to

$$(A + \epsilon I)x^\epsilon = x \tag{15.59}$$

since $Ax = 0$. Thus inverse iteration tends to amplify the null vector. It would be reasonable to expect that for $f = x + e$, where $x^\star e = 0$, the solution to (15.58) would satisfy

$$x^\epsilon = \epsilon^{-1}x + w + \mathcal{O}(\epsilon) \qquad (15.60)$$

for some w. We can prove (15.60) in the case where $A = A^\star$ as follows.

Let $V = \{y \in \mathbb{C}^n \mid x^\star y = 0\}$. Then A maps $V \to V$ and is invertible when restricted to V. The reason that the range of A is again V corresponds to the fact that $A^\star x = A x = 0$ (see section 3.4.4). Note that $A + \epsilon I$ also maps $V \to V$ invertibly for ϵ small. Define

$$w^\epsilon = (A + \epsilon I)^{-1}e. \qquad (15.61)$$

Then set $x^\epsilon = \epsilon^{-1}x + w^\epsilon$ and compute

$$(A + \epsilon I)x^\epsilon = x + (A + \epsilon I)w^\epsilon = x + e. \qquad (15.62)$$

Thus (15.60) holds, with $w \in V$ determined uniquely by $Aw = e$.

15.2.2 Rayleigh quotient iteration

Rayleigh quotient iteration (RQI) is simply inverse iteration together with the choice (15.5) to define the eigenvalue approximation. Thus

$$\begin{aligned}
(A - \lambda_k I)y^k &= x^k \\
x^{k+1} &= \|y^k\|_2^{-1}y^k \\
\lambda_{k+1} &= \lambda^R(x^{k+1}).
\end{aligned} \qquad (15.63)$$

This is very similar to Newton's method applied to the eigensystem as described in section 7.2.4, except for the use of the Rayleigh quotient instead of (7.60). We know that Newton's method is quadratically convergent, so we might wonder why we would use (15.63) instead. It turns out that Rayleigh quotient iteration is even faster: it converges cubicly (at least for normal matrices [123]).

We will not attempt to prove the cubic convergence in full detail but instead give a simple example that demonstrates this. Suppose that x and y are normalized eigenvectors of A with eigenvalues λ and μ, respectively, with $x^\star y = 0$. By normalized, we mean that $x^\star x = y^\star y = 1$. Suppose that for some k, the iterate in (15.63) satisfies

$$x^k = \alpha_k x + \beta_k y, \qquad (15.64)$$

with $|\alpha_k|^2 + |\beta_k|^2 = 1$, so that $x^{k\star}x^k = 1$. We also assume that

$$\begin{aligned}
\lambda_k = \lambda^R(x^k) = x^{k\star}A x^k &= (\alpha_k x + \beta_k y)^\star A(\alpha_k x + \beta_k y) \\
&= (\alpha_k x + \beta_k y)^\star(\alpha_k \lambda x + \beta_k \mu y) = |\alpha_k|^2\lambda + |\beta_k|^2\mu.
\end{aligned} \qquad (15.65)$$

Note that

$$\lambda - \lambda_k = (1 - |\alpha_k|^2)\lambda - |\beta_k|^2\mu = |\beta_k|^2(\lambda - \mu) \qquad (15.66)$$

and

$$\mu - \lambda_k = -|\alpha_k|^2\lambda + (1 - |\beta_k|^2)\mu = |\alpha_k|^2(\mu - \lambda). \qquad (15.67)$$

If $0 < |\alpha_k| < 1$ and $\lambda \neq \mu$, as we now assume, then $\lambda_k \neq \lambda$ and $\lambda_k \neq \mu$. Then y^k is defined by solving

$$(A - \lambda_k I)y^k = x^k = \alpha_k x + \beta_k y. \tag{15.68}$$

But the solution satisfies $y^k = \hat{\alpha}_k x + \hat{\beta}_k y$, where

$$(\lambda - \lambda_k)\hat{\alpha}_k = \alpha_k \quad \text{and} \quad (\mu - \lambda_k)\hat{\beta}_k = \beta_k. \tag{15.69}$$

In particular, $y^k \neq 0$. Therefore,

$$\begin{aligned}
\hat{\alpha}_k &= \frac{\alpha_k}{\lambda - \lambda_k} = \frac{\alpha_k}{|\beta_k|^2(\lambda - \mu)} \quad \text{and} \\
\hat{\beta}_k &= \frac{\beta_k}{\mu - \lambda_k} = \frac{\beta_k}{|\alpha_k|^2(\mu - \lambda)}.
\end{aligned} \tag{15.70}$$

Finally, $x^{k+1} = sy^k$, where $s = 1/\|y^k\|_2 > 0$. Thus $(\alpha_{k+1}, \beta_{k+1}) = s(\hat{\alpha}_k, \hat{\beta}_k)$. The value of s can be determined via

$$\begin{aligned}
s^{-2} &= \|y^k\|_2^2 = |\hat{\alpha}_k|^2 + |\hat{\beta}_k|^2 = \frac{1}{|\lambda - \mu|^2}\left(\frac{|\alpha_k|^2}{|\beta_k|^4} + \frac{|\beta_k|^2}{|\alpha_k|^4}\right) \\
&= \frac{|\alpha_k|^6 + |\beta_k|^6}{|\lambda - \mu|^2|\alpha_k|^4|\beta_k|^4} = \frac{(1 - |\beta_k|^2)^3 + |\beta_k|^6}{|\lambda - \mu|^2|\alpha_k|^4|\beta_k|^4} \\
&= \frac{1 - 3|\beta_k|^2 + 3|\beta_k|^4}{|\lambda - \mu|^2|\alpha_k|^4|\beta_k|^4} = \frac{1 - 3|\beta_k|^2 + 3|\beta_k|^4}{|\lambda - \mu|^2(1 - |\beta_k|^2)^2|\beta_k|^4}.
\end{aligned} \tag{15.71}$$

Now let us think about the component of the eigenvector y as the error. Thus we assume that β_k is small, and we can approximate (15.71) to give

$$s = \frac{|\lambda - \mu|(1 - |\beta_k|^2)|\beta_k|^2}{\sqrt{1 - 3|\beta_k|^2 + 3|\beta_k|^4}} \approx |\lambda - \mu||\beta_k|^2. \tag{15.72}$$

Thus we find that

$$|\beta_{k+1}| = s|\hat{\beta}_k| = \frac{s|\beta_k|}{|\alpha_k|^2|\lambda - \mu|} \approx |\beta_k|^3. \tag{15.73}$$

Thus the error is decreasing cubicly.

Note the symmetry between α and β in (15.70). Thus if we start closer to y than to x, RQI converges to (μ, y) instead of (λ, x). Moreover, if we start exactly in the middle, that is, $\alpha_0 = \beta_0 = 1/\sqrt{2}$, then $x^k = x^0$ and $\lambda_k = \frac{1}{2}(\lambda + \mu)$ (and $s = \frac{1}{2}|\lambda - \mu|$) for all k.

15.3 SINGULAR VALUE DECOMPOSITION

The singular value decomposition of a matrix A is closely related to the eigenvalue decomposition of $B = A^\star A$. Since B is Hermitian and positive semidefinite, we have $B = U^\star \Lambda U$, where U is unitary and Λ is a diagonal matrix with nonnegative entries. Thus we can define

$$\Sigma = \sqrt{\Lambda}, \tag{15.74}$$

that is, Σ is the diagonal matrix with diagonal entries given by the (nonnegative) square roots of the diagonal entries of Λ. Note that this holds even if A is not square. The diagonal entries of Σ are called the *singular values* of A.

The *singular value decomposition* is the representation

$$\boxed{A = V\Sigma U^{\star},} \tag{15.75}$$

where V is also unitary. Suppose that A is an $m \times n$ matrix. Then $B = A^{\star}A$ is $n \times n$ and thus so are U and Σ, as we have defined them above. However, in (15.75) the matrix Σ is $m \times n$, and V is $m \times m$. Thus we need a new way to see what Σ is since (15.74) does not suffice in the case where $m \neq n$. Since we do not need this case for further developments, we will leave the details to further reading [160].

So we now suppose that $m = n$, and then assume that all the singular values are positive. Then Σ is unambiguously defined by (15.74) and is invertible, and the representation (15.75) is equivalent to

$$V = AU\Sigma^{-1}. \tag{15.76}$$

Let us take (15.76) as the definition of V. We just need to check that it is unitary:

$$\begin{aligned}
V^{\star}V &= (\Sigma^{-1}U^{\star}A^{\star})AU\Sigma^{-1} = \Sigma^{-1}U^{\star}A^{\star}AU\Sigma^{-1} \\
&= \Sigma^{-1}U^{\star}BU\Sigma^{-1} = \Sigma^{-1}\Lambda\Sigma^{-1} = I.
\end{aligned} \tag{15.77}$$

Let us ask the question of how the singular values relate to eigenvalues of A in the case where A is diagonalizable. Suppose that $A = W^{-1}DW$, where D is a diagonal matrix (whose diagonal elements are thus the eigenvalues of A). Then

$$\begin{aligned}
B &= A^{\star}A = (W^{-1}DW)^{\star}W^{-1}DW = (W^{\star}\overline{D}(W^{-1})^{\star})W^{\star}DW \\
&= W^{\star}\overline{D}DW = W^{\star}|D|^{2}W,
\end{aligned} \tag{15.78}$$

where $|D|$ is the diagonal matrix whose entries are the complex modulus of the entries of D. Therefore, the eigenvalues of B are the same as the entries of $|D|^{2}$ (up to some ordering), and hence the singular values of A are just the complex modulus of the eigenvalues of A. Thus the singular values provide a natural generalization of eigenvalues, provided we are interested only in magnitude and not in phase. But they are also well-defined in the case where A is not diagonalizable and even when A is not square [160].

15.4 COMPARING FACTORIZATIONS

We have seen three factorizations involving unitary matrix factors, and it may be useful to see how they are related. The factorization $A = QR$ (5.54) gives the Cholesky factor of $A^{\star}A = R^{\star}R$, as noted in (5.55). Similarly, we have seen that the singular value decomposition (15.75) of A can be related to the eigen decomposition of $A^{\star}A$. On the other hand, we contrasted the Schur decomposition with QR in section 6.2. Of course, for Hermitian A, the Schur decomposition is the eigen decomposition.

15.5 MORE READING

We have left the full proof of the general form of the Perron-Frobenius theorem 8.18 to further reading [13, 168]. The text [160] was mentioned in section 15.3 and should be consulted more generally. The classic monograph [172] is still of primary interest. For the interpretation of inverse iteration as Newton's method, see [128]. The QR decomposition is commonly used to solve eigenproblems [152]. See [123] for more on Rayleigh quotient iteration.

15.6 EXERCISES

Exercise 15.1 *Use (6.26) to compute* $\rho(A)$ *via*
$$\rho(A) = \lim_{k \to \infty} \|A^k\|_\infty^{1/k}. \tag{15.79}$$
Prove that this works for diagonal A.

Exercise 15.2 *Suppose that* $A x = \lambda x$ *and* $B = A + \alpha I$. *Show that* $B x = (\lambda + \alpha) x$.

Exercise 15.3 *Prove that the Rayleigh quotient (15.6) cannot vanish for* $x \neq 0$ *for a symmetric, positive definite matrix* A. *(Hint: if* $x^T A x = 0$, *then* $(U x)^T \Lambda (U x) = 0$, *where* $A = U^T \Lambda U$ *and* U *is orthogonal.)*

Exercise 15.4 *Prove that the Rayleigh quotient (15.6) is continuous on the unit sphere* $\{ y \in \mathbb{C}^n \mid \|y\| = 1 \}$ *for any norm.*

Exercise 15.5 *Suppose that* x^k *is a sequence of vectors in* $\mathbb{C}^n$ *such that* $\lim_{k \to \infty} x^k = x$ *and that* α_k *is a sequence of scalars such that* $\lim_{k \to \infty} \alpha_k = \alpha$. *Prove that* $\lim_{k \to \infty} \alpha_k x^k = \alpha x$. *Prove also that* $\lim_{k \to \infty} \|x^k\| = \|x\|$. *If* y^k *is another sequence of vectors in* $\mathbb{C}^n$ *such that* $\lim_{k \to \infty} y^k = y$, *prove that* $\lim_{k \to \infty} (x^k)^\star y^k = x^\star y$.

Exercise 15.6 *Suppose that* $A = B^{-1} C B$. *Prove that for any integer* $k \geq 1$, $A^k = B^{-1} C^k B$. *What are the minimal assumptions required of the matrix* C *for this to be true? (Hint: use induction on* k.)

Exercise 15.7 *Consider the matrix* A *defined in (15.41) and the vectors* $X^{(n)}$ *generated by the power method. Characterize what happens when the Rayleigh quotient is applied to the vectors* $X^{(n)}$.

Exercise 15.8 *Prove that the power method for the matrix* $\begin{pmatrix} 1 & 1 \\ 0 & 1 \end{pmatrix}$ *will converge for any starting vector to the eigenvector* $\begin{pmatrix} 1 \\ 0 \end{pmatrix}$ *for any starting vector. (Hint: compute*
$$\begin{pmatrix} 1 & 1 \\ 0 & 1 \end{pmatrix}^n \begin{pmatrix} a \\ b \end{pmatrix} = \begin{pmatrix} a + nb \\ b \end{pmatrix} = n \begin{pmatrix} \frac{a}{n} + b \\ \frac{b}{n} \end{pmatrix} \tag{15.80}$$
and let $n \to \infty$.)

Exercise 15.9 *Suppose that A is Hermitian ($A^* = A$). Show that the Rayleigh quotient (15.6) is a real number. Define*

$$\lambda = \sup_{0 \neq x \in \mathbb{C}^n} \lambda^R(x) = \sup_{0 \neq x \in \mathbb{C}^n} \frac{x^* A x}{x^* x}. \qquad (15.81)$$

Prove that λ is an eigenvalue of A. Do the same for sup replaced by inf.

Exercise 15.10 *Prove (15.42).*

Exercise 15.11 *Prove (15.47). (Hint: write $A = \lambda I + J$ and use the binomial theorem to expand A^k. Here J is the matrix with 1's on the superdiagonal and 0's elsewhere.)*

Exercise 15.12 *Consider the matrix*

$$A = \begin{pmatrix} 0 & 1 & 0 & 0 \\ -1 & 0 & 0 & 0 \\ 0 & 0 & 1 & 1 \\ 0 & 0 & 0 & 1 \end{pmatrix}. \qquad (15.82)$$

Show that this matrix has three eigenvalues which all have complex modulus equal to 1. Describe how the power method for this matrix behaves for different starting vectors. (Hint: consider what happens with starting vectors x, where (1) $x_1 = x_2 = 0$, (2) $x_3 = x_4 = 0$, and (3) $x_1 = x_2 = x_3 = 0$. What is the special role of having x_4 nonzero?)

Exercise 15.13 *An alternate definition of the power method is sometimes given as*

$$y^k = A x^k, \qquad x^{k+1} = \|y^k\|_2^{-1} y^k. \qquad (15.83)$$

Note that by definition, $\|x^k\|_2 = 1$. Prove that if $x^k \to x$, then x is an eigenvector with eigenvalue $\lambda = \|y\|_2$, where $y = \lim_{k \to \infty} y^k$. In particular, $\lambda > 0$. Compare this algorithm with (15.4) for the matrix

$$A = \begin{pmatrix} -3 & 1 \\ 2 & 1 \end{pmatrix}. \qquad (15.84)$$

Exercise 15.14 *Suppose that A is any matrix whose eigenvalues satisfy (15.20). Prove that the power method will converge to the largest eigenvalue for a suitable starting vector. Give a characterization of the conditions required for the starting vector to guarantee convergence. (Hint: use the Jordan decomposition for A and apply lemma 15.5.)*

Exercise 15.15 *Let A be a real, symmetric matrix and number its eigenvalues $\lambda_1 \leq \lambda_2 \leq \cdots \leq \lambda_n$. Let $\mathcal{G}$ denote the set of all subspaces of $\mathbb{R}^n$. For $S \in \mathcal{G}$, we say $x \perp S$ if $x^T y = 0$ for all $y \in S$. Prove that*

$$\lambda_i = \sup_{S \in \mathcal{G}, \dim S = i-1} \left(\inf_{0 \neq x \perp S} \lambda^R(x) \right) = \inf_{S \in \mathcal{G}, \dim S = n-i} \left(\sup_{0 \neq x \perp S} \lambda^R(x) \right),$$
$$(15.85)$$

where $\lambda^R(x)$ denotes the Rayleigh quotient (15.6). This is known as the Courant-Fischer theorem. (Hint: see exercise 15.9.)

Exercise 15.16 *Show that RQI (15.63) can be written as fixed-point itera-tion $y^{k+1} = g(y^k)$, where*

$$g(y) = \frac{1}{\|y\|_2} \left(A - \lambda^R(y)I \right)^{-1} y. \tag{15.86}$$

Would you expect this to converge to a fixed-point $y = g(y)$. Why or why not?

Exercise 15.17 *Let A be an arbitrary complex matrix and define $B = \frac{1}{2}(A + A^\star)$, the Hermitian part of A. Prove that the Rayleigh quotient for B is the real part of the Rayleigh quotient for A. (Hint: show that $x^\star Bx = \mathcal{R}e\,(x^\star A\,x)$.)*

Exercise 15.18 *Prove that any matrix A can be written as the limit of diagonalizable matrices, i.e., that the set of diagonalizable matrices is dense. (Hint: write $A = U^\star TU$ using the Schur decomposition and perturb the diagonal entries of T to make the eigenvalues unique.)*

15.7 SOLUTIONS

Solution of Exercise 15.11. We have

$$A^k = (\lambda I + J)^k = \sum_{j=0}^{k} \binom{k}{j} \lambda^{k-j} J^j. \tag{15.87}$$

But J^j is 1 on the jth superdiagonal (indices kl such that $l = k + j$) and 0 elsewhere.

Solution of Exercise 15.13. First, if $x^k \to x$, then $y^k = A\,x^k \to A\,x := y$. Since $y^k = \|y^k\|_2 x^{k+1}$, we have

$$A\,x = y = \lim_{k \to \infty} y^k = \lim_{k \to \infty} \|y^k\|_2 x^{k+1} = \|y\|_2 x \tag{15.88}$$

(see exercise 15.5 for the last step).

Chapter Sixteen

Ordinary Differential Equations

> Bode's Law of planetary distances says that the distance d_n
> of the nth planet from its star behaves like $d_n = a + bc^n$
> for $n = 1, 2, \ldots$. For our solar system, $a = 0.4$, $b = 0.3$,
> and $c = 2$, and the model is remarkably accurate for all the
> major planets except Neptune and with the addition of the
> asteroid Ceres [129].

Ordinary differential equations (ODEs) can be used to model a remarkable range of natural phenomena [91]. For example, they can be used to predict the movement of celestial bodies [166]. They are used to model the evolution of our planetary system [70] and potentially could shed light on the validity of Bode's Law (see the introductory comment above). However, there are significant mathematical challenges to performing simulations of such systems on the required time scales. We provide a brief introduction to these issues in this and the following chapter.

An indefinite integral is the simplest form of ordinary differential equation. If $u(x) = \int_a^x f(y)\,dy$, then $u'(x) = f(x)$ for all x where the integral makes sense. Differential equations become more complex when f is allowed to depend on u as well as x (cf. (16.3)). In this case, a simple integral no longer suffices, but it is not surprising that techniques developed for numerical quadrature play a role in numerical methods for solving differential equations.

We develop the basic theory of ordinary differential equations here for two reasons. For one, it keeps the book self-contained. But the more important reason is that the techniques used are constructive and similar to ones we have used for numerical algorithms.

16.1 BASIC THEORY OF ODES

The simplest differential equation to solve is an ordinary differential equation

$$\frac{du}{dt} = f(u, t) \tag{16.1}$$

with initial value

$$u(0) = u_0, \tag{16.2}$$

where we are interested in solving on some interval $[0, T]$. Equations (16.1) and (16.2) can be cast as a single integral equation

$$u(x) = u_0 + \int_0^x f(u(t), t) \, dt. \tag{16.3}$$

Most of what we develop will be insensitive to the "type" of u, so we could think of this as having vector values in general, i.e., $u : [0, T] \to \mathbb{R}^n$, in which case $f : \mathbb{R}^{n+1} \to \mathbb{R}^n$. We could let $| \cdot |$ be a fixed norm on $\mathbb{R}^n$, so that $|u|$ and $|f(u, t)|$ indicate norms of the quantities $u \in \mathbb{R}^n$ and $f(u, t) \in \mathbb{R}^n$. However, nothing of what we pursue here changes in this case, and it requires some additional care in the notation (e.g., in integrating vector functions and the definitions of function spaces), so we prefer to stick with the case $n = 1$.

The theory for the existence and uniqueness of solutions for systems (16.1) is still a subject of research, but the basic results are classical. Suppose that the function $f \in C^0(\mathbb{R} \times [0, T])$ satisfies a Lipschitz condition

$$|f(u, t) - f(v, t)| \le L|u - v| \quad \text{for all} \quad u, v \in \mathbb{R} \text{ and } t \in [0, T]. \tag{16.4}$$

Then (16.1) has a unique solution u whose first derivative is continuous. We will derive this result as theorem 16.2 for completeness.

Let us consider a simple example that provides some insight into the range of behaviors to expect. Let $f(u, t) = -u^2$, so that f is independent of t. Then by differentiation, we can verify that the corresponding solution to (16.1) is

$$u(t) = \frac{1}{u_0^{-1} + t} = \frac{u_0}{1 + t u_0} \tag{16.5}$$

since $u'(t) = -(u_0^{-1} + t)^{-2} = -u^2$. There are two distinct regimes of behavior. If $u_0 > 0$, then u decreases algebraically to zero as $t \to \infty$, but if $u_0 < 0$, then u blows up in finite time, at $t = -1/u_0$. These two different behaviors are depicted in figure 16.1. Thus in general, solutions to (nonlinear) ordinary differential equations may not exist for all time.

16.2 EXISTENCE AND UNIQUENESS OF SOLUTIONS

There are several ways to prove the existence of smooth solutions to (16.1). We will present the Picard[1] method, which uses (16.3) as the basis for a fixed-point iteration to define a sequence of functions that converge to a solution u. The function for which the fixed-point is being sought is a mapping Φ from the function space $C^0([0, T])$ to itself, defined by

$$\Phi(v)(x) = u_0 + \int_0^x f(v(t), t) \, dt. \tag{16.6}$$

Thus u is a solution to (16.3), and, equivalently, to (16.1) if and only if $u = \Phi(u)$.

---

[1]Charles Emile Picard (1856–1941) was a mentor of Bernstein (see page 187), as well as Jacques Hadamard, Paul Painlevé (who, like Lagrange, is interred in the Pantheon), and André Weil, among others.

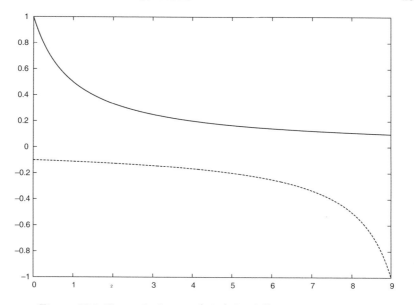

Figure 16.1 Two solutions to (16.5) for different initial values.

Lemma 16.1 *Suppose that $f \in C^0(\mathbb{R} \times [0, T])$ satisfies the Lipschitz condition (16.4) with Lipschitz constant L. Then the expression (16.6) defines a Lipschitz mapping Φ from the function space $C^0([0, T])$ to itself where*

$$\|\Phi(v) - \Phi(w)\|_{\infty, [0, T]} \leq LT\|v - w\|_{\infty, [0, T]}. \tag{16.7}$$

An immediate corollary of lemma 16.1 is the uniqueness of solutions of (16.3) and (equivalently) (16.1). If there were two solutions $u = \Phi(u)$ and $v = \Phi(v)$, then we conclude that

$$\|u - v\|_{\infty, [0, T]} = \|\Phi(u) - \Phi(v)\|_{\infty, [0, T]} \leq LT\|u - v\|_{\infty, [0, T]} \tag{16.8}$$

for any T. But (16.8) implies that $\|u - v\|_{\infty, [0, T]} = 0$ if $LT < 1$. Of course, this shows uniqueness only on the interval $[0, 1/L]$, but we can iterate this process to obtain uniqueness on the intervals $[kL, (k + 1)L]$ for any $k \geq 0$ by induction. That is, we can view u and v as solutions of a new problem starting at $t = kL$ with initial values $u(kL) = v(kL)$. This is often called the *semigroup property* of (16.1). Thus uniqueness holds on any interval $[0, T]$.

Proof. A result slightly stronger than (16.7) is true, as follows. For all $x \in [0, T]$,

$$|\Phi(v)(x) - \Phi(w)(x)| = \left| \int_0^x f(v(t), t) - f(w(t), t) \, dt \right|$$

$$\leq \int_0^x |f(v(t), t) - f(w(t), t)| \, dt \tag{16.9}$$

$$\leq L \int_0^x |v(t) - w(t)| \, dt.$$

This says that Φ is a Lipschitz mapping from L^1 to C^0. Bounding the last integrand in (16.9), we find

$$|\Phi(v)(x) - \Phi(w)(x)| \le Lx\|v - w\|_{\infty,[0,x]} \le Lx\|v - w\|_{\infty,[0,T]} \qquad (16.10)$$

for all $x \in [0, T]$. Taking the supremum of (16.10) over $x \in [0, T]$ completes the proof.   QED

The inequality (16.8) shows that Φ is a *contraction* for $T < 1/L$. This property can be used to establish the existence of solutions (exercise 16.1) as well. However, we use a slightly different approach to establish existence.

Suppose that we have an initial function u^0, e.g., $u^0(t) = u_0$ for all $t \ge 0$. Given u^n, define $u^{n+1} = \Phi(u^n)$, that is,

$$u^{n+1}(x) = u_0 + \int_0^x f(u^n(t), t)\,dt. \qquad (16.11)$$

For $n = 0$, (16.11) implies that

$$|u^1(x) - u^0(x)| = \left| \int_0^x f(u_0, t)\,dt \right| \le Mx \quad \forall x \in [0, T], \qquad (16.12)$$

where the constant M is defined by

$$M = \sup\left\{ |f(u_0, t)| \mid t \in [0, T] \right\}. \qquad (16.13)$$

Applying (16.9), we find

$$|u^{n+1}(x) - u^n(x)| \le L \int_0^x |u^n(t) - u^{n-1}(t)|\,dt \qquad (16.14)$$

for $n \ge 1$. Using (16.12) and (16.14), we conclude that

$$|u^2(x) - u^1(x)| \le \int_0^x LMt\,dt = \tfrac{1}{2}LMx^2 \quad \forall x \in [0, T]. \qquad (16.15)$$

Proceeding by induction, (16.14) implies that

$$|u^n(x) - u^{n-1}(x)| \le \frac{ML^{n-1}x^n}{n!} \quad \forall x \in [0, T] \qquad (16.16)$$

for all $n \ge 1$. Using a telescoping sum, we can write

$$
\begin{aligned}
|u^n(x) - u_0| &= \left| \sum_{i=1}^n u^i(x) - u^{i-1}(x) \right| \\
&\le \sum_{i=1}^n \frac{ML^{i-1}x^i}{i!} \le \frac{M}{L}\left(e^{Lx} - 1 \right).
\end{aligned}
\qquad (16.17)
$$

Thus the infinite series

$$u(x) = u_0 + \sum_{i=1}^\infty u^i(x) - u^{i-1}(x) \qquad (16.18)$$

converges absolutely for all $x \in [0, T]$ and satisfies

$$|u(x) - u_0| \le \frac{M}{L}\left(e^{Lx} - 1 \right). \qquad (16.19)$$

Each u^i is a C^1-function by definition (16.11) (and by induction). The uniform convergence of (16.18) implies that u is continuous (exercise 16.2). The differentiability of u is more complex to establish. We could establish bounds on the derivatives of the u^i's and show that they converge. Or we could simply show that u solves (16.3), which we need to do in any case.

We can use (16.16) to establish a rate of convergence:

$$|u(x) - u^n(x)| = \left| \sum_{i=n+1}^{\infty} u^i(x) - u^{i-1}(x) \right| \le \frac{M}{L} \sum_{i=n+1}^{\infty} \frac{(Lx)^i}{i!}$$

$$\le \frac{M}{L} \frac{(Lx)^{n+1}}{(n+1)!} \sum_{i=0}^{\infty} \frac{(Lx)^i}{i!} \le \frac{M}{L} \epsilon_{n+1} e^{Lx} \tag{16.20}$$

for any $x \in [0, T]$, where

$$\epsilon_n = \frac{(LT)^n}{n!}. \tag{16.21}$$

For any $\gamma > 0$, there is a $C_\gamma < \infty$ such that

$$\epsilon_n \le C_\gamma \gamma^n \tag{16.22}$$

(see exercise 16.3). Therefore, by the Lipschitz condition (16.4),

$$\left| u(x) - u_0 - \int_0^x f(u(t), t)\, dt \right| = \left| u(x) - u^{n+1}(x) \right.$$

$$\left. + \int_0^x f(u^n(t), t)\, dt - \int_0^x f(u(t), t)\, dt \right| \tag{16.23}$$

$$\le \left| u(x) - u^{n+1}(x) \right| + L \int_0^x \left| u^n(t) - u(t) \right| dt$$

$$\le C_\gamma (M/L) e^{Lx} \left(\gamma^{n+2} + LT \gamma^{n+1} \right)$$

for all $x \in [0, T]$. Choosing $\gamma < 1$ and letting $n \to \infty$ proves that u solves (16.3). Thus we have proved the following.

Theorem 16.2 *Suppose that $f \in C^0([0, T] \times \mathbb{R})$ satisfies the Lipschitz condition (16.4). Then there is a unique solution $u \in C^1([0, T])$ to (16.1) (and equivalently to (16.3)) that satisfies the bound*

$$|u(x) - u_0| \le \frac{M}{L} \left(e^{Lx} - 1 \right) \quad \forall x \in [0, T], \tag{16.24}$$

where M is defined by (16.13): $M = \sup \{ |f(u_0, t)| \mid t \in [0, T] \}$.

Theorem 16.2 is the best possible in a sense, in that non-Lipschitz functions f yield pathological situations. For example, the equation $u' = \sqrt{u}$ has two solutions with the initial data $u(0) = 0$, namely, $u \equiv 0$ and $u(x) = \frac{1}{4}x^2$. Moreover, we know from the example $u' = -u^2$ that solutions need not remain bounded for all T, as indicated in (16.5); here $f(u) = -u^2$ is only locally Lipschitz (cf. exercise 16.4).

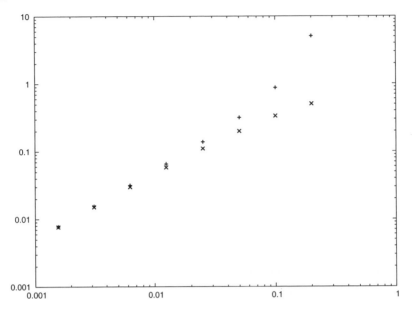

Figure 16.2 Relative errors in computing e^π by the explicit Euler method (16.26)
($\times$'s) and the implicit Euler method (16.27) (+'s). The horizontal
axis is Δt, and the vertical axis is the relative error at $T = 1$.

16.3 BASIC DISCRETIZATION METHODS

The definition of the derivative as a limit of difference quotients suggests a
method of discretization:

$$\frac{du}{dt}(t) \approx \frac{u(t + \Delta t) - u(t)}{\Delta t}, \tag{16.25}$$

where Δt is a small, positive parameter. This suggests algorithms for generating a sequence of values $u_n \approx u(n\Delta t)$ given by (for example)

$$u_n = u_{n-1} + \Delta t f(u_{n-1}, t_{n-1}) \tag{16.26}$$

or by

$$u_n = u_{n-1} + \Delta t f(u_n, t_n), \tag{16.27}$$

where $t_n = n\Delta t$.

The algorithm (16.26) is called the *explicit Euler* method, and the algorithm (16.27) is called the *implicit Euler* method. It can be shown (see
section 16.4) that both generate a sequence with the property that

$$|u(t_n) - u_n| \leq C_{f,T}\Delta t \quad \forall t_n \leq T \tag{16.28}$$

(at least provided that we solve the implicit equation (16.27) for u_n exactly).
This is illustrated in figure 16.2, which plots the relative errors as a function
of Δt for the simple case $f(u, x) = \pi u$ and $T = 1$ for both schemes. The
errors for the explicit Euler scheme (16.26) are represented by $\times$'s.

Although both methods (16.26) and (16.27) appear to be equally success-ful for our simple test problem, they are not so accurate. In chapter 17, we consider methods that produce much more accuracy with a compara-ble amount of work. However, we will see that it is not easy to find such methods.

The issue of solving the nonlinear equation in the implicit Euler method (16.27) at each step is important but not a show stopper; one uses the methods we have studied in chapter 2 (or chapter 7 in higher dimensions) for solving nonlinear equations. Moreover, we are in a situation where we have a very accurate approximation to the solution, e.g., one given by the explicit Euler method (16.26). More precisely, (16.27) is written in the form of a fixed-point iteration, so as long as $\Delta t |f_{,u}|$ remains small, fixed-point iteration will converge.

On the other hand, we might want to use a method that converges more rapidly than fixed-point iteration, or we might want to take a larger time step than the size of $|f_{,u}|$ would allow. Thus we consider an example to see what issues may arise.

16.3.1 Nonuniqueness of the time step

If we consider the implicit Euler approximation (16.27) for (16.1) with $f = -\kappa u^2$, we have to solve the quadratic equation

$$u_n + \tau u_n^2 = u_{n-1} \tag{16.29}$$

at each time step, where

$$\tau = \kappa \Delta t \tag{16.30}$$

and Δt is the size of the time step. Thus we find (cf. (2.77))

$$
\begin{aligned}
u_n^{\pm} &= \frac{-1 \pm \sqrt{1 + 4\tau u_{n-1}}}{2\tau} \\
&\approx \frac{-1 \pm \left(1 + 2\tau u_{n-1} - 2(\tau u_{n-1})^2 + \mathcal{O}\left((\tau u_{n-1})^3\right)\right)}{2\tau} \\
&= \begin{cases} u_{n-1}(1 - \tau u_{n-1}) + \mathcal{O}\left(\tau^2 u_{n-1}^3\right) & (+) \\ u_{n-1}\left(-1 - \tau u_{n-1} + (\tau u_{n-1})^2\right)/(\tau u_{n-1}) + \mathcal{O}\left(\tau^2 u_{n-1}^3\right) & (-) \end{cases} \\
&\approx \begin{cases} u_{n-1}(1 - \xi) & (+) \\ u_{n-1}(-\xi^{-1} - 1 + \xi) & (-), \end{cases}
\end{aligned}
\tag{16.31}
$$

where $\xi = \tau u_{n-1} = \kappa \Delta t u_{n-1}$.

If ξ is small, it is not hard to identify the appropriate solution

$$u_n^+ \approx u_{n-1}(1 - \xi) = u_{n-1} - \kappa \Delta t u_{n-1}^2. \tag{16.32}$$

This solution is much closer to u_{n-1} than the other solution. In particular,

$$u_n^+(\Delta t) \to u_{n-1} \tag{16.33}$$

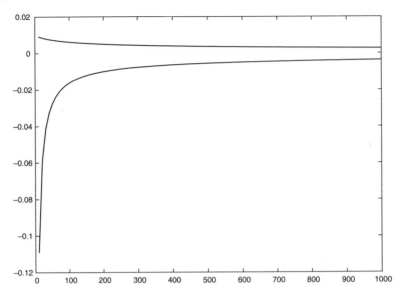

Figure 16.3 The two solutions to (16.29) as a function of τ, defined in (16.30), for $u_{n-1} = 0.01$. The horizontal axis is τ.

as $\Delta t \to 0$, whereas $u_n^-(\Delta t)$ diverges as $\Delta t \to 0$. However, there is always a second solution, as depicted in figure 16.3, and as u_n gets small, the solution (16.5) does not change very quickly. We may then become greedy and want to take τ larger. If by mistake we pick u_n negative, there is the danger that subsequent steps will remain negative, and the computation will blow up in finite time. In other problems, the unwanted behavior may be less spectacular [48] and thus more unlikely to be detected.

We might hope that a simple criterion would eliminate spurious solutions to (16.29). We can write (exercise 16.5)

$$\frac{u_{n-1} - u_n^\pm}{u_{n-1}} = 1 - \frac{-1 \pm \sqrt{1 + 4\xi}}{2\xi} \to 1 \quad \text{as } \xi \to \infty. \tag{16.34}$$

In figure 16.4, we depict the values of the expressions in (16.34) for various values of ξ. A natural constraint on the solution process might be to require

$$\left| \frac{u_n - u_{n-1}}{u_{n-1}} \right| \le K \tag{16.35}$$

for some constant K. As indicated in figure 16.4, this is satisfied for both solutions $u_n^\pm$ for reasonable values of K and ξ; cf. exercise 16.6.

16.3.2 Near uniqueness of the time step

The time-stepping equation for the implicit Euler method applied to the equation (16.1) takes the general form

$$g(u, \tau) = u - \tau f(u, \tau) - v = 0. \tag{16.36}$$

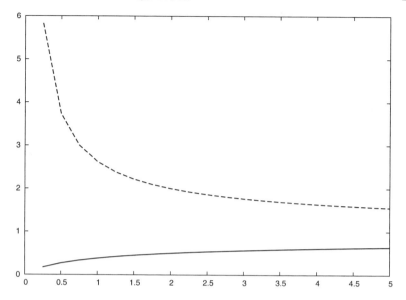

Figure 16.4 The expressions (16.34) as a function of ξ.

We can think of this as defining a curve in $\mathbb{R}^2$ given by solving for $u_n = u(\tau)$ as a function of τ, where $v = u_{n-1}$. In section 16.3.1, the solution u_n^+ corresponds to $u(\tau)$, in view of (16.33).

The derivative of u with respect to τ gives a sense of where u lies with respect to v. Differentiating (16.36) gives

$$0 = u'(\tau) - f(u(\tau), \tau) - \tau\big(u'(\tau)f_{,u}(u(\tau), \tau) + f_{,t}(u(\tau), \tau)\big). \qquad (16.37)$$

Rewriting, this becomes

$$(1 - \tau f_{,u}(u(\tau), \tau))\, u'(\tau) = f(u(\tau), \tau) - \tau f_{,t}(u(\tau), \tau). \qquad (16.38)$$

In particular,

$$u'(0) = f(v, 0). \qquad (16.39)$$

On the other hand, Newton's method for solving (16.36), starting with v as an initial guess, generates as the first step (here τ is fixed)

$$g_{,u}(v, \tau)(u - v) = -g(v, \tau). \qquad (16.40)$$

Rewriting, (16.40) becomes

$$(1 - \tau f_{,u}(v, \tau))\, (u - v) = \tau f(v, \tau). \qquad (16.41)$$

For simplicity, suppose that f is independent of t. Then (16.38) simplifies to

$$(1 - \tau f_{,u}(u(\tau)))\, u'(\tau) = f(u(\tau)), \qquad (16.42)$$

and (16.41) becomes

$$(1 - \tau f_{,u}(v))\, (u - v) = \tau f(v). \qquad (16.43)$$

Define $w = \tau^{-1}(u - v)$. Then

$$(1 - \tau f_{,u}(v))\, w = f(v). \tag{16.44}$$

Thus we conclude that Newton's method moves in the direction tangent to the solution curve (16.36). If the steps are taken small enough, then it is reasonable to hope that we will stay close to this curve and not jump to another branch as described in section 16.3.1.

16.4 CONVERGENCE OF DISCRETIZATION METHODS

We now prove the convergence result (16.28) for the explicit Euler discretization (16.26). We make a slight generalization to allow variable time steps in the spirit of the adaptive approximation in section 12.5:

$$u_n = u_{n-1} + \Delta t_n f(u_{n-1}, t_{n-1}), \tag{16.45}$$

where now the nth time point is

$$t_n = \sum_{i=1}^{n} \Delta t_i. \tag{16.46}$$

Let us write $g(t) = f(u(t), t)$ and set

$$M_n = \|g'\|_{\infty, [t_{n-1}, t_n]}. \tag{16.47}$$

We will see that the local error is bounded by

$$\epsilon_n = \Delta t_n^2 M_n \tag{16.48}$$

for all $n \geq 1$. We now show how this local error is related to global error.

16.4.1 Global error estimates

The main objective of the section will be to prove the following result.

Theorem 16.3 *Suppose that the Lipschitz estimate (16.4) holds. Then*

$$|u(t_n) - u_n| \leq \sum_{j=1}^{n} \epsilon_j e^{L(t_n - t_j)} \tag{16.49}$$

for all $n \geq 1$, where ϵ_n is defined in (16.48) and M_n is defined in (16.47).

Proof. We can write

$$u(t_n) = u(t_{n-1}) + \int_{t_{n-1}}^{t_n} f(u(t), t)\, dt$$
$$= u(t_{n-1}) + \Delta t_n f(u(t_{n-1}), t_{n-1}) + q_n, \tag{16.50}$$

where q_n is the quadrature error

$$q_n = \int_{t_{n-1}}^{t_n} f(u(t), t)\, dt - \Delta t f(u(t_{n-1}), t_{n-1}). \tag{16.51}$$

The techniques of chapter 13 can be applied to prove that

$$|q_n| \leq c\Delta t_n^2 M_n \tag{16.52}$$

for some constant $c < 1$ (exercise 16.7). Define $e_n = u(t_n) - u_n$. Subtracting (16.45) from (16.50), we have

$$e_n = e_{n-1} + \Delta t_n \left(f(u(t_{n-1}), t_{n-1}) - f(u_{n-1}, t_{n-1}) \right) + q_n. \tag{16.53}$$

Using the Lipschitz estimate (16.4), we find

$$|e_n| \leq |e_{n-1}|(1 + \Delta t_n L) + \epsilon_n. \tag{16.54}$$

Note that $u_0 = u(0)$, so

$$|e_1| \leq |q_1| \leq \Delta t_1^2 M = \epsilon_1. \tag{16.55}$$

We apply the elementary inequality

$$1 + \mu \leq e^\mu \tag{16.56}$$

(see exercise 16.10) to (16.54) to get

$$|e_n| \leq |e_{n-1}| e^{\Delta t_n L} + \epsilon_n \tag{16.57}$$

for all $n \geq 1$. By induction, we will show that this implies

$$|e_n| \leq \sum_{j=1}^{n} \epsilon_j e^{L(t_n - t_j)} \tag{16.58}$$

for all $n \geq 1$, which is the same as (16.49). Note that (16.55) implies (16.58) for $n = 1$. So assume (16.58) holds for some $n \geq 1$. Applying (16.57) to (16.58), we find

$$\begin{aligned}
|e_{n+1}| &\leq |e_n| e^{\Delta t_{n+1} L} + \epsilon_{n+1} \\
&\leq e^{\Delta t_{n+1} L} \sum_{j=1}^{n} \epsilon_j e^{L(t_n - t_j)} + \epsilon_{n+1} \\
&= \sum_{j=1}^{n} \epsilon_j e^{L(t_n + \Delta t_{n+1} - t_j)} + \epsilon_{n+1} \\
&= \sum_{j=1}^{n} \epsilon_j e^{L(t_{n+1} - t_j)} + \epsilon_{n+1} \\
&= \sum_{j=1}^{n+1} \epsilon_j e^{L(t_{n+1} - t_j)},
\end{aligned} \tag{16.59}$$

which verifies the induction step, completing the proof of (16.58) and thus the theorem.

QED

16.4.2 Interpretation of error estimates

We interpret the right-hand side of (16.49) as follows. It says that the error at time t_n is influenced by all the discretization errors $\epsilon_j = \Delta t_j^2 M_j$. But the recent errors are less important than the earlier errors. This is because the earlier errors can be amplified over time. We can make this more precise as follows. Suppose we assume that

$$\Delta t_n M_n \leq \phi(t_n) \tag{16.60}$$

for all n for some function ϕ. Thus the error terms ϵ_n defined in (16.48) satisfy $\epsilon_n \leq \phi(t_n)\Delta t_n$. Then we may view the right-hand side of (16.49) as bounded by a quadrature rule applied to an integral:

$$\sum_{j=1}^{n} \epsilon_j e^{L(t_n - t_j)} \leq \sum_{j=1}^{n} \phi(t_n)\Delta t_j e^{L(t_n - t_j)} \leq \int_0^{t_n} \phi(t) e^{L(t_n - t)} \, dt. \tag{16.61}$$

In fact, if ϕ is nonincreasing, we can prove (exercise 16.11) that

$$\sum_{j=1}^{n} \phi(t_j)\Delta t_j e^{L(t_n - t_j)} \leq \sum_{j=1}^{n} \int_{t_{j-1}}^{t_j} \phi(t) e^{L(t_n - t)} \, dt$$
$$= e^{Lt_n} \int_0^{t_n} \phi(t) e^{-Lt} \, dt \tag{16.62}$$

for all $n \geq 1$. Thus we could define

$$\phi(t) = \sup \left\{ \Delta t_n M_n \mid t_n \geq t \right\}. \tag{16.63}$$

However, the case of interest is when ϕ is increasing. If ϕ is a smooth function, then the approximation

$$|u(t_n) - u_n| \leq \sum_{j=1}^{n} \phi(t_j)\Delta t_j e^{L(t_n - t_j)} \approx e^{Lt_n} \int_0^{t_n} \phi(t) e^{-Lt} \, dt \tag{16.64}$$

remains a good guide. We see that ϕ could be exponentially increasing without having any serious impact. For example, suppose we take $\phi(t) = \delta e^{Lt}$ for some $\delta > 0$. As long as

$$\Delta t_n M_n \leq \phi(t_n) = \delta e^{Lt_n},$$

(16.64) implies

$$|u(t_n) - u_n| \leq \delta t_n e^{Lt_n}. \tag{16.65}$$

16.4.3 Discretization error example

Let us consider an example. Suppose $f(u, t) = u$, so the equation is $u' = u$ and $u(t) = e^t$ (we take $u(0) = 1$). Then the discrete solution is given by

$$u_n = u_{n-1} + \Delta t u_{n-1} = (1 + \Delta t)u_{n-1} = (1 + \Delta t)^n \tag{16.66}$$

for all $n \geq 0$. Similarly, $M_n \approx e^t$, so we take $\phi(t) = \Delta t e^t$ (note that $L = 1$ here). Thus (16.65) predicts that

$$u_n - u(t_n) \approx t_n e^{t_n} \Delta t.$$

This is easily verified (exercise 16.14). Note that this says that the relative error in the approximation

$$e_n = \frac{|u(t_n) - u_n|}{|u(t_n)|} = \frac{|u(t_n) - u_n|}{e^{t_n}} \tag{16.67}$$

is bounded by $t_n \Delta t$ and thus grows only linearly in time.

16.5 MORE READING

There are many books on the theory of ordinary differential equations, but two more recent ones which cover modern ideas of dynamical systems are [10, 104].

16.6 EXERCISES

Exercise 16.1 *The* contraction mapping principle *says that any Lipschitz function Φ with Lipschitz constant less than 1 must have a fixed point. Verify this by using (16.8) to construct a fixed point for Φ defined by (16.6). (Hint: show that for any x, the sequence $u^n(x)$ defined by fixed-point iteration (16.11) forms a Cauchy sequence.) Prove that the limit function $u(x)$ forms a $C^0([0, T])$ function.*

Exercise 16.2 *Suppose that the infinite sum*

$$v(x) = \sum_{i=1}^{\infty} v_i(x)$$

converges uniformly for $x \in [0, T]$ and that each $v_i \in C^0([0, T])$. Prove that $v \in C^0([0, T])$. (See the footnote on Seidel on page 118.)

Exercise 16.3 *Show that ϵ_n as defined in (16.21) satisfies $\epsilon_n \leq C_\gamma \gamma^n$ for any $\gamma > 0$.*

Exercise 16.4 *Show that there is a unique solution to (16.1) for some $T > 0$ for a locally Lipschitz function f. Use this to prove existence and uniqueness for $f(u) = u^2 + \sin u$. (Hint: f is locally Lipschitz if it is Lipschitz on any bounded set of u's.)*

Exercise 16.5 *Prove the equality in (16.34).*

Exercise 16.6 *Suppose that $K > 2$ and $(K-2)^{-1} \leq \xi = \tau u_{n-1}$, cf. (16.34). Prove that (16.35) holds for both solutions in (16.31).*

Exercise 16.7 *Consider the quadrature rule*

$$\int_a^b f(x)\, dx \approx Qf = (b - a)f(a). \tag{16.68}$$

Prove that

$$\left| \int_a^b f(x)\, dx - (b-a)f(a) \right| \le c(b-a)^2 \|f'\|_{\infty,[a,b]} \tag{16.69}$$

for some constant $c < 1$. (Hint: apply the Peano kernel theorem 13.5.)

Exercise 16.8 *Prove that*

$$\sum_{k=0}^{n-1} (1+\mu)^k = ((1+\mu)^n - 1)/\mu$$

for any $\mu > 0$ and any n. (Hint: this looks more familiar if you write $r = 1 + \mu$.)

Exercise 16.9 *Prove that the estimate (16.28) holds for the implicit Euler method. (Hint: repeat the argument in section 16.4 but with the quadrature rule*

$$\int_a^b f(x)\, dx \approx Qf = (b-a)f(b) \tag{16.70}$$

instead of (16.68).)

Exercise 16.10 *Prove that $1 + x \le e^x$ for $x \ge 0$. (Hint: see exercise 9.17.)*

Exercise 16.11 *Suppose that ϕ is an integrable, nonincreasing function, e.g., the step function as defined in (16.63). Prove that*

$$\Delta t_j \phi(t_j) e^{L(t_n - t_j)} \le \int_{t_{j-1}}^{t_j} \phi(t) e^{L(t_n - t)}\, dt \tag{16.71}$$

for all $j = 1, \ldots, n$.

Exercise 16.12 *Consider the function*

$$\phi(x) = \frac{1}{x^2}\left(\frac{e^x}{1+x} - 1 \right). \tag{16.72}$$

Prove that $\phi(0) = \frac{1}{2}$ and that ϕ is decreasing for $x \in [0,1]$. (Hint: consider the Taylor expansion of e^x around zero.)

Exercise 16.13 *Plot the function ϕ defined in (16.72) for $x \in [0,4]$. For what interval is it true that $\phi \le \frac{1}{2}$?*

Exercise 16.14 *Suppose that $1 > \Delta t > 0$ and that $n \le C\Delta t^{-1}$ for some constant C. Prove that*

$$e^{t_n} - (1 + \Delta t)^n \approx t_n e^{t_n} \Delta t, \tag{16.73}$$

where $t_n = n\Delta t$. (Hint: note that (16.73) is equivalent to

$$\left(\frac{1 + \Delta t}{e^{\Delta t}} \right)^n \approx 1 - cn\Delta t^2 \tag{16.74}$$

for some constant c and that $(1 + \Delta t)/e^{\Delta t} \approx 1 - \frac{1}{2}\Delta t^2$. Use (16.72).)

Exercise 16.15 *Experiment with graded meshes of the form*

$$\Delta t_n = \delta e^{ct_{n-1}},$$

for a fixed parameter $\delta > 0$, in solving $u' = u$ on $[0,1]$ with initial data $u(0) = 1$. Work with the constant c as a small parameter and consider both positive and negative values. What is the best choice to make the relative error (16.67) smallest? Suppose that we define the cost-benefit factor to be ne_n; what strategy (choice of c) minimizes this factor?

Exercise 16.16 *Experiment with graded meshes of the form $\Delta t_n = \delta e^{ct_{n-1}}$, for a fixed parameter $\delta > 0$, in solving $u' = u$ on $[0,1]$ with initial data $u(0) = 1$. Work with the constant c as a small parameter and consider both positive and negative values. What is the best choice to make the absolute error $e_n = |u(t_n) - u_n|$ smallest? Suppose that we define the cost-benefit factor to be ne_n; what strategy (choice of c) minimizes this factor?*

Exercise 16.17 *Consider using Newton's method for solving (16.29), starting the iteration with $v_0 = u_{n-1}$. More precisely, define $f(v) = v + \tau v^2 - u_{n-1}$ and apply Newton's method to solve $f(v) = 0$ (this determines $u_n = v$). Show that Newton's method converges for all $\tau > 0$ in this case and that the solution is $v = u_n^+$, assuming that $u_{n-1} > 0$. (Hint: apply exercise 2.20.)*

16.7 SOLUTIONS

Solution of Exercise 16.3. Let K be the smallest integer such that $K \geq LT/\gamma$. By definition (16.21), we have

$$\gamma^{-n}\epsilon_n = \frac{(LT/\gamma)^n}{n!} \leq \frac{K^n}{n!}. \qquad (16.75)$$

We will show that the function $\phi_K(n) - K^n/n!$ is increasing for $n < K$ and decreasing for $n > K$ and that $\phi_K(K-1) = \phi_K(K)$. Therefore,

$$\gamma^{-n}\epsilon_n \leq \frac{K^K}{K!} = C_\gamma, \qquad (16.76)$$

where the last equality is our definition of C_γ.

To prove the asserted monotonicity properties of $\phi_K(n)$, observe that

$$\frac{\phi_K(n)}{\phi_K(n-1)} = \frac{K}{n} \begin{cases} > 1 & \text{if } n < K \\ = 1 & \text{if } n = K \\ < 1 & \text{if } n > K. \end{cases} \qquad (16.77)$$

Solution of Exercise 16.12. We write

$$\phi(x) = \frac{1}{x^2}\left(\frac{e^x}{1+x} - 1\right) = \frac{e^x - 1 - x}{x^2(1+x)}. \qquad (16.78)$$

Write $\boxed{\psi(x) = e^x - 1 - x}$ so that

$$\phi(x) = \frac{\psi(x)}{x^2(1+x)}. \tag{16.79}$$

Differentiating, we find

$$\phi'(x) = \frac{\psi'(x)x^2(1+x) - \psi(x)(2x+3x^2)}{x^4(1+x)^2}. \tag{16.80}$$

Since $\psi'(x) = \psi(x) + x$, we can simplify to get

$$
\begin{aligned}
\phi'(x) &= \frac{(\psi(x)+x)x^2(1+x) - \psi(x)(2x+3x^2)}{x^4(1+x)^2} \\
&= \frac{\psi(x)(x^2(1+x) - (2x+3x^2)) + x^3(1+x)}{x^4(1+x)^2} \\
&= \frac{\psi(x)(-2 - 2x + x^2) + x^2(1+x)}{x^3(1+x)^2} \\
&= \frac{(\psi(x) - \frac{1}{2}x^2)(-2 - 2x + x^2) + \frac{1}{2}x^4}{x^3(1+x)^2} \\
&= \frac{(\psi(x) - \frac{1}{2}x^2 - \frac{1}{6}x^3)(-2 - 2x + x^2) - \frac{1}{6}x^3(2 - x - x^2)}{x^3(1+x)^2}.
\end{aligned}
\tag{16.81}
$$

Using the Taylor expansion of the exponential at zero, we find

$$\psi(x) - \tfrac{1}{2}x^2 - \tfrac{1}{6}x^3 = x^4 \sum_{k=0}^{\infty} \frac{x^k}{(k+4)!}. \tag{16.82}$$

If we define $\eta(x) = \sum_{k=0}^{\infty} \frac{x^k}{(k+4)!}$, then (16.81) simplifies to

$$\phi'(x) = \frac{x\eta(x)(-2 - 2x + x^2) - \frac{1}{6}(2 - x - x^2)}{(1+x)^2}, \tag{16.83}$$

so we see that $\phi'(0) = -\frac{1}{3}$. Moreover, since $\eta(x) > 0$ for $x \geq 0$,

$$\phi'(x) \leq \frac{-\frac{1}{6}(2 - x - x^2)}{(1+x)^2} < 0 \tag{16.84}$$

for $0 \leq x < 1$. We can extend this to $x = 1$ since

$$\phi'(1) = -3\eta(1)/4 < 0.$$

Thus ϕ is strictly decreasing on $[0, 1]$.

Solution of Exercise 16.14. We know that

$$e^{\Delta t} = 1 + \Delta t + \tfrac{1}{2}\Delta t^2 + \mathcal{O}\left(\Delta t^3\right).$$

Using (16.72), let us write

$$\frac{e^{\Delta t}}{1 + \Delta t} = 1 + \phi(\Delta t)\Delta t^2, \tag{16.85}$$

where $\phi(0) = \frac{1}{2}$ and ϕ is decreasing as a function of Δt. We need now to estimate the nth power of the left-hand side of (16.85), and so we use the estimate

$$|(1+\epsilon)^n - 1 - n\epsilon| \leq \tfrac{1}{2}\epsilon^2 n^2 e^{(n-2)\epsilon}, \tag{16.86}$$

which we prove subsequently. We now take $\epsilon = \phi(\Delta t)\Delta t^2$. Thus

$$\left|\frac{e^{n\Delta t}}{(1+\Delta t)^n} - 1 - n\phi(\Delta t)\Delta t^2\right| \leq \frac{C^2}{8}e^{C\Delta t/2}\Delta t^2 \leq K\Delta t^2, \tag{16.87}$$

since $n\Delta t \leq C$ and $\phi(\Delta t) \leq \frac{1}{2}$, where we can take

$$K = \frac{C^2}{8}e^{C/2}.$$

Multiplying (16.89) by $(1+\Delta t)^n$, we find

$$|e^{t_n} - (1+\Delta t)^n - \phi(\Delta t)t_n(1+\Delta t)^n\Delta t| \leq K\Delta t^2(1+\Delta t)^n$$
$$\leq K\Delta t^2 e^{n\Delta t} \leq K\Delta t^2 e^C, \tag{16.88}$$

where $t_n = n\Delta t$, since $e^{\Delta t} > 1 + \Delta t$ for $\Delta t > 0$, cf. (16.56). In particular,

$$0 < e^{t_n} - (1+\Delta t)^n \leq (1+\Delta t)^n\left(\tfrac{1}{2}t_n\Delta t + K\Delta t^2\right)$$
$$\leq e^{t_n}\left(\tfrac{1}{2}t_n\Delta t + K\Delta t^2\right) \leq e^C\left(\tfrac{1}{2}t_n\Delta t + K\Delta t^2\right). \tag{16.89}$$

Applying (16.89) in (16.88), we get

$$|e^{t_n} - (1+\Delta t)^n - \phi(\Delta t)t_n e^{t_n}\Delta t| \leq \widetilde{K}\Delta t^2, \tag{16.90}$$

where $\widetilde{K}$ is a constant. Since $\phi(\Delta t) \approx \frac{1}{2} + \mathcal{O}(\Delta t)$, this completes the proof.

The proof of (16.86) is as follows. Using the binomial expansion,

$$(1+\epsilon)^n - 1 - n\epsilon = \sum_{j=2}^{n}\binom{n}{j}\epsilon^j = \epsilon^2\sum_{j=2}^{n}\binom{n-2}{j-2}\frac{n(n-1)}{j(j-1)}\epsilon^{j-2}$$
$$= \epsilon^2\sum_{k=0}^{n-2}\binom{n-2}{k}\frac{n(n-1)}{(k+2)(k+1)}\epsilon^k \tag{16.91}$$
$$\leq \tfrac{1}{2}n^2\epsilon^2\sum_{k=0}^{n-2}\binom{n-2}{k}\epsilon^k = \tfrac{1}{2}n^2\epsilon^2(1+\epsilon)^{n-2}.$$

The proof of (16.86) is completed by using (16.56) (see exercise 16.10).

Chapter Seventeen

Higher-order ODE Discretization Methods

The internal title page of the book [12] reads (essentially) as follows:

An Attempt
to Test

The Theories of Capillary Action
by comparing
the theoretical and measured forms
of drops of fluid
by Francis Bashforth, B.D.
with an explanation of the method of integration
employed in constructing the tables which give the theoretical
forms of such drops
by J. C. Adams, M.A., F.R.S.

In section 16.3, we considered two simple, low-accuracy methods and proved in section 16.4 (and exercise 16.9) that they converge to the solutions of the corresponding ordinary differential equations (ODEs). There are many reasons for wanting higher accuracy methods. The most obvious ones are that we want a more accurate solution or a method that requires less work or both. In some cases, it is essential to have a higher-order method to obtain an acceptable answer. We discuss one aspect of this in section 18.1.4.

Major advances in the understanding of numerical techniques for solving ordinary differential equations occurred at the end of the 19th century and in the beginning years of the 20th century. The Adams-Bashforth[1] [12] and Runge-Heun-Kutta[2] [30] methods were established at a time when desktop

---

[1] John Couch Adams (1819–1892) was best known for his mathematical study of planetary motion. In 1843, Adams won the Cambridge mathematical Tripos and was thus named Senior Wrangler; Second Wrangler that year was Francis Bashforth. Other top Wranglers include Stokes (page 118) in 1841 and Rayleigh (page 232) in 1865.

[2] Martin Wilhelm Kutta (1867–1944) is known both for his difference method and for his work in aerodynamics. Karl L. W. M. Heun (1859–1929) is known as well for his equation related to the hypergeometric equation [114]. For information about Runge, see page 159.

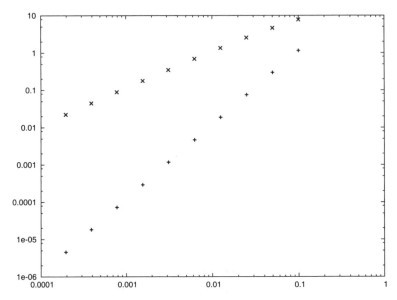

Figure 17.1 Absolute error in computing $e^{\pi} \approx 23.141$ by the explicit Euler method
(16.26) ($\times$'s) and by the centered scheme (17.1) (+'s). The horizontal
axis is Δt, and the vertical axis is the absolute error at $T = 1$.

computing devices[3] were becoming common, allowing much more extensive
computations than had been possible previously.

17.1 HIGHER-ORDER DISCRETIZATION

The main way to achieve more accuracy is to choose a more accurate approximation to the derivative (or quadrature rule, if we think of this as solving
an integral equation). The simplest higher-order method is the centered
difference

$$u_{n+1} = u_{n-1} + 2\Delta t f(u_n, t_n). \tag{17.1}$$

It requires extra work to get started because we need u_{-1}, but let us ignore
this difficulty for the moment. The application of (17.1) to the case where
$f(x, u) = \pi u$ is illustrated in figure 17.1 by the +'s, which indicate the error
as a function of Δt, for $T = 1$. The situation is much improved over the first-
order scheme (16.26) illustrated by the $\times$'s. For example, it takes $\Delta t \approx 10^{-6}$
to get an error of 10^{-4} for explicit Euler (16.26), whereas with the centered
scheme (17.1) we get an error of 10^{-6} with $\Delta t \approx 10^{-4}$.

---

[3]In 1886, the American Arithmometer Company was founded to manufacture and sell
an "adding" machine invented by William Seward Burroughs. The company was later
named the Burroughs Adding Machine Company in 1905, and it shipped its one-millionth
adding machine in 1928. Burroughs and Sperry Corporation merged in 1986 to form

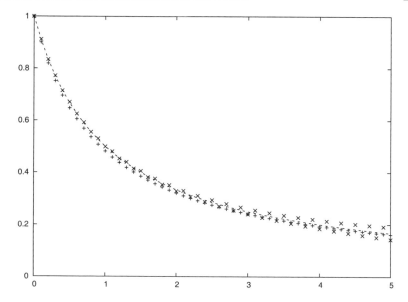

Figure 17.2 Comparison of values u_n computed via the centered scheme (17.1) ($\times$'s) with those computed via the explicit Euler scheme (16.26) (+'s) for solving $u' = -u^2$. The dotted line is the exact solution (16.5).

Let us apply a cost-benefit analysis (cf. exercise 16.15) where the cost is the number of time steps, the benefit is the size of the error, and the product of the two is the cost-benefit factor of interest. Then the centered scheme is 10^4 times more effective than the explicit Euler scheme. Of course, this number is not universal; it depends on the interval of interest and the absolute level of accuracy required, and it will be different for different problems (different f). But this shows that the benefit of using a high-order scheme can be astronomical.

Although the centered scheme is much more accurate for a fixed T, it has some disturbing qualities as T increases for fixed Δt. In figure 17.2, the centered scheme (17.1) is compared to the explicit Euler scheme (16.26) for the problem

$$u' = -u^2.$$

Initially, the centered scheme is more accurate, as indicated in figure 17.3, but at later times, the error begins to increase (and "bounce"). The data in figure 17.3 are the relative absolute errors e_n defined by

$$e_n = \frac{|u_n - u(t_n)|}{|u(t_n)|}, \tag{17.2}$$

which are worse for $t_n \geq 3$ for the scheme (17.1). Note that for $t_n \geq 3$, the

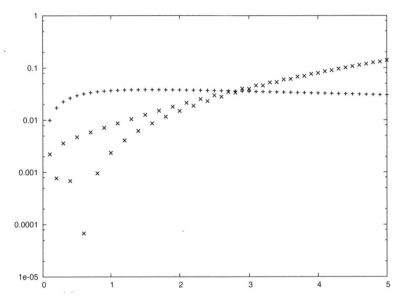

Figure 17.3 Comparison of relative absolute errors (17.2) for the centered scheme
(17.1) (×'s) and the explicit Euler scheme (16.26) (+'s) for solving
$u' = -u^2$.

absolute value masks the bounce in the error (alternation between positive and negative signs) seen in figure 17.2.

As we integrate longer (see figure 17.4), the centered scheme becomes much less accurate, and eventually the approximate solution becomes negative, at which point the character of the equation changes dramatically The discretization scheme locks onto a solution that blows up in finite time.

The success of a discretization method for a nonlinear problem depends on Δt being small enough. For our model problem $u' = -u^2$ and the explicit Euler scheme, we see that $u_1 = u_0 - \Delta t(u^0)^2 < 0$ if $\Delta t > 1/u_0$. Thus in one step we cross from a region where the solution tends to zero into one where it blows up in finite time (see section 16.3.1).

17.1.1 An unstable scheme

We have seen that the centered scheme provides a more accurate method as Δt is decreased for fixed T, but at a cost. Our first reaction may be to seek a more accurate scheme, but we will see that things can actually get worse.

The book [86, chapter 8, section 1.4] derives what might be called a forward differentiation formula

$$\frac{du}{dt}(t_{n-1}) \approx P'(t_{n-1}) = \frac{1}{\Delta t}\sum_{i=0}^{k} a_i^k u_{n-i}, \qquad (17.3)$$

where P is a polynomial interpolating the values u_j at $t_j = j\Delta t$ and the coefficients a_i^k are just the derivatives of the Lagrange basis functions for

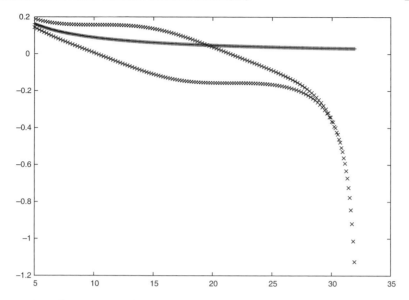

Figure 17.4 Comparison of values u_n computed via the centered scheme (17.1) ($\times$'s) with those computed via the explicit Euler scheme (16.26) ($+$'s) for solving $u' = -u^2$, integrated until $T = 32$.

this interpolation process, scaled by Δt. (The a_i^k thus correspond to the derivatives at -1 of the Lagrange basis functions for the points

$$-k, 1-k, \ldots, -1, 0;$$

cf. exercise 17.2). This leads to the scheme

$$a_0^k u_n = -\sum_{i=1}^{k} a_i^k u_{n-i} + \Delta t f(u_{n-1}, t_{n-1}). \tag{17.4}$$

We could choose a different interpolation scheme involving $k+1$ points, but if we choose P to be of degree k, we provide maximum accuracy Δt^k. For low values of k, we do not get anything new. For $k = 1$, this is the explicit Euler scheme, and for $k = 2$, this is the centered difference scheme (17.1) (exercise 17.3). However, for $k = 3$, the scheme is new and takes the form

$$\tfrac{1}{3} u_n = -\tfrac{1}{2} u_{n-1} + u_{n-2} - \tfrac{1}{6} u_{n-3} + \Delta t f(u_{n-1}, t_{n-1}). \tag{17.5}$$

Unfortunately, we will see that this scheme is unconditionally unstable. In figure 17.5, we see what happens when (17.5) is applied to the solve the equation $u' = u$ on $[0, 1]$. Shown are the results of two time steps, $\Delta t = 0.1$ and $\Delta t = 0.05$. Unfortunately, the results are worse for the *smaller* value of Δt, at least near $t = 1$. One might suspect that round-off error is the cause of the problem, but the same behavior is found if rational arithmetic is used (exercise 17.5).

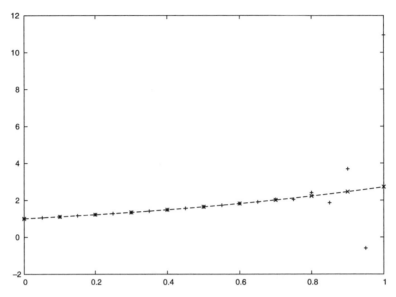

Figure 17.5 Result of using (17.5) to solve $u' = u$ on $[0, 1]$ with two different values for the time step $\Delta t = 0.1$ ($\times$'s) and $\Delta t = 0.05$ ($+$'s). The horizontal axis is t, and the vertical axis is the value of u.

17.1.2 Improved Euler

So far we have seen some obstacles in deriving high-order schemes. We now consider a scheme that is second-order accurate and does not exhibit the oscillation problem of the centered scheme:

$$
\begin{aligned}
\tilde{u}_n &= u_{n-1} + \Delta t f(u_{n-1}, t_{n-1}) \\
u_n &= u_{n-1} + \tfrac{1}{2}\Delta t \left(f(u_{n-1}, t_{n-1}) + f(\tilde{u}_n, t_n) \right).
\end{aligned}
\tag{17.6}
$$

The algorithm (17.6) is often called the *improved Euler* scheme. We see that the explicit Euler scheme is embedded in it, in the sense that the formula for $\tilde{u}_n$ comes directly from that method. But this value is then used in what appears to be the trapezoidal rule applied to (16.3), except that $\tilde{u}_n$ is used in the evaluation to make the computation explicit.

In figure 17.6, we compare the improved Euler scheme to the explicit Euler scheme. We see that the latter is much more accurate, and it does not suffer the error degradation in time that the centered scheme does.

It is not immediately obvious why (17.6) should be second-order accurate since it is based on using a first-order scheme to define the important ingredient $\tilde{u}_n$. Thus a careful examination is required to establish the order of accuracy rigorously.

The improved Euler scheme (17.6) may be viewed as an elementary example of various classes of schemes. We can view it as a simple type of *predictor-corrector* scheme; the first equation in (17.6) computes the (lower-order, in general) predictor $\tilde{u}_n$, and the second, corrector equation in (17.6)

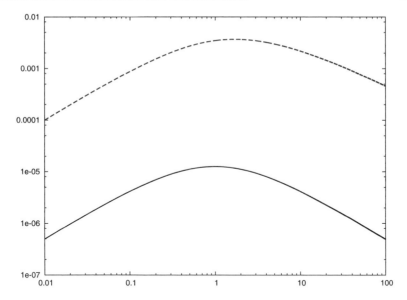

Figure 17.6 Relative absolute errors (17.2) for the explicit Euler scheme (16.26)
(upper curve) and the improved Euler scheme (17.6) (lower curve) for
solving $u' = -u^2$, using $\Delta t = 0.01$. The horizontal axis is time.

uses the predictor to compute the final value u_n. On the other hand, the
improved Euler scheme (17.6) was one of the original schemes analyzed by
Runge [30] and forms a basic example of the Runge-Kutta schemes.

17.2 CONVERGENCE CONDITIONS

We have seen several different schemes for solving ODEs with different behaviors. We now try to understand how to predict these behaviors based on
abstract properties of the scheme. We will see that two conditions that arise
naturally are required for a convergent difference approximation: stability
and consistency. We derive these by considering a very simple situation.

17.2.1 Constant solutions

If we have an ODE of the form (16.1) with $f(u, t) = 0$, then we find u is
constant. This could happen at any time; $u(t)$ could enter a regime where
$f(u, t) = 0$, and it would stay there forever. Although we might not expect
the numerical approximation to be constant in such a situation, we might
be interested in knowing whether or not the approximate solution is growing
or not.

Having $f = 0$ simplifies analysis of the numerical schemes. In this case

Scheme	a_0	a_1	a_2	a_3
Euler	1	-1		
Centered	1	0	-1	
Unstable	$\frac{1}{3}$	$\frac{1}{2}$	-1	$\frac{1}{6}$

Table 17.1 Coefficients a_i for various discretization schemes. The coefficients are the same for all three (explicit, implicit, improved) Euler schemes.

they all take the form

$$\sum_{i=0}^{k} a_i u_{n-i} = 0, \tag{17.7}$$

where $a_0 \neq 0$ and $a_k \neq 0$. For the explicit Euler scheme (16.26), $k = 1$; for the schemes (17.1) and (17.5), $k = 2$ and 3, respectively. The improved Euler scheme (17.6) is a bit more complex in appearance, but when $f = 0$, it reduces to the explicit Euler scheme (16.26). When $f = 0$, the implicit Euler scheme (16.27) also matches the explicit Euler scheme (16.26). We collect these values of a_i for the various schemes in table 17.1.

For the explicit Euler scheme, we find that $u_n = u_{n-1}$ when $f = 0$, so we have a constant approximate solution which matches perfectly the behavior of the exact solution. But for the higher-order schemes, we need a more subtle analysis to see what is going on. There are several ways to do this, but the most direct is to observe that there are solutions to (17.7) of the form $u_n = \xi^n$ if and only if

$$0 = \sum_{i=0}^{k} a_i \xi^{n-i} = \xi^{n-k} \sum_{i=0}^{k} a_i \xi^{k-i} = \xi^{n-k} p(\xi), \tag{17.8}$$

where the *characteristic polynomial* p for the difference stencil (17.8) is defined by

$$p(\xi) = \sum_{i=0}^{k} a_i \xi^{k-i}. \tag{17.9}$$

Since we assume that $a_k \neq 0$, then $\xi = 0$ is not a root. Thus we have proved the following result.

Lemma 17.1 *There is a solution to the equation (17.7) of the form $u_n = \xi^n$ if and only if $p(\xi) = 0$, where p is the characteristic polynomial p as defined by (17.9).*

If all roots satisfy $|\xi| < 1$, then all solutions $u_n = \xi^n$ will decay exponentially to zero. But this is too strong a design criterion; it turns out that at least one root must not lead to decaying solutions.

17.2.2 Consistency

To be convergent, it is natural to assume that the difference approximation applied to a constant approximation gets the right answer, namely, zero. That is, since $u' = 0$ for constant u, we may expect this as well for the discrete approximation. Thus we expect that all schemes must have $p(1) = 0$, i.e.,

$$\sum_{i=0}^{k} a_i = 0 \qquad (17.10)$$

(cf. exercise 17.7). Thus we are forced to consider schemes whose characteristic polynomials have a root ξ for which $\xi = 1$.

For the explicit Euler scheme this is the only root. With the centered scheme (17.1), $p(\xi) = \xi^2 - 1$ and the roots are $\xi = \pm 1$. The root $\xi = -1$ is the cause of the oscillations we see in figures 17.2, 17.3, and 17.4.

17.2.3 Unbounded discrete solutions

It is clear that if there is a solution of $p(\xi) = 0$ with $|\xi| > 1$, then an explosion can result even with no driving force (i.e., $f = 0$). We find a solution to $p(\xi) = 0$ for (17.5) where

$$-\xi = \frac{5 + \sqrt{29}}{4} \approx 2.5963. \qquad (17.11)$$

This explains the extreme blowup shown in figure 17.5.

There is a more subtle way in which schemes can be unbounded when they should be constant. Suppose that there is a root ξ of p in (17.9) such that $|\xi| = 1$. Suppose further that this is a double root of p, so that also $p'(\xi) = 0$. Then

$$0 = p'(\xi) = \sum_{i=0}^{k} a_i(k - 1)\zeta^{k-i-1}. \qquad (17.12)$$

Thus a sequence of the form $u_n = n\xi^n$ is also a solution to (17.7), and unfortunately it is not bounded. Thus we are led to define a root condition as follows.

Definition 17.2 *We say that a difference stencil of the form (17.7) satisfies the* root condition *if*

- *none of the roots $p(\xi) = 0$ satisfy $|\xi| > 1$*

- *all the roots $p(\xi) = 0$ satisfying $|\xi| = 1$ are simple.*

We have shown that whenever the root condition is violated, there are unbounded solutions of (17.7).

17.2.4 Zero stability

The first stability concept we will consider is *zero stability*, or 0-stability. Informally, this condition states that there are no unbounded solutions of (17.7). Thus it is natural to define zero stability just in terms of definition 17.2. But to know that this is a sufficient condition for boundedness, we must prove the following.

Theorem 17.3 *The difference method (17.7) has bounded solutions if and only if the root condition definition 17.2 is satisfied.*

Proof. We have already seen that a root $p(\xi) = 0$ with $|\xi| > 1$ leads to unbounded solutions, as well as to multiple roots for which $|\xi| = 1$.

Now we need to show that if the root condition (definition 17.2) is satisfied, then the solutions must remain bounded. Let us start by noting that u_n depends linearly on $u_{n-1}, \ldots, u_{n-k}$. Think of $v^{(n)} \in \mathbb{R}^k$ as the vector $v = (u_n, \ldots, u_{n-k+1})$. Then $v^{(n)} = A v^{(n-1)} = A^n v^{(0)}$, where the matrix A is defined by

$$A = \begin{pmatrix} -a_1/a_0 & -a_2/a_0 & \cdots & -a_{k-1}/a_0 & -a_k/a_0 \\ 1 & 0 & \cdots & 0 & 0 \\ 0 & 1 & \cdots & 0 & 0 \\ \vdots & \vdots & \vdots & \vdots & \vdots \\ 0 & 0 & \cdots & 1 & 0 \end{pmatrix}. \tag{17.13}$$

By exercise 17.8, we have

$$\begin{aligned} (-1)^{k-1}\det(A - \lambda I) &= (-a_1/a_0 - \lambda)\lambda^{k-1} + \sum_{i=2}^{k}(-a_i/a_0)\lambda^{k-i} \\ &= -\lambda^k - \sum_{i=1}^{k}(a_i/a_0)\lambda^{k-i} \\ &= -p(\lambda)/a_0. \end{aligned} \tag{17.14}$$

Therefore, the roots of p are the eigenvalues of A. Thus we see that the term "characteristic polynomial" for the difference stencil is consistent with the usual notion of the characteristic polynomial of the matrix A.

Let $\lambda_1, \ldots, \lambda_k$ be the eigenvalues of A that satisfy $|\lambda_j| = 1$. By assumption, these are simple eigenvalues, so there are eigenvectors x^j such that $A x^j = \lambda_j x^j$, $j = 1, \ldots, k$.

Let $\lambda_1, \ldots, \lambda_k$ denote the eigenvalues of A such that $|\lambda_j| = 1$. Using the Jordan decomposition, we can write

$$A = S^{-1} \begin{pmatrix} D & 0 \\ 0 & T \end{pmatrix} S, \tag{17.15}$$

where S is nonsingular, D is the diagonal matrix with entries λ_j, and T is upper-triangular, with the remaining eigenvalues of A on the diagonal of

T. Thus the diagonal entries of T are all less than 1 in complex modulus. Therefore (exercise 17.12),

$$v^{(n)} = A^n v^{(0)} = S^{-1} \begin{pmatrix} D^n & 0 \\ 0 & T^n \end{pmatrix} S v^{(0)}. \tag{17.16}$$

By lemma 6.13, we find that $\|T^n\|_\infty \to 0$ as $n \to \infty$ (see exercise 17.13).

QED

17.2.5 Absolute stability regions

For reasons we will explain subsequently, we now consider complex-valued solutions to ODEs. Solutions to the model problem

$$u' = \lambda u \tag{17.17}$$

are bounded for all complex λ with nonpositive real part, that is, $\mathcal{R}e\,\lambda \leq 0$. It is thus of interest to ask for what λ a particular scheme generates bounded solutions. This is similar to the question we asked for $\lambda = 0$ in section 17.2.4, so we might call this λ-stability. Since we have a linear f in our test problem ($f(u,t) = \lambda u$), we can easily work with implicit schemes as well as explicit ones.

Definition 17.4 *The region of absolute stability (or λ-stability) is the set of values of λ and Δt for which a numerical scheme for solving (17.17) has the property that u_n can be bounded in terms of $u_{n-1}, u_{n-2}, \ldots, u_{n-k}$ for some fixed k.*

Although in general the absolute stability regions could describe complicated relations between Δt and λ, in all the cases we will consider, the regions take the form of a set S in the complex plane such that the product $\lambda \Delta t$ resides in S.

We should explain why complex values of λ might be of interest. Consider a system of equations such as

$$\begin{pmatrix} u_1 \\ u_2 \end{pmatrix}' = B \begin{pmatrix} u_1 \\ u_2 \end{pmatrix}, \quad \text{where } B = \begin{pmatrix} 0 & 1 \\ -1 & 0 \end{pmatrix}, \tag{17.18}$$

which corresponds to the second-order equation $u_1'' = -u_1$. We can diagonalize the system using a similarity transformation that converts the matrix B to diagonal form. If we call v_j the transformed variables, then the equations become $v_j' = (-1)^j i v_j$ since the eigenvalues of B are $\pm i$ (where i is the imaginary unit). In many cases, the behavior of a linear system reduces to that of a diagonal system with the (complex, in general) eigenvalues on the diagonal. Thus even if we are interested only in systems with real coefficients, we need to consider the behavior of schemes for (17.17) for complex λ for essentially the same reason we are forced to consider complex eigenproblems for real matrices.

For the explicit Euler scheme (16.26), the sequence of approximates satisfies $u_n = u_{n-1} + \lambda \Delta t u_{n-1} = (1 + \lambda \Delta t) u_{n-1}$. Thus $|u_n| \leq |u_{n-1}|$ iff

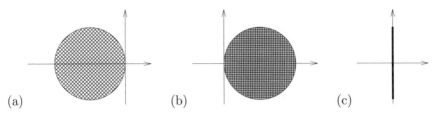

Figure 17.7 Absolute stability regions for the (a) explicit Euler scheme (16.26), (b) implicit Euler scheme (16.27), and (c) centered scheme (17.1). The region for the implicit Euler scheme is the complement of the hashed disk.

$|1 + \lambda \Delta t| \leq 1$. The set of values of the product $\lambda \Delta t$ satisfying this condition is the unit disk in the complex plane centered at -1, as shown in figure 17.7(a).

The region of absolute stability for the implicit Euler scheme (16.27) is similar. The approximates are defined by $u_n = u_{n-1} + \lambda \Delta t u_n$, so that absolute stability (i.e., $|u_n| \leq |u_{n-1}|$) holds iff $|1 - \lambda \Delta t|^{-1} \leq 1$. Thus the region of absolute stability is the set of $\lambda \Delta t$ in the complement of the unit disk in the complex plane centered at $+1$, as depicted in figure 17.7(b) (it is the complement of the hashed region). Note that this includes values of λ for which solutions to the model problem $u' = \lambda u$ would *not* be bounded.

The centered scheme (17.1) applied to the model problem (17.17) yields $u_n = u_{n-2} + 2\lambda \Delta t u_{n-1}$. Now we cannot express u_n simply in terms of u_{n-1} but rather must form a matrix equation similar to (17.13). We can thus write

$$\begin{pmatrix} u_n \\ u_{n-1} \end{pmatrix} = A \begin{pmatrix} u_{n-1} \\ u_{n-2} \end{pmatrix}, \quad \text{where } A = \begin{pmatrix} 2\lambda \Delta t & 1 \\ 1 & 0 \end{pmatrix}. \tag{17.19}$$

The eigenvalues μ of A satisfy

$$0 = \det(A - \mu I) = -\mu(2\lambda \Delta t - \mu) - 1 = \mu^2 - (2\lambda \Delta t)\mu - 1, \tag{17.20}$$

so that

$$\mu_{\pm} = z \pm \sqrt{z^2 + 1}, \tag{17.21}$$

where $z = \lambda \Delta t$. We are interested in the set of z such that both

$$|\mu_+(z)| \leq 1 \quad \text{and} \quad |\mu_-(z)| \leq 1. \tag{17.22}$$

Let us start by observing that for purely imaginary $z = it$, we find

$$|\mu_{\pm}(it)| = \left| it \pm \sqrt{-t^2 + 1} \right| = 1 \quad \text{as long as } |t| \leq 1, \tag{17.23}$$

since $\sqrt{-t^2 + 1}$ is real in this case. In general, we have

$$\mu_+ \mu_- = -1, \tag{17.24}$$

either by multiplying out the expressions in (17.21) or by observing that the determinant of A in (17.19), which is -1, must be the same as the product of its eigenvalues. But then

$$|\mu_+| = 1/|\mu_-|, \tag{17.25}$$

k	a_0	a_1	$2a_2$	$3a_3$	$4a_4$	$5a_5$	$6a_6$	ρ'
1	1	-1						NA
2	$3/2$	-2	1					0.33333
3	$11/6$	-3	3	-1				0.42640
4	$25/12$	-4	6	-4	1			0.56086
5	$137/60$	-5	10	-10	5	-1		0.70871
6	$147/60$	-6	15	-20	15	-6	1	0.86338

Table 17.2 Coefficients of the 0-stable BDF schemes of degree k. The number ρ' is the largest modulus of the roots of the characteristic polynomial (17.9) (and equivalently, the eigenvalues of the matrix (17.13)) excluding the common root (eigenvalue) $\xi = 1$.

and the only way to satisfy $\max\{|\mu_+|, |\mu_-|\} \leq 1$ is to have both $|\mu_+| = 1$ and $|\mu_-| = 1$, as we found for $z = it$ and $t \in [-1, 1]$ in (17.23). But the equations $|\mu_+| = |\mu_-| = 1$ and $\mu_+\mu_- = -1$ imply we can write (exercise 17.14)

$$\mu_\pm = \pm \cos\theta + i \sin\theta. \qquad (17.26)$$

Adding the plus and minus terms in (17.21), we have

$$2z = \mu_+ + \mu_- = 2i \sin\theta. \qquad (17.27)$$

That is, we find that $\max\{|\mu_+|, |\mu_-|\} \leq 1$ implies that $z = it$ for $t \in [-1, 1]$, as we found in (17.23). Thus the absolute stability region contains only an interval on the imaginary axis: the set $\{it \mid |t| \leq 1\}$, depicted in figure 17.7(c) as the dark interval on the vertical axis.

17.3 BACKWARD DIFFERENTIATION FORMULAS

Another way to increase the accuracy in (16.28) is to use a *backward differentiation formula* (BDF)

$$\frac{du}{dt}(t_n) \approx P'(t_n) = \frac{1}{\Delta t} \sum_{i=0}^{k} a_n u_{n-i} = f(u_n, t_n), \qquad (17.28)$$

where the coefficients $\{a_i \mid i = 0, \dots k\}$ are chosen so that (17.28) is exact for polynomials of degree k. The BDF for $k = 1$ is the same as the implicit Euler scheme. BDF methods arise by inverting the relation (13.58) to get

$$hD = \log(I + \Delta) \qquad (17.29)$$

and then expanding via the Taylor series $\log(1 + x) = -\sum_{j=1}^{\infty}(-x)^j/j$:

$$hD = \sum_{j=1}^{\infty} \frac{(-1)^{j+1}}{j}\Delta^j, \qquad (17.30)$$

which is valid as an expression for operators on polynomials (exercise 17.15). By truncating the infinite series (17.30) at a finite point, we get BDF formulas:

$$\sum_{i=0}^{k} a_i u_{n-i} = \sum_{j=1}^{k} \frac{(-1)^j}{j} \Delta^j u_n, \qquad (17.31)$$

where we define Δu_n to be the sequence whose nth entry is $u_n - u_{n-1}$. The higher powers are defined by induction: $\Delta^{j+1} u_n = \Delta(\Delta^j u_n)$. (There is multiple notation abuse here: Δu_n really means the nth element of the sequence $\Delta\{u.\}$, where $\{u.\}$ denotes the full sequence.) For example, $\Delta^2 u_n = u_n - 2u_{n-1} + u_{n-2}$, and in general Δ^j has coefficients given by Pascal's triangle. We thus see that $a_0 \neq 0$ for all $k \geq 1$.

In table 17.2 we give the coefficients for the first few instances of the BDF formulas. We see that $a_0 = \sum_{i=1}^{k} 1/i$, $a_1 = -k$, and for $j \geq 2$, ja_j is an integer conforming to Pascal's triangle.

Given this simple definition of the general case of a BDF, it is hard to imagine what could go wrong regarding stability. Unfortunately, the condition that $|\xi| \leq 1$ for roots of $p_k(\xi) = 0$ restricts k to be 6 or less for the BDF formulas. Presumably, the same feature that makes the forward difference formula fail at $k = 3$ is at work. We simply cannot compute such accurate approximations to the derivative by looking so exclusively in one direction in a stable way.

One can compute the roots of $p_k(\xi) = 0$ by forming the matrix (17.13) and computing its eigenvalues. We find they are all simple up to order $k = 7$ (and higher), but there is a complex pair with $|\xi| \approx 1.0222$ for $k = 7$. For smaller values of k, all the eigenvalues other than the required $\xi = 1$ satisfy $|\xi| < 1$.

17.4 MORE READING

The careful reader will notice that we have not proved that a stable, consistent numerical method converges to the solution of an ordinary differential equation when the mesh is refined. This is our stopping point; we leave to further study the general formulation of such results for different classes of difference methods. There are many basic books on the numerical solution of ordinary differential equations, such as the classic by Henrici[4] [79] and the more recent [29, 75], as well as advanced books such as [31, 74, 76].

---

[4]Peter Henrici (1923–1987) was a student of Eduard Stiefel and spent most of his career at ETH in Switzerland, but he was at UCLA for several years where he was a colleague of Hestenes and the thesis advisor of Gilbert Strang, his first student and recipient of the first Henrici prize in 2007.

17.5 EXERCISES

Exercise 17.1 *Consider the ODE $u'(t) = f(t)$ with $u(0) = 0$. Then $u(T) = \int_0^T f(t)\,dt$. Consider the explicit Euler scheme (16.26) for solving the ODE. Interpret this as a quadrature rule for computing the integral. What rule is it?*

Exercise 17.2 *Show that the coefficients a_i^k in (17.3) are the derivatives of the basis functions for Lagrange interpolation for the points*
$$-k\Delta t, (1-k)\Delta t, \ldots, -\Delta t, 0.$$
Show that the scaling with respect to Δt is correct. (Hint: first you need to show that the coefficients are independent of n. Then just do a scaling of the t variable by Δt.)

Exercise 17.3 *Show that the coefficients a_i^k in (17.3) for $k = 1$ correspond to the explicit Euler scheme, and for $k = 2$ to the centered scheme (17.1).*

Exercise 17.4 *Show that the coefficient a_0^k in (17.3) is never zero. (Hint: show that the basis function ϕ_0 has a nonzero derivative at -1. Observe that $\phi_0(-1) = 0$ is a simple zero.)*

Exercise 17.5 *Show that round-off error is not the cause of the instability observed for (17.5). (Hint: try writing the equations in terms of rational numbers.)*

Exercise 17.6 *Examine the behavior of the improved Euler scheme (17.6) on the test problem $u' = -u^2$ and compare its accuracy with that of the explicit Euler scheme (16.26). (Hint: the values $\tilde{u}_n$ are not the same as in (16.26) because they restart at each step with the result of (17.6) at the previous time step.)*

Exercise 17.7 *Show that any scheme of the form (17.4) must satisfy*
$$\sum_{i=0}^{k} a_i = 0 \tag{17.32}$$
in order that $u_n \to u(n\Delta t)$ as $\Delta t \to 0$ uniformly for $n\Delta t \leq T$, where u is the solution to (16.1). (Hint: show that we must have $\lim_{\Delta t \to 0} \sum_{i=0}^{k} a_i u_{n-i} = u'(n\Delta t)$. Apply this to the special case in which u is constant.)

Exercise 17.8 *Prove that*
$$\det \begin{pmatrix} \alpha_1 & \alpha_2 & \cdots & \alpha_{k-1} & \alpha_k \\ 1 & -\lambda & \cdots & 0 & 0 \\ . & . & \cdots & . & . \\ 0 & 0 & \cdots & 1 & -\lambda \end{pmatrix} = (-1)^{k-1} \sum_{i=1}^{k} \alpha_i \lambda^{k-i}. \tag{17.33}$$
(Hint: by induction.)

Exercise 17.9 *Verify the values claimed for the BDF coefficients, namely, that* $a_0 = \sum_{i=1}^{k} 1/i$, $a_1 = -k$, *and for* $j \geq 2$, ja_j *is an integer conforming to Pascal's triangle.*

Exercise 17.10 *Verify directly that the roots* $\mu_\pm$ *in (17.21) satisfy*

$$|\mu_\pm(it)| > 1$$

for one of the choices of roots provided that $|t| > 1$. *Also show that for* r *real, then* $\pm\mu_\pm(r) > 1$.

Exercise 17.11 *Suppose that* $\lambda_1 \neq \lambda_2$ *are two eigenvalues of a symmetric matrix* A *with corresponding eigenvectors* x^j, $j = 1, 2$. *Prove that* $(x^1)^\star x^2 = 0$. *(Hint: suppose that* $\lambda_1 \neq 0$ *and write* $(x^1)^\star x^2 = (1/\lambda_1)(A\,x^1)^\star x^2$.)

Exercise 17.12 *Suppose that an* $m \times m$ *matrix* A *can be written in the block form*

$$A = \begin{pmatrix} V & 0 \\ 0 & W \end{pmatrix}, \tag{17.34}$$

where V *is a* $k \times k$ *matrix and* W *is an* $(m-k) \times (m-k)$ *matrix. Prove that*

$$A^n = \begin{pmatrix} V^n & 0 \\ 0 & W^n \end{pmatrix} \tag{17.35}$$

for any integer $n \geq 1$.

Exercise 17.13 *Suppose that an* $m \times m$ *matrix* A *can be written in the block form*

$$A = \begin{pmatrix} V & 0 \\ 0 & W \end{pmatrix}, \tag{17.36}$$

where V *is a* $k \times k$ *matrix and* W *is an* $(m-k) \times (m-k)$ *matrix. Prove that*

$$\|A\|_\infty \leq \max\{\|V\|_\infty, \|W\|_\infty\}. \tag{17.37}$$

(Hint: write m-vectors x *in block form:* $x = (y, z)^T$, *where* y *is a* k-*vector.)*

Exercise 17.14 *Suppose that* $\mu_\pm$ *are complex numbers such that* $|\mu_+| = |\mu_-| = 1$ *and* $\mu_+\mu_- = -1$. *Prove that* $\mu_\pm = \pm\cos\theta + i\sin\theta$ *for some value of* $\theta \in \mathbb{R}$. *(Hint: start by writing* $\mu_\pm = \cos\theta_\pm + i\sin\theta_\pm$.)

Exercise 17.15 *Verify that (17.30) is valid as an expression for operators on polynomials, i.e., for any polynomial* P *and any* x,

$$hP'(x) = \sum_{j=1}^{\infty} \frac{(-1)^{j+1}}{j} \Delta^j P(x) = \sum_{j=1}^{k} \frac{(-1)^{j+1}}{j} \Delta^j P(x),$$

where k *is the degree of* P. *(Hint: show that* $\Delta^j P \equiv 0$ *for* $j > k$.)

Exercise 17.16 *Prove that there are eigenvectors of the form*
$$(\xi^{k-1}, \xi^{k-2}, \ldots, \xi, 1) \tag{17.38}$$
for the matrix (17.13) when ξ is a root of (17.9).

Exercise 17.17 *Suppose that the polynomial (17.9) has multiple roots ξ satisfying $|\xi| = \rho(A)$, the spectral radius of the matrix (17.13). Prove that A is defective (not diagonalizable). (Hint: if A were diagonalizable, could there be solutions like (17.12) that grow faster than $\rho(A)^n$?)*

Exercise 17.18 *Consider the polynomial $p(\xi) = (\xi - 1)^2$. Determine the matrix B corresponding to (17.13) for this polynomial. Show that the only eigenvalue of B is $\xi = 1$ and that B has only one eigenvector. Prove that*
$$B^k = \begin{pmatrix} k+2 & -(k+1) \\ k+1 & -k \end{pmatrix} = \begin{pmatrix} 1 & 0 \\ 0 & 1 \end{pmatrix} + k \begin{pmatrix} 1 & -1 \\ 1 & -1 \end{pmatrix} \tag{17.39}$$
for all integers $k \geq 1$.

17.6 SOLUTIONS

Solution of Exercise 17.5. If we multiply (17.5) by 24 and write it expressly for the case $f(u, t) = u$, we find
$$8u_n = (-12 + 24\Delta t)u_{n-1} + 24u_{n-2} - 4u_{n-3}. \tag{17.40}$$
Thus for $\Delta t = 1/12$ or $1/24$ we have an expression with integer coefficients. Take, e.g., $\Delta t = 1/24$. Then (17.40) becomes
$$8u_n = -11u_{n-1} + 24u_{n-2} - 4u_{n-3}. \tag{17.41}$$
Writing $u_n = p_n/q_n$, we find
$$\begin{aligned} p_i &= -11p_{i-1}q_{i-2}q_{i-3} + 24q_{i-1}p_{i-2}q_{i-3} - 4q_{i-1}q_{i-2}p_{i-3} \\ q_i &= 8q_{i-1}q_{i-2}q_{i-3}. \end{aligned} \tag{17.42}$$
To begin the algorithm, we can use the approximation $e^x \approx 1 + x + x^2/2 + x^3/6$ with $x = \Delta t, 2\Delta t$ to provide rational starting data for u_1 and u_2. To reduce the growth in coefficients, we can divide both p_n and q_n by their greatest common divisor (GCD). Using octave with this code produces integer results for p_n and q_n (use "format bank" to verify this), and by plotting p_n/q_n we see the same results (to graphical accuracy) as for the corresponding algorithm (17.5) performed in floating-point. The main loop in the octave code to verify this looks like

```
pp=-11*p(i-1)*q(i-2)*q(i-3)
    +24*q(i-1)*p(i-2)*q(i-3)
     -4*q(i-1)*q(i-2)*p(i-3);
qq=8*q(i-1)*q(i-2)*q(i-3);
g=gcd(pp,qq);
p(i)=pp/g;
q(i)=qq/g;
u(i)=-(3/2)*u(i-1)+3*u(i-2)-0.5*u(i-3)+3*dt*u(i-1);
```

where $\mathtt{dt} = \Delta t$.

Solution of Exercise 17.8. For $k = 2$, this is evident. For general k, we expand along the first column and use induction:

$$
\det \begin{pmatrix} \alpha_1 & \alpha_2 & \alpha_3 & \cdots & \alpha_{k-1} & \alpha_k \\ 1 & -\lambda & 0 & \cdots & 0 & 0 \\ 0 & 1 & -\lambda & \cdots & 0 & 0 \\ \vdots & \vdots & \vdots & \vdots & \vdots & \vdots \\ 0 & 0 & 0 & \cdots & 1 & -\lambda \end{pmatrix} = \alpha_1 \det \begin{pmatrix} -\lambda & 0 & \cdots & 0 & 0 \\ 1 & -\lambda & \cdots & 0 & 0 \\ \vdots & \vdots & \vdots & \vdots & \vdots \\ 0 & 0 & \cdots & 1 & -\lambda \end{pmatrix}
$$

$$
- \det \begin{pmatrix} \alpha_2 & \alpha_3 & \cdots & \alpha_{k-1} & \alpha_k \\ 1 & -\lambda & \cdots & 0 & 0 \\ \vdots & \vdots & \vdots & \vdots & \vdots \\ 0 & 0 & \cdots & 1 & -\lambda \end{pmatrix}
$$

$$
= \alpha_1(-\lambda)^{k-1} - (-1)^{k-2} \sum_{i=1}^{k-1} \alpha_{i+1}\lambda^{k-1-i}
$$

$$
= \alpha_1(-\lambda)^{k-1} - (-1)^{k-2} \sum_{j=2}^{k} \alpha_j\lambda^{k-j} = (-1)^{k-1} \sum_{j=1}^{k} \alpha_j\lambda^{k-j}.
$$

$$(17.43)$$

Solution of Exercise 17.10. For $|t| > 1$,

$$
\mu_\pm(it) = i\left(t \pm \sqrt{t^2 - 1}\right), \tag{17.44}
$$

and thus $|\mu_\pm(it)| > 1$ for one of the choices of roots (e.g.,

$$
\mu_- = t - \sqrt{t^2 - 1} \tag{17.45}
$$

if $t < -1$). On the other hand, if $z = r$ is purely real, then

$$
\mu_\pm(r) = r \pm \sqrt{r^2 + 1} \tag{17.46}
$$

is also real. For $r > 0$, then

$$
\mu_+^2 = \left(r + \sqrt{r^2 + 1}\right)^2 > 1 + r > 1; \tag{17.47}
$$

if $r < 0$, then

$$
\mu_-^2 = \left(r - \sqrt{r^2 + 1}\right)^2 < -1 + r < -1. \tag{17.48}
$$

Chapter Eighteen

Floating Point

> "To summarize, although there are practically occurring matrices for which partial pivoting yields a moderately large, or even exponentially large, growth factor, the growth factor is almost invariably found to be small. Explaining this fact remains one of the major unsolved problems in numerical analysis." N. Higham in [82].

It is beyond the scope of this book to consider the effect of floating-point in detail. We have noted its effect in certain areas where some rigorous estimates can be given. In this chapter we look at some central topics and point out areas where further research is needed.

One recurring concern in numerical computation has been the accumulation of floating-point errors when a very large number of steps are involved. One of the simplest computations of this type is the summation of n numbers with n large. We will analyze this computation in great detail and show that the floating-point error can be bounded rigorously by a reasonable upper bound. However, we will also see that in typical computations, these upper bounds are too pessimistic and fall short of predicting the observed floating-point error.

Another topic is perturbation theory for systems of linear equations. We derive the standard theory and show how it can motivate the iterative improvement algorithm. But we also point out limitations of the standard theory in predicting actual errors in solving systems of linear equations, which are again very pessimistic compared to observed computational errors.

18.1 FLOATING-POINT ARITHMETIC

In section 1.5, we introduced a simple model of floating-point arithmetic. This provides a way to analyze algorithms which are executed in floating-point arithmetic on a digital computer. The algorithms we have considered involve real numbers at an abstract level, but we are forced to work with a finite approximation of them in actual computations. Including an analysis of the effect of floating-point allows us to prove theorems about the actual computations, not just about their theoretical counterparts executed using real numbers. The latter is of interest but not sufficient to guarantee success of the approximated computations done using floating-point arithmetic.

Cancellation is a major source of error. Note that $f\ell(a-b) = (a-b)(1+\delta)$ does not mean that $f\ell(f\ell(a+e)-b)$ is at all close to $a+e-b$. *Floating-point arithmetic is not associative.* Cancellation amplifies errors that have already occurred. It can easily be that $f\ell(a + e) = a$, and if $b = a$, we get zero for the result instead of e. However, we will see that in other cases, the effect of floating-point arithmetic is much more subtle.

Floating-point arithmetic is quite complicated and does not lend itself to a simple representation. A floating-point hardware standard has been adopted by the Institute of Electrical and Electronics Engineers (IEEE), and most hardware follows this at the moment. However, these specifications simply provide bounds on the behavior of floating-point. But it is possible to use such a model to guarantee the success of computations using this standard [82].

We should note that there is no absolute guarantee that on every occasion a computer will not make a mistake, say, because of some external force. Indeed, it used to be common that parity errors occurred in computers, that is, a piece of memory would be corrupted by a 1-bit error. Current technology typically can correct such errors and detect 2-bit errors (which are far more unlikely). We ignore such unlikely errors; they would be even more unlikely to recur.

The model of floating-point arithmetic used in this book is a standard way of representing an upper bound of the inaccuracies of floating-point. We emphasize that it is just a model and that other more accurate models are possible. One flaw of this model is that, even if some error bound is shown to be the worst case for the model, it may not be the worst case for floating-point arithmetic for any set of data. And even if there is the possibility of a worst-case scenario, it is in many cases extremely unlikely to occur.

18.1.1 Summation

A common subproblem in many algorithms is the computation of a sum of n numbers for large n. This appears in the computation of inner products in the conjugate gradients algorithm (chapter 9), and in all the iterative methods the computation of a matrix-vector product is of this type. Even in Gaussian elimination (chapter 3) or in a direct method such as the Cholesky method, computation of inner products is a central issue. Consider computing the sum

$$A_n = \sum_{i=1}^{n} a_i. \tag{18.1}$$

Let $\tilde{A}_n$ denote the result obtained by floating-point arithmetic. The equation (1.31) implies that we can write

$$\tilde{A}_k = (1 + \delta_k)(\tilde{A}_{k-1} + a_k), \tag{18.2}$$

where $|\delta_k| \leq \epsilon$ (see section 1.5) and where we define $\tilde{A}_1 = A_1 = a_1$. Define $e_k = \tilde{A}_k - A_k$. Subtracting A_k from (18.2), we have

$$
\begin{aligned}
e_k = \tilde{A}_k - A_k &= \tilde{A}_{k-1} - A_{k-1} + \delta_k(\tilde{A}_{k-1} + a_k) \\
&= (1 + \delta_k)(\tilde{A}_{k-1} - A_{k-1}) + \delta_k A_k = (1 + \delta_k)e_{k-1} + \delta_k A_k.
\end{aligned}
\tag{18.3}
$$

Subtracting e_{k-1} from both sides of (18.3) gives

$$
e_k - e_{k-1} = \delta_k e_{k-1} + \delta_k A_k.
\tag{18.4}
$$

Then by summing (18.4), we have (since $e_1 = 0$)

$$
A_n - \tilde{A}_n = e_n = \sum_{i=2}^{n} A_i \delta_i + \sum_{i=2}^{n} e_{i-1} \delta_i = \sum_{i=2}^{n} A_i \delta_i + \mathcal{O}(\epsilon^2)
\tag{18.5}
$$

for $n \geq 2$. We postpone for a moment making this a rigorous estimate for $|A_n - \tilde{A}_n|$. It provides a useful guide in practice, but even if we ignore the $\mathcal{O}(\epsilon^2)$ term, it can still be hard to interpret. The individual errors δ_i are likely to be distributed around zero, not biased in predictable ways, so if the partial sums A_i are slowly varying in i, there can be substantial cancellation.

One prediction that (18.5) makes is that the error will be smaller if we add the numbers starting with the smallest a's first and moving to the larger a's at the end since this makes the partial sums A_i smaller. Let us consider an example, the slowly diverging series

$$
A_n = \sum_{i=1}^{n} \frac{1}{i} \approx \gamma + \log n,
\tag{18.6}
$$

where $\gamma = 0.57721 \cdots$ is Euler's constant; cf. (13.79). A thought experiment explains why the order of summation might matter. If we start with the largest numbers, then we eventually (for n large enough) get to the point where $1/i$ is below the level of round-off error for the sum A_i accumulated so far, i.e., $i^{-1} < \epsilon A_i$. Thus the result of the summation process A_n will quit changing for some finite n and give the false impression that this is a convergent series. On the other hand, if we start with the smallest terms, the results do not stabilize (exercise 18.1) for any n, although they might eventually overflow.

The errors for these two orders of summation are shown in table 18.1. We see that the errors are often smaller if we sum the smallest terms first, but both errors are smaller than (18.5) predicts, as we explore in section 18.5. Moreover, the errors are not uniformly smaller for the "smallest first" approach. The error difference for the two methods as a function of n is depicted in figure 18.1, and we see that it looks somewhat like a random process. In these computations, $\epsilon \approx 2.2 \times 10^{-16}$ (exercise 18.2). We explain in section 18.1.3 how we were able to compute the errors for these sums.

In exercise 18.3, a more complicated behavior is examined with the slowly converging series

$$
\frac{\pi}{4} \approx A_n = \sum_{i=1}^{n} \frac{(-1)^{i+1}}{2i - 1}.
\tag{18.7}
$$

n	$(A_n - \tilde{A}_n^{>})/A_n$	$(A_n - \tilde{A}_n^{<})/A_n$
10^3	$+2.4 \times 10^{-16}$	$+5.9 \times 10^{-16}$
10^4	$+3.4 \times 10^{-15}$	-3.6×10^{-15}
10^5	$+7.6 \times 10^{-15}$	$+1.6 \times 10^{-15}$
10^6	$+5.1 \times 10^{-14}$	-3.3×10^{-15}
10^7	$+1.5 \times 10^{-13}$	-6.8×10^{-15}
10^8	$+7.1 \times 10^{-14}$	$+2.4 \times 10^{-14}$
10^9	-2.8×10^{-14}	$+8.4 \times 10^{-14}$

Table 18.1 Normalized errors for two methods for summing the series (18.6), largest-first ($\tilde{A}_n^{>}$) versus smallest-first ($\tilde{A}_n^{<}$), are shown for various values of n.

Now we see a new behavior that is order-dependent. If we add the odd and even terms separately, we get two diverging sums whose difference is the quantity of interest. Eventually, the answer will be smaller than ϵ times these two diverging sums, and it will have no significant digits. Thus we can see that the order of computation could provide a variety of results for a single problem. On the other hand, the estimate (18.5) says that as long as the partial sums A_i/A_n remain of reasonable relative size, the computation of a sum of numbers will not produce a very large error.

Lemma 18.1 *The error in the expression (18.5) satisfies*

$$\left| A_n - \tilde{A}_n - \sum_{i=2}^{n} A_i \delta_i \right| \leq 2n\epsilon^2 \sum_{i=1}^{n} (n + 1 - i)|A_i|, \qquad (18.8)$$

provided that $n \leq 1/2\epsilon$.

Proof. Define $\sigma_k = \sum_{i=2}^{k} A_i \delta_i$ for $k \geq 2$ and set $\sigma_1 = 0$ and $\beta_k = e_k - \sigma_k$. Note that $e_1 = 0$. Then (18.5) says that

$$e_n = \sigma_n + \sum_{i=2}^{n} e_{i-1} \delta_i, \qquad (18.9)$$

and thus

$$\beta_k = \sum_{i=2}^{k} (\beta_{i-1} + \sigma_{i-1}) \delta_i. \qquad (18.10)$$

Define $S = \sum_{i=2}^{n} |\sigma_i|$. Then we claim that for all k,

$$|\beta_k| \leq ((1 + \epsilon)^k - 1)S. \qquad (18.11)$$

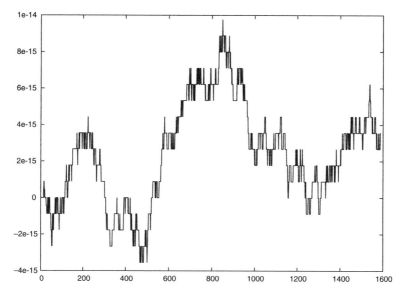

Figure 18.1 The vertical axis is the difference $\tilde{A}_n^> - \tilde{A}_n^<$ between two methods (largest-first minus smallest-first) for summing the series (18.6) for various values of n. The horizontal axis is n.

The proof is by induction using (18.10) (see exercise 18.6),

$$\begin{aligned}
|\beta_{k+1}| &\leq \epsilon\left(S + \sum_{i=2}^{k+1} |\beta_{i-1}|\right) \\
&\leq \epsilon S\left(1 + \sum_{i=2}^{k+1}((1+\epsilon)^{i-1} - 1)\right) \\
&= \epsilon S + S\left((1+\epsilon)^{k+1} - 1 - \epsilon\right) - kS \\
&\leq S\left((1+\epsilon)^{k+1} - 1\right).
\end{aligned} \tag{18.12}$$

Applying exercise 18.7 shows that

$$|\beta_k| \leq 2k\epsilon S. \tag{18.13}$$

From the definition of S and the σ_i's, we see that

$$S = \sum_{i=2}^{n} |\sigma_i| \leq \sum_{i=2}^{n}\sum_{k=2}^{i} \epsilon|A_k| = \epsilon\sum_{k=2}^{n}(n+1-k)|A_k|, \tag{18.14}$$

where we have used an elementary form of Fubini's theorem to reverse the order of the summations.   QED

18.1.2 Summation application

The most general application of the estimates in lemma 18.1 is to say that

$$\left|A_n - \tilde{A}_n\right| \approx \epsilon\sum_{i=2}^{n} |A_i| \tag{18.15}$$

since we do not know anything a priori about the distribution of the δ_i's. Here we are implicitly assuming that the term that is quadratic in ϵ is much smaller than the right-hand side of (18.15). Let us apply this idea to some examples.

We can apply the estimates in lemma 18.1 to the summation problem (18.6). Let $A_k^<$ denote the partial sums progressing from the smallest to the largest terms. Then using estimate in (18.6), we have

$$A_k^< = \sum_{i=n+1-k}^{n} \frac{1}{i} \approx \log \frac{n}{n-k}. \tag{18.16}$$

Since all the partial sums are positive, we thus find

$$\sum_{k=1}^{n} |A_k^<| = \sum_{k=1}^{n} A_k^< \approx \sum_{k=1}^{n} \log \frac{n}{n-k} = \log \left(\prod_{k=1}^{n} \frac{n}{n-k} \right)$$
$$= \log \left(\frac{n^n}{n!} \right) \approx \log \frac{e^n}{\sqrt{2\pi n}} = n - \tfrac{1}{2} \log(2\pi n) \approx n, \tag{18.17}$$

by Stirling's formula (see page 198). Similarly, if we let $A_k^>$ denote the partial sums progressing from the largest to the smallest terms, we find (exercise 18.8) that

$$\sum_{k=1}^{n} |A_k^>| = \sum_{k=1}^{n} A_k^> \approx \sum_{k=1}^{n} \gamma + \log k = \gamma n + \log(n!)$$
$$\approx \gamma n + \tfrac{1}{2} \log(2\pi n) + n \log \frac{n}{e} \approx n \log n. \tag{18.18}$$

Thus we see that the partial sums for the smallest-to-largest algorithm for computing (18.6) are smaller than those for the largest-to-smallest algorithm, but only by a factor of $\log n$. We compare the results of the two algorithms computed using octave in table 18.1 and figure 18.1. The errors are smaller for $A_n^<$, but neither grows like n as (18.17) and (18.18) would predict.

Now let us consider a more rapidly converging sum where there is a more distinct difference in the way that the sums are done:

$$\sum_{k=1}^{n} \frac{1}{k^2} \approx \frac{\pi^2}{6} = 1.644934 \cdots . \tag{18.19}$$

Now the partial sums are quite order-dependent. First,

$$A_k^< = \sum_{i=n+1-k}^{n} \frac{1}{i^2} \leq \int_{n-k}^{n} \frac{dx}{x^2} = \frac{1}{n-k} - \frac{1}{n} = \frac{k}{n(n-k)}. \tag{18.20}$$

For $k < n$, this estimate is quite tight since we also have

$$A_k^< \geq \int_{n+1-k}^{n+1} \frac{dx}{x^2} = \frac{1}{n+1-k} - \frac{1}{n+1} = \frac{k}{(n+1)(n+1-k)}. \tag{18.21}$$

However, for $k = n$, the estimate (18.20) fails, and we need to substitute

$$A_n^< = A_n \leq \frac{\pi^2}{6}.$$

Summing (18.20), we find

$$\sum_{k=1}^{n} |A_k^{\leq}| = \sum_{k=1}^{n} A_k^{\leq} \leq \frac{\pi^2}{6} + \sum_{k=1}^{n-1} \frac{k}{n(n-k)} = \frac{\pi^2}{6} + \frac{1}{n} \sum_{k=1}^{n-1} \frac{k}{n-k}$$

$$= \frac{\pi^2}{6} + \frac{1}{n} \sum_{i=1}^{n-1} \frac{n-i}{i} = \frac{\pi^2}{6} + \sum_{i=1}^{n-1} \frac{1}{i} - \frac{n-1}{n} \leq 2 + \log n \tag{18.22}$$

(see exercise 18.10). From lemma 18.1,

$$|A_n - A_n^{\leq}| \leq (2 + \log n)\epsilon + \mathcal{O}\left(\epsilon^2\right). \tag{18.23}$$

On the other hand, $A_k^{>} \geq 1$ for all k, so that

$$\sum_{k=1}^{n} A_k^{>} \geq n. \tag{18.24}$$

Thus the errors for the two methods could be quite different. In particular, (18.23) says that we can take $A_n^{\leq}$ as an estimate for A_n in evaluating the errors in $A_n^{>}$ computationally. In table 18.2 we give the differences in the results of the two algorithms, together with a more precise estimate of the individual errors. To estimate these errors, we need a formula for A_n, which we do not have explicitly. However, it is easy to see that

$$A_n = \sum_{k=1}^{n} \frac{1}{k^2} = \frac{\pi^2}{6} - \sum_{k=n+1}^{\infty} \frac{1}{k^2}. \tag{18.25}$$

With a small correction, we can recognize the latter sum as the trapezoidal rule approximating the integral

$$\sum_{k=n+1}^{\infty} \frac{1}{k^2} + \frac{1}{2n^2} \approx \int_{n}^{\infty} \frac{dx}{x^2} = \frac{1}{n}. \tag{18.26}$$

But we can also invoke the Euler-Maclaurin formula (13.77) to get an even more accurate approximation:

$$\sum_{k=n+1}^{\infty} \frac{1}{k^2} \approx \frac{1}{n} - \frac{1}{2n^2} + \frac{1}{6n^3}. \tag{18.27}$$

The next term in the Euler-Maclaurin expansion (13.77) would contribute a term of order n^{-5}. For simplicity, in table 18.2 we have just listed the estimates

$$\tilde{A}_n - \frac{\pi^2}{6} + \frac{1}{n} - \frac{1}{2n^2} + \frac{1}{6n^3}, \tag{18.28}$$

where by $\tilde{A}_n$ we mean one of the methods computed in floating-point. We refer to these as $\tilde{A}_n^{>}$ and $\tilde{A}_n^{\leq}$ in table 18.2.

We see that the algorithm starting with the smallest summands appears to have at most a 1-bit error for the values of n listed, whereas the algorithm starting with the largest entries has an error that grows with n. However, the

n	$\tilde{A}_n^> - \tilde{A}_n^<$	(18.28) for $\tilde{A}_n^>$	(18.28) for $\tilde{A}_n^<$
10^3	$+1.7764 \times 10^{-15}$	$+1.5543 \times 10^{-15}$	-2.2204×10^{-16}
10^4	$+5.5511 \times 10^{-15}$	$+5.5511 \times 10^{-15}$	0.0
10^5	$+1.5987 \times 10^{-14}$	$+1.6209 \times 10^{-14}$	$+2.2204 \times 10^{-16}$
10^6	$+4.3743 \times 10^{-14}$	$+4.3521 \times 10^{-14}$	-2.2204×10^{-16}
10^7	-9.7189×10^{-13}	-9.7189×10^{-13}	0.0
10^8	$+9.8635 \times 10^{-10}$	$+9.8635 \times 10^{-10}$	0.0
10^9	-8.0137×10^{-9}	-8.0137×10^{-9}	0.0

Table 18.2 The difference between two methods for summing the series (18.19), largest-first $(\tilde{A}_n^>)$ minus smallest-first $(\tilde{A}_n^<)$, for various values of n.

error does not grow linearly with n as might have been expected from (18.24). The unpredictable nature of the distribution of the δ_i's leads to a slower growth rate due to cancellations in the floating-point error. Unfortunately, we do not have a more precise model to predict this behavior in more detail.

We have seen that it is possible to derive rigorous error expressions for the summation problem but that these still fall short of predicting the error behavior in several cases. Moreover, the error behavior is quite data-dependent. Doing the same level of analysis for more complex algorithms remains a topic of research.

18.1.3 Better summation algorithms

There are also simple techniques [82] to reduce the errors in large sums. Probably the simplest is called *double precision accumulation* which simply stores the current accumulated value as a double-precision variable, essentially reducing the value of ϵ substantially. However, there is a much more sophisticated technique due to Kahan,[1] which can be viewed as a type of residual correction algorithm (cf. section 18.2.2).

The gist of the Kahan summation algorithm is to keep an error estimate e_k in addition to the running sum s_k. Then the error estimate is added to the next summand before adding it to the running sum:

$$
\begin{aligned}
y_k &= a_k + e_{k-1} \\
s_k &= s_{k-1} + y_k \\
d_k &= s_{k-1} - s_k \\
e_k &= d_k + y_k.
\end{aligned}
\tag{18.29}
$$

The terms d_k and y_k are temporaries and do not need to be stored from one iteration to the next, and only one storage location for the e's is needed as well. But there need to be two locations for the s's, the old and the

---

[1]William Morton (a.k.a. Velvel) Kahan (1933-) was a primary architect of the IEEE floating-point standard and won the Turing award in 1989.

new values. Moreover, the computations of d_k and e_k can be collapsed into one line, eliminating d_k completely. But it is critical that the difference, $s_{k-1} - s_k$, be computed (in floating-point) before adding y_k.

It is possible to show that the errors using (18.29) satisfy

$$A_n - \tilde{A}_n = \sum_{i=1}^{n} \delta_i a_i, \qquad (18.30)$$

where $|\delta_i| \leq 2\epsilon + \mathcal{O}\left(n\epsilon^2\right)$ [82]. The algorithm (18.29) was used to estimate the exact sum in table 18.1. The algorithm was applied with both orders, and the results were identical in all cases reported.

18.1.4 Solving ODEs

Algorithms for solving ordinary differential equations compute quantities very closely related to the simple sums (18.1). The requirement of using finite-precision arithmetic means that the best error behavior we could expect for the algorithm (16.26) is

$$|u(t_n) - u_n| \leq C_{f,T}\Delta t + n\epsilon \quad \forall t_n \leq T, \qquad (18.31)$$

where ϵ measures the precision error that occurs at each step in (16.27).

Let us suppose that $t_n = T$. It is useful to rewrite (18.31) using the fact that $n = T/\Delta t$ as

$$|u(T) - u_n| \leq C_{f,T}\Delta t + \frac{T\epsilon}{\Delta t}, \qquad (18.32)$$

which shows that the error reaches a minimum and cannot be reduced by reducing Δt. This occurs when $C_{f,T}\Delta t = T\epsilon/\Delta t$, that is, $\Delta t = \sqrt{T\epsilon/C_{f,T}}$, and the best accuracy is $\sqrt{C_{f,T}T\epsilon}$. We can turn this around to say that it occurs at the time $T = C_{f,T}\Delta t^2/\epsilon$.

Choosing a more accurate difference method helps avoid the onset of round-off error since we can use a larger Δt to get the desired accuracy. The centered difference method (17.1) instead satisfies an error estimate of the form

$$|u(t_n) - u_n| \leq \hat{C}_{f,T}\Delta t^2 + n\epsilon = \hat{C}_{f,T}\Delta t^2 + \frac{t_n\epsilon}{\Delta t} \qquad (18.33)$$

for all $t_n \leq T$. Of course, the constant $\hat{C}_{f,T}$ may be different, but it is typically not different by an order of magnitude. The critical value for Δt occurs at the much smaller level $\sqrt[3]{t_n\epsilon/\hat{C}_{f,T}}$. However, there is still a limitation in theory. In practice, very high-order schemes are used [70, 109, 166] primarily to achieve sufficient accuracy for longtime integration, but this also has the side effect of diminishing round-off error.

18.2 ERRORS IN SOLVING SYSTEMS

Iterative methods for solving linear and nonlinear systems are inherently self-correcting with regard to floating-point error. However, direct methods for solving linear systems do not have such a self-correction aspect, and it was an early concern whether it would be possible to accurately compute with such methods. We address some of these questions here.

There are various issues to analyze for direct methods. The central point is that we compute only approximate factorizations. It would be of interest to understand in detail how floating-point errors affect the resulting factors, but this issue is not yet well understood. It appears that the triangular factors are far better behaved than current analytical techniques would predict [160].

18.2.1 Condition number

The condition number of a matrix A has appeared in different contexts previously (cf. sections 9.3.4 and 15.1.4). For a given norm, it is defined by

$$\kappa(A) = \|A\| \|A^{-1}\|. \tag{18.34}$$

It is defined only for invertible matrices A. We show here how it quantifies the stability of solving a system of equations. Note that $\kappa(I) = 1$ for the identity matrix I, and indeed $\kappa(A) \geq 1$ for any matrix A (exercise 18.11). Moreover, $\kappa(A)$ is invariant under a simple scaling $A \to tA$ for any scalar t (exercise 18.12).

One characterization of a matrix factorization in floating-point is that it produces the exact factors of a perturbed system $\tilde{A} = \tilde{L}\tilde{U}$, where $\tilde{L}$ and $\tilde{U}$ are the computed factors including round-off. We will examine how this might affect the resulting solution of a perturbed system. However, the best-known rigorous bounds are quite pessimistic, and a more involved analysis like (18.5) for summation might be more revealing. However, such results are lacking.

Suppose we write $X(t)$ for the solution of

$$A(t)X(t) = F(t), \tag{18.35}$$

where we think of t as a perturbation parameter as might arise from floating-point or other errors. Here we imagine that there can be independent errors in A and F, and we want to see how this causes changes in X. We are interested in what happens for small t, so we differentiate (18.35) to find

$$A(t)X'(t) + A'(t)X(t) = F'(t), \tag{18.36}$$

or equivalently,

$$X'(t) = A^{-1}(t)F'(t) - A^{-1}(t)A'(t)X(t). \tag{18.37}$$

This equation says that the change in X has two parts. The first part says something obvious: if you make an error in F, it will cause an error in X, and the relationship involves the operator A in the obvious way. The second part is more complicated; it depends on the interaction between the perturbation in A and the solution vector X, all multiplied by the inverse of A. We begin by giving a standard estimate for both terms and then return to point out why this may be pessimistic for the second term.

We take norms in (18.37) to find

$$\|X'(t)\| \le \|A^{-1}F'\| + \|A^{-1}A'X(t)\|, \tag{18.38}$$

which we simplify to get

$$\frac{\|X'(t)\|}{\|X(t)\|} \le \frac{\|A^{-1}F'\|}{\|X(t)\|} + \frac{\|A^{-1}A'X(t)\|}{\|X(t)\|} \le \frac{\|A^{-1}F'\|}{\|X(t)\|} + \|A^{-1}A'\|$$

$$\le \|A^{-1}\| \left(\frac{\|F'\|}{\|X(t)\|} + \|A'\| \right) = \kappa(A) \left(\frac{\|F'\|}{\|A\|\|X(t)\|} + \frac{\|A'\|}{\|A\|} \right)$$

$$\le \kappa(A) \left(\frac{\|F'\|}{\|F(t)\|} + \frac{\|A'\|}{\|A\|} \right),$$

$$\tag{18.39}$$

where we used the estimate $\|F(t)\| = \|A(t)X(t)\| \le \|A(t)\|\|X(t)\|$ in the last step. The estimate (18.39) has a simple interpretation: the relative error in X is bounded by the sum of the relative errors in A and F, multiplied by the condition number of A.

Although this does provide a rigorous upper bound (see exercise 18.14), it shows that the interaction between perturbations in A and X is complicated. It could well be that for certain solutions X, the term $A^{-1}(A')X$ would be quite a bit smaller than in others. For example, if X is slowly varying (nearly constant) and if A^{-1} tends to smooth things out (cf. the matrix M in (4.22) which is the inverse of the matrix A in (4.20)), then both $A^{-1}F'$ and $A^{-1}(A')X$ could be much smaller than in other situations. Using norms to bound things gives a worst-case estimate (see exercise 18.16).

On the other hand, solving equations $HX = F$ with the Hilbert matrix (4.15) can be quite different. The inverse of the Hilbert matrix (4.15) grows exponentially with n (exercise 18.17), and since the errors A' and F' in (18.37) are not naturally correlated, we expect the error X' to be quite large, as experiments show (exercise 18.18). In this case, the estimate (18.39) is unfortunately not overly pessimistic.

18.2.2 A posteriori estimates and corrections

The residual error $R = F - AX$ left after applying some algorithm to "solve" $AX = F$ can (1) give an estimate of the error and (2) be used to correct it. More precisely, let the function f denote the solution process, so that $Y = f(A, F)$ is the result of such an algorithm, for which we know (e.g., from (18.39)) that, for some parameter μ,

$$\|f(A, Aw) - w\| \le \mu\|w\| \tag{18.40}$$

for any vector w. In particular, this implies that

$$\|Y - X\| \le \mu\|X\|. \tag{18.41}$$

Define the residual $R = F - AY$. (There is a small point in that there will be some floating-point error in computing R by this formula, but we are interested in the case where this is much smaller than the overall error parameter μ; thus we ignore this error here.) Then we "solve" $A E = R$, or more precisely, define $\hat{E} = f(A, R)$. Then, by (18.40),

$$\|\hat{E} - E\| = \|f(A, R) - E\| \leq \mu\|E\|. \tag{18.42}$$

We expect that $Y + \hat{E}$ is a better approximation to X since

$$Y + E = X. \tag{18.43}$$

(To prove (18.43), multiply by A: $A Y + A E = F - R + R = F = A X$.) Therefore,

$$Y + \hat{E} = Y + E - (E - \hat{E}) = X - (E - \hat{E}). \tag{18.44}$$

To quantify this, we just estimate:

$$\begin{aligned} \|(Y + \hat{E}) - X\| &= \|E - \hat{E}\| && \text{[by (18.44)]} \\ &\leq \mu\|E\| && \text{[by (18.42)]} \\ &= \mu\|X - Y\| && \text{[by (18.43)]} \\ &\leq \mu^2\|X\| && \text{[by (18.41)]}. \end{aligned} \tag{18.45}$$

If $\mu < 1$, (18.45) implies that $Y + \hat{E}$ is a more accurate approximation to X than Y is. This algorithm is known as *iterative improvement*. If the level of accuracy in (18.45) is not enough, the process can be repeated by making the assignment $Y \leftarrow Y + \hat{E}$ and repeating the calculations ($R = F - A Y$ and $\hat{E} = f(A, R)$) and estimates. After k applications of the solution algorithm $f(A, \cdot)$, the error satisfies

$$\|Y - X\| \leq \mu^k\|X\|. \tag{18.46}$$

This implies that the error would eventually go to zero, which does not necessarily happen in floating-point. This is true for all the iterative methods considered so far. Once the change in the iteration is the size of round-off error, it can bounce around unpredictably.

18.2.3 Pivoting

The intent of pivoting in Gaussian elimination is to reduce the error in the resulting $\tilde{A} = \tilde{L}\tilde{U}$. One cause of error is simply the size of the factors. Pivoting can reduce the size of the factors substantially. However, there is a well-known example that shows that partial pivoting is not sufficient to control the growth of factors [82]. Consider the $n \times n$ matrix A given by

$$A = \begin{pmatrix} 1 & 0 & 0 & \cdots & 0 & 0 & 1 \\ -1 & 1 & 0 & \cdots & 0 & 0 & 1 \\ -1 & -1 & 1 & \cdots & 0 & 0 & 1 \\ \vdots & \vdots & \vdots & \vdots & \vdots & \vdots \\ -1 & -1 & -1 & \cdots & -1 & 1 & 1 \\ -1 & -1 & -1 & \cdots & -1 & -1 & 1 \end{pmatrix}. \tag{18.47}$$

That is, A has entries equal to -1 below the diagonal, 1's on the diagonal and in the rightmost column, and 0 elsewhere. For this matrix, the standard partial pivoting algorithm will do no pivoting, as the diagonal term in the standard order is at each step at least as large as the entries below it. However, the entries in the right-hand column grow exponentially, so that if $A = LU$, then $\|U\|_\infty = 2^{n-1}$. In particular, it is easy to see that L and A agree on and below the diagonal and that U has 1's on the diagonal except for the nth entry, the last column is $u_{in} = 2^{i-1}$, and there are 0's elsewhere. If for example (see page 36), $n = 10^5$, then $\|U\|_\infty \approx 10^{3010}$, even though there are entries in U of order unity. On the other hand, experience shows that such catastrophic growth is very uncommon [82, 160].

18.3 MORE READING

The book by Trefethen and Bao [160] reports interesting experiments regarding the effect (or lack thereof) of floating-point computation on matrix factorization. The comprehensive text by Higham [82] should be consulted for a more detailed understanding of the effects of floating-point arithmetic. There is a recent handbook on floating-point arithmetic [119].

18.4 EXERCISES

Exercise 18.1 *Write a code to compute the slowly diverging series*

$$A_n = \sum_{i=1}^{n} \frac{1}{i}$$

and compare with (18.5), which estimates the error in terms of the partial sums. What is the worst order? The best order? See if the results "converge" for one order and diverge for the other. (Hint: you may need to work in a programming system that allows you to specify "single" precision, such as "float" in C.)

Exercise 18.2 *You can estimate the size of ϵ in our floating-point model by $\epsilon = \inf \{x > 0 \mid f\ell(f\ell(1 + x) - 1) > 0\}$. On a computer with binary-based arithmetic, you can estimate ϵ by replacing x in the set above by $x_k = 2^{-k}$. For the first k such that $f\ell(f\ell(1 + x_k) - 1) = 0$, $\epsilon \approx x_{k-1}$.*

Exercise 18.3 *Write a code to compute the slowly converging series*

$$\frac{\pi}{4} \approx \sum_{i=1}^{n} \frac{(-1)^{i+1}}{2i - 1}$$

and compare with the estimate (18.5), which estimates the error in terms of the partial sums. What is the worst order? The best order? What if you sum the odd and even terms separately?

Exercise 18.4 *Analyze the algorithms in section 1.5 for computing solutions to (1.32) for various values of b. Establish conditions on b that guarantee the success of each of the four algorithms.*

Exercise 18.5 *Perform some computational experiments with the numerical methods discussed in chapter 17 to see whether the effects of round-off error can be easily discerned.*

Exercise 18.6 *Prove that*

$$\epsilon \sum_{i=1}^{k} (1 + \epsilon)^i = (1 + \epsilon)^{k+1} - 1 - \epsilon. \tag{18.48}$$

(Hint: multiply the sum by $((1 + \epsilon) - 1)$ and see how it telescopes.)

Exercise 18.7 *Prove that*

$$(1 + \epsilon)^n - 1 \leq 2n\epsilon, \tag{18.49}$$

provided that $n \leq 1/2\epsilon$. (Hint: expand the left-hand side in a binomial series and bound the terms, or use the fact that $\log(1 + \epsilon) \leq \epsilon$ for $\epsilon > 0$.)

Exercise 18.8 *Justify all the steps in (18.18).*

Exercise 18.9 *Backward error analysis expresses*

$$\tilde{A}_n = \sum_{i=1}^{n} a_i(1 + \gamma_i). \tag{18.50}$$

This represents the computed sum as the exact sum with modified summands. Develop an analysis of the size of the γ_i's in terms of the δ_i's in (18.2).

Exercise 18.10 *Prove that*

$$\frac{\pi^2}{6} + \sum_{i=1}^{n-1} \frac{1}{i} - \frac{n-1}{n} \leq 2 + \log n. \tag{18.51}$$

(Hint: see (13.81) for an estimate of the sum when $n > 3$. For $n = 1, 2$ make a direct evaluation.)

Exercise 18.11 *Prove that $\kappa(A) \geq 1$ for any matrix A, where $\kappa(A) = \|A\|\|A^{-1}\|$ is the condition number of A. Show that this holds for any choice of norm. (Hint: note that $1 = \|AA^{-1}\|$ and apply (6.3).)*

Exercise 18.12 *Prove that $\kappa(tA) = \kappa(A)$ for any matrix A and scalar t, where $\kappa(A) = \|A\|\|A^{-1}\|$ is the condition number of A. Show that this holds for any choice of norm. Here tA is the matrix with entries ta_{ij} if $A = (a_{ij})$. (Hint: note that $1 = \|AA^{-1}\|$ and apply (6.3).)*

Exercise 18.13 *Estimate the term of order ϵ^2 in (18.23) and determine how it affects the subsequent statements regarding using $A_k^{<}$ as an "exact" sum.*

Exercise 18.14 *The following is a more conventional version of (18.39). Suppose that* $AX = F$ *and* $\tilde{A}\tilde{X} = \tilde{F}$. *Prove that*

$$\frac{\|X - \tilde{X}\|}{\|X\|} \leq \kappa(A)\left(1 - \kappa(A)\frac{\|A - \tilde{A}\|}{\|A\|}\right)^{-1}\left(\frac{\|F - \tilde{F}\|}{\|F\|} + \frac{\|A - \tilde{A}\|}{\|A\|}\right),$$

(18.52)

where $\kappa(A) = \|A\|\|A^{-1}\|$ *is the condition number of* A.

Exercise 18.15 *Experiment with different ways of computing Stirling's formula (12.80). Find a way to control the size of the numerator and denominator. (Hint: write each as a product of n terms and write the quotient of the product as the product of the quotients.)*

Exercise 18.16 *Experiment computationally with solving* $Ax = f$, *where* A *is the* $n \times n$ *matrix (4.20). Start with a given* x *and compute* $f = Ax$ *by matrix-vector multiplication. Then use a standard routine (e.g.,* y=A\f *in* octave*) to "solve"* $Ay = f$ *and compare* x *to* y *(e.g., monitor the value of* $\|x - y\|/\|x\|$*) for different* x *and different values of* n*). Also compute the size of the condition number* $\kappa(A) = \|A\|\|A^{-1}\|$ *and monitor the value of* $\|x - y\|/\kappa(A)\|x\|$*. Compare the choices where* x *is all 1's* $(x_i = 1$ *for all* $i = 1, \ldots, n)$ *and* x *is random (e.g.,* x=rand(n,1) *in* octave*).*

Exercise 18.17 *Show that the inverse of the* $n \times n$ *Hilbert matrix* H *in (4.14) and (4.15) has entries of order one and exponentially large. In particular, show that*

$$(H^{-1})_{11} = \sum_{i=1}^{n} \frac{1}{j^2} \approx \frac{\pi}{\sqrt{6}}$$

(18.53)

$$(H^{-1})_{nn} = u_{nn}^{-2} > 2^{2n} \quad (for\ n > 2),$$

where $H = U^\star U$ *is the Cholesky factorization of* H *given by (4.16). (Hint: to estimate the size of* u_{nn}*, see exercise 4.5.)*

Exercise 18.18 *Experiment computationally with solving* $Hx = f$, *where* H *is the* $n \times n$ *Hilbert matrix in (4.14) and (4.15). Start with a given* x *and compute* $f = Hx$ *by matrix multiplication. Then use a standard routine (e.g.,* y=H\f *in* octave*) to "solve"* $Hy = f$ *and compare* x *to* y *(e.g., monitor the value of* $\|x - y\|/\|x\|$*) for different* x *and different values of* n*). Also compute the size of the condition number* $\kappa(H) = \|H\|\|H^{-1}\|$ *and monitor the value of* $\|x - y\|/\kappa(H)\|x\|$*. Compare the choices where* x *is all 1's* $(x_i = 1$ *for all* $i = 1, \ldots, n)$ *and* x *is random (e.g.,* x=rand(n,1) *in* octave*).*

18.5 SOLUTIONS

Solution of Exercise 18.7. Let $r = n\epsilon$. Recall that $r \leq \frac{1}{2}$. Write

$$(1 + \epsilon)^n - 1 = \sum_{i=1}^{n} \binom{n}{i} \epsilon^i = \sum_{i=1}^{n} \frac{n!}{(n-i)!i!} \epsilon^i$$

$$= \sum_{i=1}^{n} \prod_{k=0}^{i-1} \frac{n-k}{i-k} \epsilon \leq \sum_{i=1}^{n} (n\epsilon)^k \qquad (18.54)$$

$$= \sum_{i=1}^{n} r^k = \frac{r - r^{n+1}}{1 - r} \leq \frac{r}{1 - r} \leq 2r.$$

Alternatively, since $\log(1 + \epsilon) \leq \epsilon$ for $\epsilon > 0$,

$$(1 + \epsilon)^n = e^{n \log(1+\epsilon)} \leq e^{n\epsilon}.$$

For $x \leq 1$, $e^x \leq 1 + 2x$ since $e^1 = e = 2.718\cdots < 3$ and the exponential function is strictly increasing. Therefore, $(1+\epsilon)^n \leq 1+2\epsilon n$, provided $\epsilon n \leq 1$.

Solution of Exercise 18.14. Subtract $AX = F$ and $\tilde{A}\tilde{X} = \tilde{F}$ to get

$$A(X - \tilde{X}) = F - A\tilde{X} = F - \tilde{F} + (\tilde{A} - A)\tilde{X}. \qquad (18.55)$$

Multiplying by A^{-1} and taking norms gives

$$\|X - \tilde{X}\| \leq \|A^{-1}\| \left(\|F - \tilde{F}\| + \|\tilde{A} - A\|\|\tilde{X}\| \right). \qquad (18.56)$$

We have $\|F\| \leq \|A\|\|X\|$, so that $\|X\|^{-1} \leq \|A\|/\|F\|$. Therefore

$$\frac{\|X - \tilde{X}\|}{\|X\|} \leq \kappa(A) \left(\frac{\|F - \tilde{F}\|}{\|F\|} + \frac{\|A - \tilde{A}\|}{\|A\|} \frac{\|\tilde{X}\|}{\|X\|} \right)$$

$$\leq \kappa(A) \left(\frac{\|F - \tilde{F}\|}{\|F\|} + \frac{\|A - \tilde{A}\|}{\|A\|} \left(1 + \frac{\|X - \tilde{X}\|}{\|X\|} \right) \right). \qquad (18.57)$$

Therefore,

$$\frac{\|X - \tilde{X}\|}{\|X\|} \left(1 - \kappa(A) \frac{\|A - \tilde{A}\|}{\|A\|} \right) \leq \kappa(A) \left(\frac{\|F - \tilde{F}\|}{\|F\|} + \frac{\|A - \tilde{A}\|}{\|A\|} \right). \qquad (18.58)$$

Chapter Nineteen

Notation

> "During the years 1831-80 the strange figure of Benjamin Peirce (A.B. 1829) completely dominated the situation. His great natural mathematical talent and originality of thought, combined with a total inability to put anything clearly, produced upon his contemporaries a feeling of awe that amounted almost to dread." [37]

We follow standard notation in general, but we use some notation that is different from what is sometimes used.

We will use the notation $]a, b[$ to denote the open interval $a < x < b$. Similarly, $]a, b]$ ($a < x \leq b$) and $[a, b[$ ($a \leq x < b$) denote the corresponding half-open intervals.

We use a tall vertical line as the separator in our notation for a set. Thus $]a, b] = \{x \mid a < x \leq b\}$.

When the infimum of a set is known to be attained, we will often write $\min\{\cdots\}$ for $\inf\{\cdots\}$, and similarly for max and sup.

We use the notation "argmin" to denote the point at which a minimum takes place. Thus $\operatorname{argmin} \phi(r)$ is the value of r_0 (if it exists), where $\phi(r_0) = \min \phi(r)$, where the minimum is taken over some set S. We can similarly define $\operatorname{argmin} \{\phi(r) \mid r \in S\}$ in the same way.

We use the notation $f := g$ to mean "f is defined to be g." Sometimes the definition comes first, so we write $g =: f$ in that case.

We use the notation $x \leftarrow y$ in an algorithm to mean that (the value of) y is assigned to (the value of) x. This notation is used instead of an equal sign ($=$), which is used in many programming languages.

We generally use capital letters for matrices but frequently denote their entries by lowercase. Thus the entries in A are denoted by a_{ij}, and those of B by b_{ij}. We use the notation $B = A^\star$ for the conjugate transpose: $b_{ij} = \overline{a_{ji}}$, where $\overline{z}$ denotes the complex conjugate of z. The same notation applies to vectors as well: $v^\star$ performs the complex conjugate and switches from a column vector to a row vector (or conversely).

We write the transpose as A^{T} or v^{T} for a matrix or vector, respectively. In particular, we often write a (column) vector as $v = (a, b, c, d, e, f, g)^{\mathrm{T}}$ to save space. Of course, for real matrices, $A^{\mathrm{T}} = A^\star$, and similarly for real vectors.

An expression of the form $g(x) = \mathcal{O}\left(f(x)\right)$ means that $|g(x)| \leq C|f(x)|$ for some constant $C < \infty$.

For a function f, we use the expression $f \equiv 0$ to mean that f is "identically zero." That is, we mean that $f(x) = 0$ for all x in the domain of f.

We use $\mathbb{R}$ to denote the field of real numbers, and $\mathbb{C}$ to denote the field of complex numbers. For a complex number $z = r + it$, we write $r = \mathcal{R}e\, z$ and $t = \mathcal{I}m\, z$. The notation ℓ_p is introduced in section 5.1.1 to denote $\mathbb{F}^n$ endowed with the p-norm, where $\mathbb{F}^n$ denotes either $\mathbb{R}^n$ or $\mathbb{C}^n$.

We define $\text{sign}(t)$ to be 1 when $t \geq 0$ and -1 when $t < 0$.

We use the notation "$\log x$" for the natural logarithm ($e^{\log x} = x$).

We use the notation $\mathcal{P}_n$ to denote the space of polynomials of degree n in one variable; $\mathcal{P}_\infty$ denotes the space of all such polynomials, cf. (13.51).

We use the notation $\rho(A)$ for the spectral radius of A, the modulus of the largest eigenvalue of A, cf. (6.8).

We use the notation $\kappa(A)$ for the condition number of A, cf. (9.74), section 15.1.4 and exercise 18.11. The condition number depends on the particular norm being used.

Bibliography

[1] P.-A. Absil, R. Mahony, and R. Sepulchre. *Optimization Algorithms on Matrix Manifolds*. Princeton University Press, 2007.

[2] Victor S. Adamchik and David J. Jeffrey. Polynomial transformations of Tschirnhaus, Bring and Jerrard. *SIGSAM Bulletin*, 37(3):90–94, 2003.

[3] A. C. Aitken. Studies in practical mathematics. I. the evaluation, with application, of a certain triple product matrix. *Proceedings of the Royal Society of Edinburgh*, 57:172–181, 1937.

[4] A. C. Aitken. *Gallipoli to the Somme: Recollections of a New Zealand Infantryman*. Oxford, 1963.

[5] Adrian Albert. *Modern Higher Algebra*. University of Chicago Press, 1937.

[6] P. S. Aleksandrov, N. I. Akhiezer, B. V. Gnedenko, and A. N. Kolmogorov. Sergei Natanovich Bernstein (obituary). *Russian Mathematical Surveys*, 24(3):169–176, 1969.

[7] Steven C. Althoen and Renate McLaughlin. Gauss-Jordan reduction: A brief history. *American Mathematical Monthly*, 94(2):130–142, 1987.

[8] Ned Anderson and Åke Björck. A new high order method of *regula falsi* type for computing a root of an equation. *BIT Numerical Mathematics*, 13:253–264, Sept. 1973. 10.1007/BF01951936.

[9] V. I. Arnold. Remarks on eigenvalues and eigenvectors of Hermitian matrices, Berry phase, adiabatic connections and quantum Hall effect. *Selecta Mathematica, New Series*, 1(1):1–19, 1995.

[10] Vladimir I. Arnold. *Ordinary Differential Equations*. Springer Verlag, 2006.

[11] Sheldon J. Axler. *Linear Algebra Done Right*. Springer Verlag, 1997.

[12] Francis Bashforth and John Couch Adams. *An Attempt to Test the Theories of Capillary Action*. Cambridge University Press, 1883.

[13] A. Berman and R. J. Plemmons. *Nonnegative Matrices in the Mathematical Sciences*. Society for Industrial and Applied Mathematics, Philadelphia, PA, 1994.

[14] K. Yusuf Billah and Robert H. Scanlan. Resonance, Tacoma Narrows Bridge failure, and undergraduate physics textbooks. *American Journal of Physics*, 59(2):118–124, 1991.

[15] George David Birkhoff. General mean value and remainder theorems with applications to mechanical differentiation and quadrature. *Transactions of the American Mathematical Society*, 7(1):107–136, 1906.

[16] J. L. Bona, W. G. Pritchard, and L. R. Scott. Solitary-wave interaction. *Physics of Fluids*, 23:438–441, 1980.

[17] Jonathan Borwein and Adrian S. Lewis. *Convex Analysis and Nonlinear Optimization: Theory and Examples*. Springer, 2nd edition, 2005.

[18] John Boyd. *Chebyshev and Fourier Spectral Methods*. Dover, 2nd edition, 2001.

[19] Stephen P. Boyd. *Convex Optimization*. Cambridge University Press, 2004.

[20] Ernst Breitenberger. Gauss's geodesy and the axiom of parallels. *Archive for History of Exact Sciences*, 31:273–289, Sept. 1984. 10.1007/BF00327704.

[21] S. C. Brenner and L. R. Scott. *The Mathematical Theory of Finite Element Methods*. Springer-Verlag, 3rd edition, 2008.

[22] Richard Brent. *Algorithms for Minimization Without Derivatives*. Dover Publications, 2002.

[23] C. Brezinski. The life and work of André Cholesky. *Numerical Algorithms*, 43:279–288, Nov. 2006. 10.1007/s11075-006-9059-x.

[24] C. Brezinski and M. Gross-Cholesky. La vie et les travaux d'André Cholesky. *Bulletin de la Société des Amis de la Bibliothèque de l'Éc. Polytechnique*, 39:7–32, 2005.

[25] Sergey Brin and Lawrence Page. The anatomy of a large-scale hypertextual web search engine. *Computer Networks and ISDN Systems*, 30(1-7):107–117, 1998.

[26] Ezra Brown. Square roots from 1; 24, 51, 10 to Dan Shanks. *College Mathematics Journal*, 30(2):82–95, Mar. 1999.

[27] L. Brutman. On the Lebesgue function for polynomial interpolation. *SIAM Journal on Numerical Analysis*, 15(4):694–704, 1978.

[28] Richard L. Burden and J. Douglas Faires. *Elementary Numerical Analysis*. Brooks/Cole, 8th edition, 2005.

[29] J. C. Butcher. *The Numerical Analysis of Ordinary Differential Equations: Runge-Kutta and General Linear Methods*. John Wiley & Sons, 1987.

[30] J. C. Butcher. A history of Runge-Kutta methods. *Applied Numerical Mathematics*, 20:247–260, March 1996.

[31] J. C. Butcher. *Numerical Methods for Ordinary Differential Equations*. John Wiley & Sons, 2nd edition, 2008.

[32] Florian Cajori. Historical note on the Newton-Raphson method of approximation. *American Mathematical Monthly*, 18(2):29–32, 1911.

[33] V. Chellaboina and W. M. Haddad. Is the Frobenius matrix norm induced? *IEEE Transactions on Automatic Control*, 40(12):2137–2139, 1995.

[34] E. W. Cheney. *Introduction to Approximation Theory*. American Mathematical Society, Providence, RI, 2nd edition, 2000.

[35] Lindsay N. Childs. *A Concrete Introduction to Higher Algebra*. Springer Verlag, 3rd edition, 2009.

[36] William James Cody, Jr. and William Waite. *Software Manual for the Elementary Functions*. Prentice-Hall, 1980.

[37] Julian Coolidge. Mathematics 1870–1928. In Samuel Eliot Morison, editor, *The Tercentennial History of Harvard College and University 1636–1936*, pages 248–257. Harvard University Press, 1930.

[38] P. D. Crout. A short method for evaluating determinants and solving systems of linear equations with real or complex coefficients. *Transactions of the American Institute of Electrical Engineers*, 60:1235–1240, 1941.

[39] Germund Dahlquist and Åke Björck. *Numerical Methods*. Dover, 2003.

[40] Harold Davenport. Dirichlet. *Mathematical Gazette*, 43(346):268–269, 1959.

[41] Philip J. Davis and Philip Rabinowitz. *Methods of Numerical Integration*. Academic Press, 2nd edition, 1984.

[42] Carl de Boor. Divided differences. *Surveys in Approximation Theory*, 1:46–69, 2005.

[43] Carl de Boor and Allan Pinkus. Proof of the conjectures of Bernstein and Erdös concerning the optimal nodes for polynomial interpolation. *Journal of Approximation Theory*, 24:289–303, Dec. 1978.

[44] D. W. Decker, H. B. Keller, and C. T. Kelley. Convergence rates for Newton's method at singular points. *SIAM Journal on Numerical Analysis*, 20(2):296–314, 1983.

[45] James W. Demmel. *Applied Numerical Linear Algebra*. Society for Industrial and Applied Mathematics, Philadelphia, PA, 1997.

[46] P. Deuflhard and G. Heindl. Affine invariant convergence theorems for Newton's method and extensions to related methods. *SIAM Journal on Numerical Analysis*, 16(1):1–10, 1979.

[47] Peter Deuflhard and Andreas Hohmann. *Numerical Analysis: A First Course in Scientific Computation*. Walter de Gruyter & Co., Hawthorne, NJ, USA, 1995.

[48] Luca Dieci and Donald Estep. Some stability aspects of schemes for the adaptive integration of stiff initial value problems. *SIAM Journal on Scientific and Statistical Computing*, 12:1284–1303, 1991.

[49] E. Dikmen, A. Novoselsky, and M. Vallieres. Shell model calculations of ^{108}Sb in the sdgh shell. *Physical Review C*, 66(5):057302, Nov 2002.

[50] S. M. Djouadi. Comments on "Is the Frobenius Matrix Norm Induced?". *IEEE Transactions on Automatic Control*, 48(3):518–518, 2003.

[51] M. H. Doolittle. Method employed in the solution of normal equations and the adjustment of a triangulation. *U.S. Coast and Geodetic Survey Report*, pages 115–120, 1878.

[52] M. Dowell and P. Jarratt. A modified *regula falsi* method for computing the root of an equation. *BIT Numerical Mathematics*, 11:168–174, June 1971. 10.1007/BF01934364.

[53] Tobin A. Driscoll, Kim-Chuan Toh, and Lloyd N. Trefethen. From potential theory to matrix iterations in six steps. *SIAM Review*, 40(3):547–578, 1998.

[54] William Dunham. *Euler: The Master of Us All*. Mathematics Association of America, Washington, D.C., 1999.

[55] Todd F. Dupont and L. Ridgway Scott. The end-game for Newton iteration. Research Report UC/CS TR-2010-10, University of Chicago, Department of Computer Science, 2010.

[56] Alan Edelman. Large dense numerical linear algebra in 1993: the parallel computing influence. *International Journal of High Performance Computing Applications*, 7(2):113–128, 1993.

[57] Paul Erdös. Problems and results on the theory of interpolation. II. *Journal Acta Mathematica Hungarica*, 12(1-2):235–244, 1961.

[58] D. K. Fadeev and V. N. Fadeeva. *Computational Methods of Linear Algebra*. W. H. Freeman, 1963.

[59] Richard William Farebrother. A memoir of the life of M. H. Doolittle. *Bulletin of the Institute of Mathematics and Its Application*, 23(6/7):102, 1987.

[60] P. C. Fenton. A. C. Aitken (1895–1967). *Gazette of the Australian Mathematical Society*, Mar. 1995.

[61] C. T. Fike. *Computer evaluation of mathematical functions*. Prentice-Hall, 1968.

[62] George Forsythe. Notes, 128. *Mathematical Tables and Other Aids to Computation*, 5(36):255–258, 1951.

[63] George E. Forsythe. Solving linear algebraic equations can be interesting. *Bulletin of the American Mathematical Society*, 59:299–329, 1953.

[64] David Fowler and Eleanor Robson. Square root approximations in old Babylonian mathematics: YBC 7289 in context. *Historia Mathematica*, 25:366–378, Nov. 1998.

[65] L. Fox, H. D. Huskey, and J. H. Wilkinson. Notes on the solution of algebraic linear simultaneous equations. *Quarterly Journal Mechanics and Applied Mathematics*, 1(1):149–173, 1948.

[66] Herman H. Goldstine. *A History of Numerical Analysis from the 16th through the 19th century*. Springer, 1977.

[67] Gene H. Golub and Charles F. Van Loan. *Matrix Computations*. Dover, 3rd edition, 1996.

[68] Gene H. Golub and Dianne P. O'Leary. Some history of the conjugate gradient and Lanczos algorithms: 1948–1976. *SIAM Review*, 31(1):50–102, 1989.

[69] Ronald Gowing. *Roger Cotes–Natural Philosopher*. Cambridge University Press, 1983.

[70] K. R. Grazier, W. I. Newman, W. M. Kaula, and J. M. Hyman. Dynamical evolution of planetesimals in the outer solar system. *Icarus*, 140:341–352, 1999.

[71] Anne Greenbaum. *Iterative methods for solving linear systems*. Society for Industrial and Applied Mathematics, Philadelphia, PA, 1997.

[72] David Alan Grier. *When Computers Were Human*. Princeton University Press, 2005.

[73] Wolfgang Hackbusch. *Iterative Solution of Large Sparse Systems of Equations*. Springer, 1993.

[74] E. Hairer, Christian Lubich, and G. Wanner. *Geometric Numerical Integration: Structure-Preserving Algorithms for Ordinary Differential Equations*. Springer, 2nd edition, 2006.

[75] E. Hairer, Syvert P. Nørsett, and G. Wanner. *Solving Ordinary Differential Equations I: Nonstiff Problems*. Springer, 2002.

[76] E. Hairer and G. Wanner. *Solving Ordinary Differential Equations II: Stiff and Differential-Algebraic Problems*. Springer, 2004.

[77] Anders Hald. *A History of Mathematical Statistics*. John Wiley & Sons, 1998.

[78] John Fraser Hart et al. *Computer Approximations*. Krieger, 1978.

[79] Peter Henrici. *Discrete Variable Methods in Ordinary Differential Equations*. John Wiley & Sons, 1962.

[80] Peter Henrici. *Elements of Numerical Analysis*. John Wiley & Sons, 1964.

[81] V. Hernández, J. E. Román, A. Tomás, and V. Vidal. A survey of software for sparse eigenvalue problems. Technical Report SLEPc Technical Report STR-6, Universidad Politecnica de Valencia, http://www.grycap.upv.es/slepc, 2005.

[82] Nicholas J. Higham. *Accuracy and Stability of Numerical Algorithms*. Society for Industrial and Applied Mathematics, Philadelphia, PA, 2nd edition, 2002.

[83] Nicholas J. Higham. The numerical stability of barycentric Lagrange interpolation. *IMA Journal of Numerical Analysis*, 24(4):547–556, 2004.

[84] Alston S. Householder. *Principles of Numerical Analysis*. McGraw-Hill, 1953.

[85] I. M. L. Hunter. An exceptional talent for calculative thinking. *British Journal of Psychology*, 53(3):243–258, 1962.

[86] E. Isaacson and H. B. Keller. *Analysis of Numerical Methods*. Dover, 1994.

[87] Nathan Jacobson. *Lectures in Abstract Algebra II. Linear algebra*. Springer Verlag, 1953.

[88] Ioan James. *Remarkable Mathematicians: From Euler to von Neumann*. Cambridge University Press, 2003.

[89] K. R. James and W. Riha. Convergence criteria for successive overrelaxation. *SIAM Journal on Numerical Analysis*, 12(2):137–143, 1975.

[90] Glen Jeh and Jennifer Widom. Scaling personalized web search. In *WWW '03: Proceedings of the 12th international conference on World Wide Web*, pages 271–279, New York, NY, USA, 2003. Association for Computing Machinery, New York, NY.

[91] D. Kalman. *Uncommon Mathematical Excursions: Polynomia and Related Realms*. Mathematical Association of America, 2008.

[92] Shen Kangshen, John N. Crossley, and Anthony W.-C. Lun. *The Nine Chapters of the Mathematical Art*. Oxford, 1999.

[93] Daniel Kehlmann. *Measuring the World*. Vintage, 2007.

[94] C. T. Kelley. *Iterative Methods for Linear and Nonlinear Equations*. Society for Industrial and Applied Mathematics, Philadelphia, PA, 1995.

[95] C. T. Kelley. *Iterative Methods for Optimization*. Society for Industrial and Applied Mathematics, Philadelphia, PA, 1995.

[96] Theodore A. Kilgore. A characterization of the Lagrange interpolating projection with minimal Tchebycheff norm. *Journal of Approximation Theory*, 24:273–288, Dec. 1978.

[97] D. A. Knoll and D. E. Keyes. Jacobian-free Newton-Krylov methods: A survey of approaches and applications. *Journal of Computational Physics*, 193(2):357–397, 2004.

[98] Gina Bari Kolata. Geodesy: Dealing with an enormous computer task. *Science*, 200(4340):421–466, 1978.

[99] Nick Kollerstrom. Thomas Simpson and "Newton's method of approximation:" an enduring myth. *British Journal for the History of Science*, 25(3):347–354, 1992.

[100] Vladimir Ivanovich Krylov. *Approximate Calculation of Integrals*. Macmillan Press, 1962.

[101] J. L. Lagrange. *Analytical Mechanics*. Springer, 2001.

[102] F. M. Larkin. Root-finding by fitting rational functions. *Mathematics of Computation*, 35(151):803–816, 1980.

[103] Imre Latakos. *Proofs and Refutations*. Cambridge University Press, 1976.

[104] Norman Lebovitz. *Ordinary Differential Equations*. Brooks/Cole, 2002.

[105] W. Ledermann. Issai Schur and his school in Berlin. *Bulletin of the London Mathematical Society*, 15(2):97–106, 1983.

[106] Hou-Biao Li and Ting-Zhu Huang. On a new criterion for the H-matrix property. *Applied Mathematics Letters*, 19:1134–1142, 2006.

[107] Ren-Cang Li. Near optimality of Chebyshev interpolation for elementary function computations. *IEEE Transactions on Computers*, 53(6):678–687, 2004.

[108] Elliott H. Lieb and Michael Loss. *Analysis*. American Mathematical Society, Providence, RI, 2nd edition, 2001.

[109] A. Logg. Multi-adaptive Galerkin methods for ODEs I. *SIAM Journal on Scientific Computing*, 24(6):1879–1902, 2003.

[110] A. Logg. Automating the finite element method. *Archives of Computational Methods in Engineering*, 14(2):93–138, 2007.

[111] G. G. Lorentz. *Approximation of Functions*. Chelsea, New York, 2nd edition, 1986.

[112] G. G. Lorentz and K. L. Zeller. Birkhoff interpolation. *SIAM Journal on Numerical Analysis*, 8(1):43–48, 1971.

[113] L. A. Lyusternik, O. A. Chervonenkis, and A. R. Yanpol'skii. *Handbook for Computing Elementary Functions*. Pergamon Press, 1965.

[114] Robert S. Maier. On reducing the Heun equation to the hypergeometric equation. *Journal of Differential Equations*, 213:171–203, 2005.

[115] L. Maligranda. Why Hölder's inequality should be called Rogers' inequality. *Mathematical Inequalities and Applications*, 1(1):69–83, 1998.

[116] Roy Mathias. Proof of two matrix theorems via triangular factorizations. Technical report, University of Birmingham.

[117] Frank McSherry. A uniform approach to accelerated PageRank computation. In *WWW '05: Proceedings of the 14th International Conference on the World Wide Web*, pages 575–582, New York, NY, USA, 2005. Association for Computing Machinery, New York, NY.

[118] Gérard Meurant. *The Lanczos and Conjugate Gradient Algorithms*. Society for Industrial and Applied Mathematics, Philadelphia, PA, Philadelphia, PA, 2006.

[119] Jean-Michel Muller, Nicolas Brisebarre, Florent de Dinechin, Claude-Pierre Jeannerod, Vincent Lefèvre, Guillaume Melquiond, Nathalie Revol, Damien Stehlé, and Serge Torres. *Handbook of Floating-Point Arithmetic*. Birkhäuser Boston, 2010. ACM G.1.0; G.1.2; G.4; B.2.0; B.2.4; F.2.1., ISBN 978-0-8176-4704-9.

[120] Jorge Nocedal and Stephen Wright. *Numerical Optimization*. Springer, 2nd edition, 2006.

[121] J. Tinsley Oden and Leszek F. Demkowicz. *Applied Functional Analysis*. CRC Press, 1996.

[122] A. M. Ostrowski. *Solution of Equations and Systems of Equations*. Academic Press, 1966.

[123] B. N. Parlett. The Rayleigh quotient iteration and some generalizations for nonnormal matrices. *Mathematics of Computation*, 28(127):679–693, 1974.

[124] Beresford Parlett. Very early days of matrix computations. *SIAM News*, 36(9), 2003.

[125] Paul C. Pasles. *Benjamin Franklin's Numbers: An Unsung Mathematical Odyssey*. Princeton University Press, 2008.

[126] David A. Patterson and John L. Hennessy. *Computer Organization and Design: The Hardware/Software Interface*. Morgan Kaufmann, 3rd edition, 2007.

[127] Giuseppe Peano. Resto nelle formule di quadratura, espresso con un integrale definito. *ATTI della Reale Accademia Dei Lincei-Rendiconti*, 22:562–569, 1913.

[128] G. Peters and J. H. Wilkinson. Inverse iteration, ill-conditioned equations and Newton's method. *SIAM Review*, 21(3):339–360, 1979.

[129] Clifford A. Pickover. *Archimedes to Hawking*. Oxford, 2008.

[130] M. J. D. Powell. *Approximation Theory and Methods*. Cambridge University Press, 1981.

[131] William H. Press, Saul A. Teukolsky, William T. Vetterling, and Brian P. Flannery. *Numerical Recipes*. Cambridge University Press, 3rd edition, 2007.

[132] Liqun Qi. Eigenvalues and invariants of tensors. *Journal of Mathematical Analysis and Applications*, 325(2):1363 – 1377, 2007.

[133] Liqun Qi, W. Sun, and Y. Wang. Numerical multilinear algebra and its applications. *Frontiers of Mathematics in China*, 2(4):501–526, 2007.

[134] Alfio Quarteroni, Riccardo Sacco, and Fausto Saleri. *Numerical Mathematics*. Springer, 2nd edition, 2007.

[135] Th. M. Rassias, H. M. Srivastava, and A. Yanushauskas. *Topics in Polynomials of One and Several Variables and Their Applications*. River Edge, 1993.

[136] Edgar Reich. On the convergence of the classical iterative method of solving linear simultaneous equations. *Annals of Mathematical Statistics*, 20(3):448–451, 1949.

[137] J. R. Rice. *The Approximation of Functions*, volume 1. Addison-Wesley, 1969.

[138] J. R. Rice. *The Approximation of Functions*, volume 2. Addison-Wesley, 1969.

[139] Ralph Tyrell Rockafellar. *Convex Analysis*. Princeton University Press, 1996.

[140] R. Roy. The work of Chebyshev on orthogonal polynomials. In Th. M. Rassias, H. M. Srivastava, and A. Yanushauskas, editors, *Topics in Polynomials of One and Several Variables and Their Applications*, pages 495–512. River Edge, 1993.

[141] Walter Rudin. *Principles of Mathematical Analysis*. McGraw-Hill, 3rd edition, 1976.

[142] Walter Rudin. *Real and Complex Analysis*. McGraw-Hill, 3rd edition, 1986.

[143] Hans Schneider. Olga Taussky-Todd's influence on matrix theory and matrix theorists. *Linear & Multilinear Alg.*, 5:197–224, 1977.

[144] L. R. Scott, T. W. Clark, and B. Bagheri. *Scientific Parlallel Computing*. Princeton University Press, 2005.

[145] L. R. Scott and Xie Dexuan. Parallel linear stationary iterative methods. In Petter Bjørstad and Mitchell Luskin, editors, *Parallel solution of partial differential equations*, pages 31–55. Springer–Verlag, 2000.

[146] J. R. Silvester. Determinants of block matrices. *Mathematical Gazette*, 84(501):460–467, 2000.

[147] Simon J. Smith. Lebesgue constants in polynomial interpolation. *Annales Mathematicae et Informaticae*, 33:109–123, 2006.

[148] Arnold Sommerfeld. *Mechanics*. Academic Press, 1952.

[149] Blair K. Spearman and Kenneth S. Williams. Characterization of solvable quintics $x^5 + ax + b$. *American Mathematical Monthly*, 101(10):986–992, 1994.

[150] Neal Stephenson. *The Baroque Cycle, Volume 1: Quicksilver*. Harper Perennial, 2004.

[151] G. W. Stewart. *Matrix Algorithms, Volume I: Basic Decompositions*. Philadelphia : Society for Industrial and Applied Mathematics, 1998.

[152] G. W. Stewart. *Matrix Algorithms, Volume II: Eigenvalue Problems.* Philadelphia : Society for Industrial and Applied Mathematics, 2001.

[153] G. Strang and K. Borre. *Linear Algebra, Geodesy, and GPS.* Wellesley Cambridge Press, 1997.

[154] V. Szebehely, D. Saari, J. Waldvogel, and U. Kirchgraber. Eduard L. Stiefel (1909–1978). *Celestial Mechanics and Dynamical Astronomy,* 21:2–4, Jan. 1980. 10.1007/BF01230237.

[155] René Taton. Evariste Galois and his contemporaries. *Bulletin of the London Mathematical Society,* 15(2):107–118, 1983.

[156] David J. Thomas and Judith M. Smith. Joseph Raphson, F.R.S. *Notes and Records of the Royal Society of London,* 44(2):151–167, 1990.

[157] John Todd. Numerical analysis at the National Bureau of Standards. *SIAM Review,* 17(2):361–370, 1975.

[158] H. R. Tolley and Mordecai Ezekiel. The Doolittle method for solving multiple correlation equations versus the Kelley-Salisbury "iteration" method. *Journal of the American Statistical Association,* 22(160):497–500, 1927.

[159] Lloyd N. Trefethen. Is Gauss quadrature better than Clenshaw-Curtis? *SIAM Review,* 50(1):67–87, 2008.

[160] Lloyd N. Trefethen and David Bau, III. *Numerical Linear Algebra.* Society for Industrial and Applied Mathematics, Philadelphia, PA, 1997.

[161] A. M. Turing. Rounding-off errors in matrix processes. *Quarterly Journal of Mechanics and Applied Mathematics,* 1(1):287–308, 1948.

[162] Herbert Westren Turnbull. *University of St. Andrews James Gregory Tercentenary.* St. Andrews: The University, 1939.

[163] Ian Tweddle. The prickly genius–Colin Maclaurin (1698–1746). *Mathematical Gazette,* 82(495):373–378, 1998.

[164] Robert A. van de Geijn and Enrique S. Quintana-Ortí. *The Science of Programming Matrix Computations.* www.lulu.com, 2008.

[165] Henk A. van der Vorst. *Iterative Krylov Methods for Large Linear Systems.* Cambridge University Press, 2003.

[166] F. Varadi, B. Runnegar, and M. Ghil. Successive refinements in long-term integrations of planetary orbits. *Astrophysical Journal,* 592:620–630, 2003.

[167] Richard S. Varga. *Gershgorin and His Circles.* Springer, 2000.

[168] Richard S. Varga. *Matrix Iterative Analysis*. Springer, 2nd edition, 2000.

[169] John H. Welsch. Algorithm 280: Abscissas and weights for Gregory quadrature. *Communincations of the ACM*, 9(4):271, 1966.

[170] E. T. Whittaker and G. Robinson. *The Calculus of Observations: A Treatise on Numerical Mathematics*. Blackie, 1942.

[171] E. T. Whittaker and G. Robinson. *The Calculus of Observations: An Introduction to Numerical Analysis*. Dover, 4th edition, 1967.

[172] J. H. Wilkinson. *The Algebraic Eigenvalue Problem*. Oxford, 1965.

[173] Ragnar Winther. Some superlinear convergence results for the conjugate gradient method. *SIAM Journal on Numerical Analysis*, 17(1):14–17, 1980.

[174] Tjalling J. Ypma. Historical development of the Newton-Raphson method. *SIAM Review*, 37(4):531–551, 1995.

Index